GUIDE TO BIOMETRICS

Ruud M. Bolle Jonathan H. Connell
Sharath Pankanti Nalini K. Ratha
Andrew W. Senior

GUIDE TO BIOMETRICS

With 131 Illustrations

Springer

Ruud M. Bolle
e-mail: bolle@us.ibm.com

Jonathan H. Connell
e-mail: jconnell@us.ibm.com

Sharath Pankanti
sharat@us.ibm.com

Nalini K. Ratha
ratha@watson.ibm.com

IBM T.J. Watson research Center
19 Skyline Drive
Hawthorne, NY 10598

Library of Congress Cataloging-in-Publication Data
Guide to Biometrics / Ruud M. Bolle ... [et al.].
p. cm.
ISBN 0-387-40089-3
1. Biometric identification. 2. Pattern recognition systems. 3. Identification—Automation. I. Bolle, Ruud. II. Series.
TK7882.B56G85 2003
006.4—dc21 2003052962

ISBN 0-387-40089-3

Bolle et. al.: Guide to Biometrics

First Indian Reprint 2009

ISBN 978-81-8489-160-7

This edition is manufactured in India for sale only in India, Pakistan, Bangladesh, Nepal and Sri Lanka and any other country as authorized by the publisher.

This edition is published by Springer (India) Private Limited,
A part of Springer Science+Business Media, Registered Office: 906-907,
Akash Deep Building, Barakhamba Road, New Delhi – 110 001, India.

Printed in India by Rakmo Press (P) Ltd., New Delhi.

Contents

List of Figures

List of Tables

Foreword

Starting with fingerprints more than a hundred years ago, there has been ongoing research in biometrics. Within the last forty years face and speaker recognition have emerged as research topics. However, as recently as a decade ago, biometrics itself did not exist as an independent field. Each of the biometric-related topics grew out of different disciplines. For example, the study of fingerprints came from forensics and pattern recognition, speaker recognition evolved from signal processing, the beginnings of face recognition were in computer vision, and privacy concerns arose from the public policy arena.

One of the challenges of any new field is to state what the core ideas are that define the field in order to provide a research agenda for the field and identify key research problems. Biometrics has been grappling with this challenge since the late 1990s. With the maturation of biometrics, the separate biometrics areas are coalescing into the new discipline of biometrics. The establishment of biometrics as a recognized field of inquiry allows the research community to identify problems that are common to biometrics in general. It is this identification of common problems that will define biometrics as a field and allow for broad advancement.

Guide to Biometrics provides the groundwork for defining biometrics as a discipline and the basis for advancing biometrics as a field. The selection of topics shows that biometrics as a field includes technical, application, and policy perspectives. The publication of this volume, the first scientific monograph on biometrics, is an important step forward in the development of biometrics as a recognized field of study. For the first time, we have a book which presents in one place an introduction to biometrics in general, the properties of specific biometrics, issues associated with biometrics, measures of performance for systems biometrics, selection criteria for biometrics, and mathematical techniques for analysis of biometric systems. In their selection of topics, the authors have chosen to discuss areas that are common to the field of biometrics, rather than concentrate on the properties of individual biometrics. This approach facilitates an appraisal and review of biometrics as a discipline.

The authors, Ruud Bolle, Jonathan Connell, Sharath Pankanti, Nalini Ratha, and Andrew Senior, are recognized experts in biometrics. Their research interests include biometrics in general, system analysis, evaluations, performance statistics, fingerprints, face processing, and audio and video processing. The authors' wide and diverse areas of expertise and experience in biometrics give them the knowledge and perspective to write the first technical book on biometrics.

This book will become required reading for the next generation of biometric scientists and engineers as they prepare to advance the frontiers of biometrics.

Dr. Jonathon Phillips
Program Manager,
Human ID at a Distance (HumanID)
Bethesda, MD
March, 2003

Preface

Biometrics is a constantly evolving field which has engendered a viable industry holding great promise for the future. In recent years we have seen growing maturity as technologies have improved and understanding of usage has increased with expanding experience. This experience has changed the focus of biometrics technologies from simply "replacing passwords" to being fundamental components of secure systems, but components whose use and integration demands very careful planning. This involves the consideration of many issues, such as recognition accuracy, total cost of ownership, acquisition and processing speed, intrinsic and system security, privacy and legal requirements, as well as interface usability and user acceptance.

Biometrics has survived despite a great amount of hype and wild predictions of its imminent ubiquity. We will attempt to untangle the surrounding hyperbole and myths by describing the state of the art of biometrics research, theory, and practical implementation. Biometrics is not just a fledgling technology area described by a scattering of more-or-less successful trials; biometrics is about solving serious and important problems.

Biometrics is also a fascinating area of research with many aspects: there are legal and social issues, along with perhaps more tractable aspects such as ergonomics, security, data integrity, and large system integration, including fault tolerance and system recovery. Currently, however, pattern recognition is the main discipline that concerns itself with biometrics, with increasing interest being shown by practitioners in other fields as biometrics becomes more widespread. Within the pattern recognition area, there are a host of disciplines that form pieces of the biometric puzzle. The areas of image processing, computer vision, signal processing, speech recognition, VLSI, and machine learning are all relevant to developing recognition technologies for biometrics.

This is an exciting period for the field of biometrics. Sensors are rapidly dropping in price, computing power is plentiful, and the technological infrastructure is already in place. It appears to be only a matter of time until everyday authentication protocols controlling secure access with biometric identifiers become commonplace.

Overview and Goals

Because biometrics involves pattern recognition, it brings along by necessity the ideas of statistics and probability theory in the design of matchers, but also in the analysis of accuracy, because, as we shall see, no biometric system can be error-free. There is always some chance of falsely authenticating (accepting) an imposter or falsely rejecting a genuine user. Because of this inherent uncertainty, the computer security community may not feel that biometrics is in their charter. It should be recognized that biometrics is developing serious theoretical work, testing procedures, and statistical analyses of the test results. This warrants, we feel, due attention from areas like computer security.

In our consulting position at the IBM Corporation we are constantly asked the question, *Which biometric is the best?* Given the range of services offered by IBM, this question comes from all corners of the information technology industry. The question is a good one, but one which we always answer with, "*It depends on ....*" followed by an extensive list of factors to be considered. On the one hand, this book attempts to crystallize our experience and record a comprehensive expression of the many contingencies in our answer. In this way we hope to make it possible for readers to begin to answer this question for themselves. On the other hand, its goal is to define biometric authentication, describe the state of the art, and point out the many problems still to be solved.

Audience

The book is divided into four parts for structured reading. The material progresses from easily accessible overviews to deep analyses of particular problems. The parts are meant to be fairly independent and the reader is encouraged to read just those parts that are directly relevant, skimming through introductory material or skipping over the heavy math as appropriate.

This book is oriented toward a broad group of readers. Students wishing to gain an overview of the topic can read the early sections of the books. Business people charged with selecting or recommending biometric systems can read the middle. Graduate students wishing to delve deeper into the performance issues can scrutinize the end. The book can also serve as a reference guide: we have tried to make it as up to date as possible and provided a number of pointers to advanced research.

Organization and Features

Part I *Basics*

This part provides an introduction to the field of biometrics and presents the basic terminology. Core biometrics concepts are outlined in Chapter 1. Chapter 2 then discusses general authentication protocols for verification, identification, and screening and shows how these protocols can be augmented with biometrics. Chapter 3 discusses the most frequently used biometrics (finger, face, voice, iris, hand, signature), while some emerging and less common alternatives are described in Chapter 4.

Part II *Performance and Selection*

This part explains the various issues involved in selecting a particular biometric as the basis for an authentication system. Chapters 5 and 6 give the definition of the fundamental measurable aspects which affect system accuracy, while Chapter 7 provides a summary of realistic error rates as found in the literature. Chapter 8 takes all these factors into consideration in the choice of a particular biometric for a given application and debunks a number of myths and misconceptions about biometrics.

Part III *System Issues*

This part looks beyond the selection of a particular biometrics and examines how biometrics is embedded in a systems context. Chapter 9 discusses the methodology behind *training* biometrics systems, a crucial topic since a biometrics system is only as good as the quality of its database. Chapter 10 looks at two applications in depth, detailing the logistics of enrollment and verification as well as issues involving exception handling. Chapter 11 addresses the state of the art of *integration* of disparate biometric evidence and points out research issues that are still very much open problems. The threat model analysis in Chapter 12 shows possible attack points on the overall system. We investigate these, paying special attention to vulnerabilities that may be introduced because of the use of biometrics. Finally, Chapter 13 lists biometric APIs, standards, and public databases, all of which are important considerations for ensuring compatible and commensurate systems.

Part IV *Mathematical Analyses*

This part looks in more detail at some key issues involved in evaluating and selecting biometric systems. The focus of Chapter 9 is the mysterious notion of the *"intrinsic error rate,"* the ultimate discriminatory power associated with each biometric. *Error parameter estimation* is a topic that requires careful attention and has some unique pitfalls, as will be seen in Chapter 15. Comparing matchers for the same biometrics but from different vendors is analyzed in depth in Chapter 16.

Finally, the last chapter offers a discussion of biometrics in practical applications and as a research topic, and presents a number of the authors' experience-based opinions and gut-level conjectures. Please note that the views expressed here and throughout the book are in no way the official position of the IBM corporation; the authors are solely responsible for any misrepresentations or inaccuracies.

Acknowledgments

The authors would like to acknowledge that this book is a result of the "Matchbox" First-of-a-Kind (FOAK) project and was made possible by the combined support of IBM Research, Sales & Distribution, and Global Services. We thank Norman Haas for his valuable suggestions and help in organizing the content of this book. We are grateful to John McKeon at IBM Global Services for being a great advocate of biometrics, and to Donna Fuentes who put in a tremendous amount of effort in the arduous process of finalizing what should be the content of this book. Youri Zoutman helped us enormously checking the copy edits.

IBM Thomas J. Watson Research Center
19 Skyline Drive
Hawthorne, NY
March 17, 2003

Ruud M. Bolle
Jonathan H. Connell
Sharath Pankanti
Nalini K. Ratha
Andrew S. Senior

Part I

Basics of Biometrics

1

Introduction

Reliable authorization and authentication are becoming necessary for many everyday actions (or applications), be it boarding an aircraft, performing a financial transaction, or picking up a child from daycare. Authorization is almost always vested in a single individual or in a small group of individuals. Identity verification becomes a challenging task when it has to be automated with high accuracy and hence with low probability of break-ins and reliable non-repudiation. The user should not be able to deny having carried out the transaction and should be inconvenienced as little as possible, which only makes the task more difficult.

Figure 1.1: Distinct personal characteristics: physiological (fingerprint) and behavioral characteristics (signature).

Recognizing people is a fundamental activity at the heart of our society and culture, since for many activities (*applications*), ensuring the identity and authenticity of people is a prerequisite. Biometric identification, or *biometrics*, refers to identifying an individual based on his or her distinguishing characteristics. More precisely, *biometrics* is the science of identifying, or verifying the identity of, a person based on physiological or behavioral characteristics. Physiological biometrics, like fingerprints or hand geometry, are physical characteristics generally measured at some point in time. Behavioral biometrics, like signature or voice, on the other hand, consist of the way some action is carried out and extend over time. Figure 1.1 illustrates two such characteristics. Behavioral biometrics are learned or acquired over time and are dependent on one's state of mind or even subject to deliberate alteration. We also use the term *a biometric* to refer to a specific mode of recognizing people (thus fingerprint and face recognition are two *biometrics*) or examples of the specific characteristics being recognized (an iris image and a signature are both *biometrics*).

Loosely speaking, physiological biometrics are rich enough that a one-time sample may suffice for comparing biometric identifiers. For behavioral biometrics, any given sample may give no information about a person's identity, but it is the temporal variation (behavioral influence) of the signal that contains the information.

Because identification cards may be lost, forged, or misplaced and passwords may be forgotten or compromised, the conventional methods of identification, based on unique ID cards and ID numbers, or based on exclusive knowledge (such as passwords or even social security numbers) are not sufficient. It is clear that any reliable positive person identification *must* entail biometric identification. There is simply no way around it.

It is being accepted by government and industry alike that automated biometric authentication will become a necessary fact of life. Just as other methods of computerized authentication will become more pervasive, the use of automated face, finger, voice, etc., recognition will become more widespread. But the publicity surrounding biometrics has been misinterpreted in various ways to make it appear that there are no remaining challenges in automatically identifying people. Consequently, biometrics is surrounded by much expectation, but also by myths and misunderstandings. This is a heavy burden for an emerging technology to bear, and there is a risk that biometrics usage could die a premature death because of its failure to live up to the great expectations created by its strongest proponents.

This book is intended as a guide for introducing the reader to the principal issues and practical considerations of the field, while debunking the common myths about biometrics in the popular media, as recited by both its detractors or champions. Biometrics is a fledgling technology that may pose threats to privacy; however, technology solutions that guarantee privacy are being developed by biometric researchers. Biometrics, as an area of research and technology, is an area that deserves attention and needs nurturing. This book explains and explores the many facets of biometrics; more advanced mathematical aspects of biometrics are also included in this book.

1.1 Person authentication, in general

Person authentication is not a new problem of course, and society has adopted ways to verify the identity of a person, i.e., to authenticate the person. There are three *traditional* modes of authentication [136] (Figure 1.2):

1. *Possessions:* Physical possessions such as keys, passports, and smartcards.

2. *Knowledge:* Pieces of information that are supposed to be kept secret, only known to the rightful person, i.e., passwords and pass phrases.

   Knowledge can include more-or-less personal and private information that might not be secret, such as a user ID, *mother's maiden name*, and *favorite color*.

3. *Biometrics:* Physiological and behavioral appearances or characteristics of individuals that distinguish one person from the next. These are characteristics of the human body and human actions that differentiate people from each other.

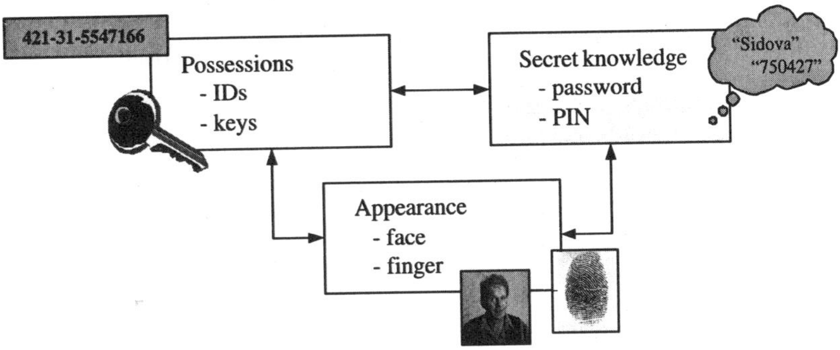

Figure 1.2: The three basic ways a person can prove identity.

Formally, this includes any personal trait that has biometric characteristics: uniqueness, permanence, etc., the properties as discussed later in this section.

The three modes of authentication can be used in combination, especially in automated authentication. A password plus a user ID combine public knowledge with secret knowledge, an ATM card (money card) is a possession that requires knowledge to carry out a transaction, a passport is a possession with a face picture and signature biometrics.

In biometrics we distinguish two authentication methods:

1. **Verification** is based on a unique identifier which singles out a particular person (e.g., some ID number) and that person's biometrics, and thus is based on a combination of authentication modes.

2. **Identification**, on the other hand, is based *only* on biometric measurements. It compares these measurements to the entire database of enrolled individuals instead of just a single record selected by some identifier.

Biometric identification can be viewed as "pure" biometric authentication and is much harder to design and implement because of the biometric database search capabilities that are needed. For example, in performing many biometric matches each against only one entry from the database ("1:1" matches), such systems need to be able to search efficiently through large collections of biometric representations and find matching biometrics. Verification systems, on the other hand, perform just one or a few 1:1 biometric comparisons.

The distinction between physiological and behavioral biometrics in our definition of biometrics is useful. But we should also keep in mind other attributes that are necessary to make a biometric practical. These include the five properties described by Clarke [40]:

1. *Universality:* Every person should have the biometric characteristic.

2. *Uniqueness:* No two persons should be the same in terms of the biometric characteristic.

3. *Permanence:* The biometric characteristic should be invariant over time.

4. *Collectability:* The biometric characteristic should be measurable with some (practical) sensing device.

A property perhaps less inherently connected with a particular biometric is—

5. *Acceptability:* The particular user population and the public in general should have no (strong) objections to the measuring/collection of the biometric.

It is the combination of all these attributes that determines the effectiveness of a biometric, and therefore, the effectiveness of a biometric people authentication system using a particular biometric in a particular application [96]. There is no biometric that satisfies any of these properties absolutely, nor one which has all to a completely satisfactory level simultaneously, especially if *acceptability* is taken into account. This means that *any* biometric authentication solution is the result of many compromises.

Biometrics can provide protection to applications in a number of ways. The ones of most concern in this book are—

1. **Physical access control** of, for example, an airport. Here the airport infrastructure, or travel infrastructure in general, is the application.

2. **Logical access control** of, for example, a bank account; i.e., the application is the access to and the handling of money.

3. **Ensuring uniqueness** of individuals. Here the focus is typically on preventing double enrollment in some application, for example, a social benefits program.

1.2 A quick overview of biometrics

There are many issues involved in designing and commissioning a practical biometric system. These range from the legal issues of the purposes to which biometrics can be put, and social issues such as user acceptance, to more practical issues of ergonomics, physical and data security, rights management, and back-up procedures when the system fails. In this book we deal with all of these issues, but at the heart of any such system is biometric sensing, comparison, and decision-making, and it is to these biometric functions that we shall devote most of our attention.

1.2.1 Biometric identifiers

The left half of Table 1.1 lists the six biometrics most commonly used in today's automated authentication systems. These are discussed in more depth in this book, both from the perspective of the matching technology and achievable error rates. These biometric identifiers are considered most "mature" and are generally expected to find widespread use in the shorter term. On the right, biometric identifiers that are discussed to a lesser extent in this book (principally in Chapter 4) are listed.

The retina biometric is often confused with the iris and sometimes called the "eye biometric." In reality, the iris and retina require quite different biometric sample acquisition.

Physiological	Behavioral
Face	Signature
Fingerprint	Voice
Hand geometry	
Iris	

Physiological	Behavioral
DNA	Gait
Ear shape	Keystroke
Odor	Lip motion
Retina	
Skin reflectance	
Thermogram	

Table 1.1: The six most commonly used biometrics (left). Some other biometric identifiers that are either used less frequently, or that are still in the early stages of research (right).

To obtain a retina image, the user is inconvenienced to a greater extent because an image of inside the eye is needed (the image a medical doctor examines for diagnosing various diseases). For iris images less intrusive sensing suffices. Although the retina biometric is relatively mature, we consider it to be a "niche biometric" because of these sensing issues. The fact that a retina image could reveal certain medical conditions makes the retina biometric perhaps a bit controversial.

Much recent work exploring biometrics has been motivated by the Human Identification at a Distance [44] program, which calls for the development of automated biometric identification technologies to detect, recognize, and identify humans at great distances. The 2004 plans include the development and demonstration of a human identification system that operates out to 150 meters (500 ft.) using visible imagery.

1.2.2 Biometric subsystems

Any biometric authentication system can be viewed as a pattern recognition system, as shown in Figure 1.3. Such a system consists of *biometric readers*, or sensors; *feature extractors* to compute salient attributes from the input signals; and *feature matchers* for comparing two sets of biometric features.

An authentication system consists of two subsystems: one for *enrollment* and one for *authentication* (see Figure 1.3). During enrollment, biometric measurements are captured from a subject, relevant information from the raw measurements is gleaned by the feature extractor, and this information is stored in the database. Along with the machine representation of the biometric features, some form of ID for the subject (such as a unique number) is linked to the representation along with other data such as the person's name. These pieces of information may be consolidated into a physical token, such as an ATM card, issued to the user.

The task of the authentication module of Figure 1.3 is to recognize a subject at a later stage, and is either identification of one person from among many, or verification that a person's biometric matches a claimed identity.

1. For *identification:* The system acquires the biometric sample from the subject, extracts features from the raw measurements, and searches the entire database for

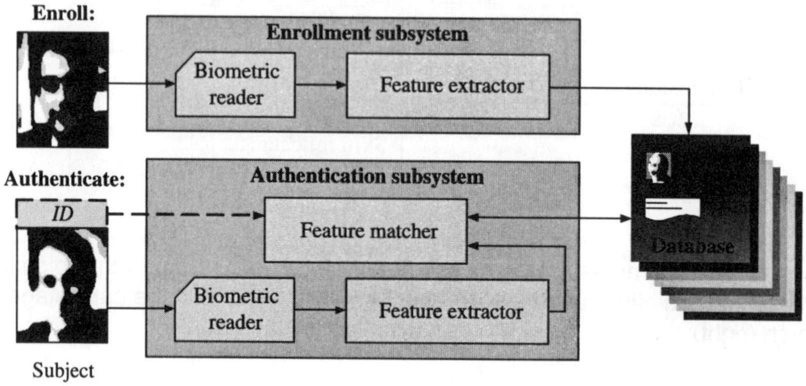

Figure 1.3: Architecture of a typical biometric authentication system: Enrollment subsystem, authentication subsystem and database.

matches using the extracted biometric features.

2. For *verification:* A subject presents some form of identifier (user ID, ATM card) and a biometric. The system senses the biometric measurements, extracts features, compares the input features to the features enrolled in the system database under the subject's ID. The system then either determines that the subject is who he claims to be or rejects the claim. The presentation of the unique identifier in Figure 1.3 is indicated by the dashed input arrow.

In some situations, a single system operates as both an identification and a verification system with a common database.

1.2.3 System performance and design issues

Design of a biometric recognition system can essentially be reduced to the design of a pattern recognition system that satisfies the basic design specifications. Pattern recognition system designers have adopted a sequential phase-by-phase modular architecture (see Figure 1.4).

Four basic design specifications of biometric systems, and pattern recognition systems in general, are—

System accuracy: In a verification system, when enrolled subjects present their biometric identifier and (correct) identity to the system, does it always make the correct decision? How about an intruder posing under a different (false) identity? Does the system correctly reject the intruder? Similar questions can be asked about identification systems. The accuracy of a biometric system cannot be measured exactly, and can only be estimated. These error rates include the chance of accepting an intruder

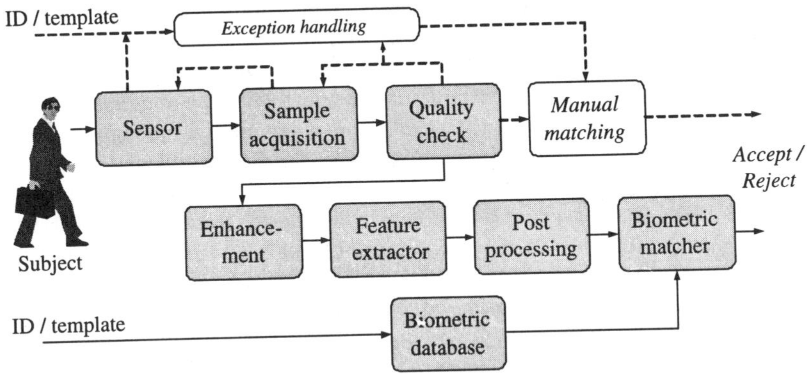

Figure 1.4: The building blocks of a biometric authentication system: The shaded boxes are automated processes, the white boxes are manual processes.

(the *False Accept Rate* or FAR) and the probability of rejecting a genuine individual (the *False Reject Rate* or FRR). These error rates are often just estimated for the portion of the user population that is not subject to exception handling.

Computational speed: How fast the system makes the decisions is of course an important parameter of a biometric system. For some systems it is particularly important to know if it the system is *scalable* from small populations to large populations.

Exception handling: Any biometric system will need an "exception handling" procedure that involves a manual matching process, as indicated by the white boxes in Figure 1.4. A subject may simply choose not to use the biometric authentication system, a subject may be among the portion of the population for which the biometric cannot be enrolled, or a subject may have a "bad biometric day." These are the Failure to Use (*FTU*), the Failure to Enroll (*FTE*), and the Failure to Acquire (*FTA*) events. A Failure to Acquire, for example, is detected through the feedback loops in the sequential architecture (see Figure 1.4).

An important factor in deciding on biometric authentication is the exception rate of an application, which is of course hard to estimate *a priori*. Somehow in the design, however, it should be specified what exception rate is acceptable and how the exception process is implemented.

System cost: This includes the costs of all the components of the authentication system. It includes the one-time costs as well as the recurring costs of routine operation and system maintenance; further included are the costs of user and operating personnel training. The exception rate is, of course, a large cost factor.

Given the speed, accuracy, and cost performance specifications of an end-to-end identification system, the following design questions need to be addressed: (i) How to achieve a low exception handling rate? (ii) How to acquire the input data/measurements (biometric)? (iii) What internal representation (feature set) of the input data is invariant and amenable for automatic feature matching? (iv) Given the input data, how to extract the internal representation from it? (v) Given two input samples in the selected internal representation, how to define a matching metric that translates the intuition of "similarity" among the patterns? (vi) How to implement the matching metric? Additionally, for reasons of efficiency, the designer may also need to address issues involving (vii) organization of the input samples (representations) into a database, and (viii) effective methods of searching a database for a given input biometric. Building a biometric system now means satisfying all these design specifications, if possible.

There are at least two additional system specifications that cannot be defined very precisely because biometric data are particularly sensitive:

Security: The fact that the decisions made by biometric systems can be used as positive proofs/denials of an individual's authorization and/or presence at a sensor raises serious questions about integrity of the overall biometric systems. In what ways can the system be compromised? What can one do once a system/identity has been compromised? How can one fight attacks on the integrity of the system?

Privacy: Anonymity is one of the most cherished values of any free society. Civil liberty organizations consider the automatic identification capabilities of biometrics to be dehumanizing. They believe that biometrics could potentially be used as a tool abetting/sustaining totalitarian regimes because biometrics allow linking otherwise distinct (pseudo)identities using distinctive characteristics and hence are a threat to individual anonymity. Conventional security technology has to be applied in a thorough fashion to address data confidentiality.

1.2.4 Competing system design issues

Given all this, a biometric person authentication system has to satisfy requirements that are often contradictory. Most importantly, a biometric authentication system has to guarantee security, which implies accuracy, without compromising too much on the convenience of its

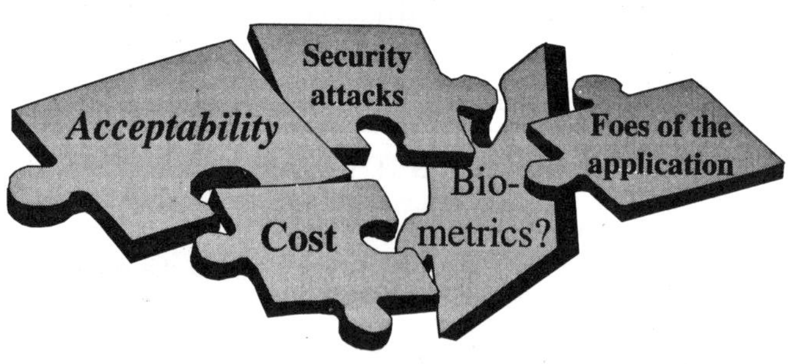

Figure 1.5: Designing a biometric authentication system involves many pieces of a complicated puzzle.

users, *and* it has to do this cost effectively. The specification of accuracy of such a system is typically in terms of *quantitative* parameters such as error rates, which, of course, need to be as low as possible. There are more *qualitative* system specifications like levels of security, convenience and privacy that need to be satisfied in addition to a system's accuracy.

However, the biometrics literature is often confusing and obscure when it comes to error numbers of the different biometrics. Only sparse, incomplete and often contradictory information is available. When faced with the task of designing a biometric system, one is immediately confronted with many difficult questions, such as: Which biometric is best for a given application? How are the error numbers that are reported for the different biometrics to be interpreted? What exactly is security when it comes to trusting people instead of information? Are new security holes created because of the use of biometrics? It's an ongoing list of questions, and the many facets of biometric authentication indeed form a complicated puzzle (Figure 1.5). Additionally, biometrics is surrounded by multitudes of myths and misconceptions, of which some are catalogued in Section 8.6.

This book is intended to introduce and explain the pieces of the puzzle in an impartial way (as far as that is possible) and to look at the many issues that arise when building biometric authentication systems. Therefore, we first introduce the various biometric functions in a little more detail.

1.3 Biometric identification

Biometric identification is based on biometric characteristic ("credentials"), and is based *only* on the biometric credentials. Figure 1.6 shows the basic building blocks of a biometric identification system. First, such a system has access to a biometric database (right) containing biometric samples or representations of biometric samples (hereafter called *templates*, where a template may contain representations of multiple biometric samples).

A biometric identification system further has the capability of *searching* the biometric database to determine if there are any database entries that resemble the input subject sample. This function is performed by the middle block of Figure 1.6. One by one, the database templates are matched to the input sample. The output of the matching phase is some candidate list of subject identifiers from the database that resemble the input biometric.

Such a biometric identification system can be used in two different modes:

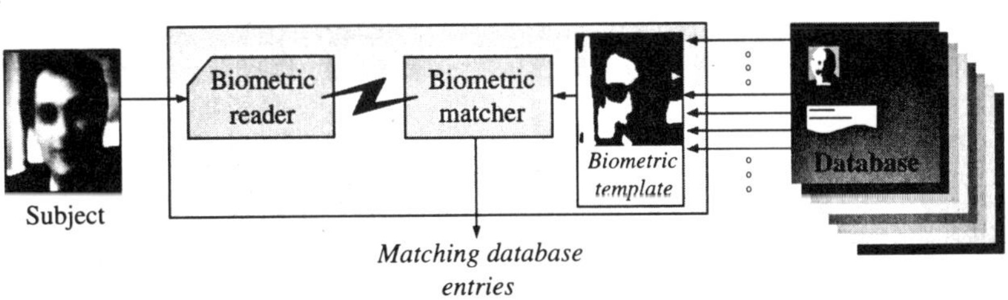

Figure 1.6: Biometric identification is done *only* on the basis of biometrics and involves database search.

Positive identification: This refers to determining that a given individual is in a (member) database. The errors that can be made are a *False Accept* and a *False Reject*. A subject is falsely accepted, causing intruders to enter the system, or a legitimate subject is denied service, i.e., falsely rejected.

These are the same errors that can occur in a biometric verification system (see below). In fact, positive identification is functionally similar to verification.

Negative identification: This amounts to determining that a subject is *not* in some negative databases. This database could, for instance, be some "most wanted" database. We also call negative identification "*screening*," because the input subject is in effect *screened* against the biometric database.

This is a very different biometric system where *False Negative* errors and *False Positive* errors are made, i.e., missing a match and false detection of a match.

Biometric identification systems may return multiple candidate matches. It is required for positive identification that the list of candidate matches is of size 1, or at least that the candidate list can be quickly narrowed down to size 1 by some other match mechanism. For negative identification, it is desired that the candidate list be small so it can be examined by human operators.

1.4 Biometric verification

Biometric verification differs from identification in that the presented biometric is only compared with a single enrolled biometric entity. There may still be a large enrolled population, but the user presents a token which indicates one biometric template from the database for comparison. This may be done in one of two ways, both of which are illustrated in Figure 1.7.

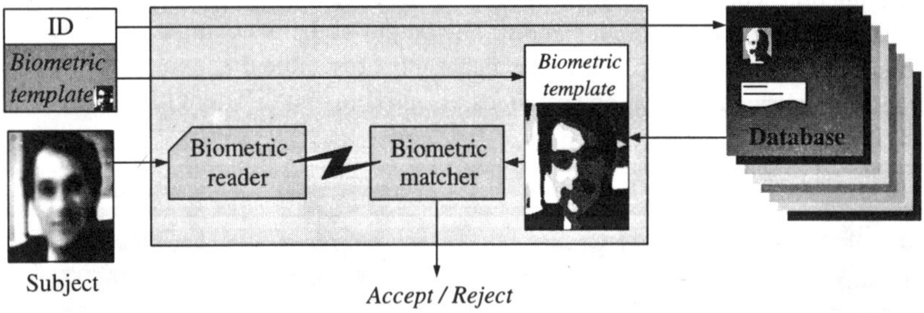

Figure 1.7: Biometric verification is based on an input biometric and an enrolled template from a single fragment of a distributed database, or by indexing a unique ID in a central database.

Like an identification system, a verification system has access to a biometric database (right). This database contains biometric templates, associated with subjects. However, unlike in a biometric identification system, a distinct identifier (ID number) is associated with each biometric template. Hence, from the database, the biometric template associated with some subject is easily retrieved using the unique ID. The input to the verification system is a biometric input sample of the subject in addition to some identifier ID associated with the identity that the subject claims to be. The output of the matching phase is an *Accept/Reject* decision. There are two possible database configurations (Figure 1.7):

Centralized database: A central database stores the biometric information of the enrolled subjects. The user presents some identity token (swipes a card, types a user ID) stating an identity, which allows the retrieval of the corresponding biometric template. This is compared with the newly presented biometric sample.

This includes applications where one subject or a small group of subjects is authorized to use some application, such as a laptop, PDA, or cell phone.

Distributed database: The database stores the biometric information in a distributed fashion (e.g., smartcards). There is no need to maintain a central copy of the biometric database.

A subject presents some biometric device containing a single biometric template directly to the system, for instance, by swiping a mag-stripe card or presenting a smartcard. The biometric system can compare this to the newly presented biometric sample from the user and confirm that the two match. Typically some additional information, such as an ID or name, is also presented for indexing the transaction, and the presentation must be made through a secure protocol.

In practice, many systems may use both kinds of database — a distributed database for off-line day-to-day verification and a central database for card-free, on-line verification or simply to allow lost cards to be reissued without reacquiring the biometric.

1.5 Biometric enrollment

Biometric enrollment (Figure 1.8) is the process of registering subjects in biometric databases:

Positive enrollment (enrollment for verification and positive identification): The purpose of enrollment is to construct a database of eligible subjects or members. It has to be somehow determined what makes a subject eligible to be enrolled, and all enrollees must be checked against these criteria.

Biometric samples and other credentials are stored in the database, which in case of a verification system might be a distributed database. Each subject is enrolled with a biometric template. The subject is issued an ID number or some possession that contains the biometric template, as indicated in Figure 1.8.

Negative enrollment (enrollment for negative identification): The collection of a database of subjects that are ineligible for some application, where the database is centralized by definition. It has to be determined for what reasons a subject would be ineligible, like membership of a most-wanted list. (The rules *themselves* that are used to exclude

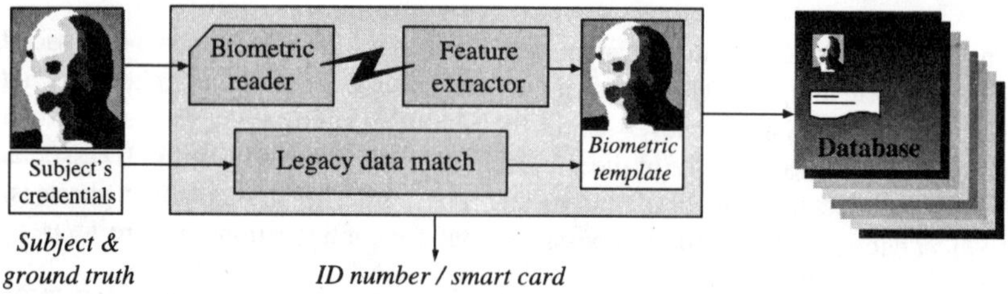

Figure 1.8: Enrolling a subject into a positive or negative biometric database.

individuals from an application are application-dependent, and are not a topic of this book.)

Biometric samples and other credentials are stored in negative identification databases. This may be done in an involuntary or possibly even covert fashion; i.e., without the subject's cooperation or even knowledge.

Enrollment is based on information about the user or subject population in the form of "ground truth," i.e., credentials such as birth certificates, passports, etc., in legacy databases and criminal data documented in government databases. This is indicated in Figure 1.8 as the block that matches an input subject to these so-called "seed documents" and other such legacy data. Determining the legacy data match most likely will involve manual labor, and hence human visual matching. This of course is a potential source of error, because of the imperfections of humans and their procedures.

1.6 Biometric system security

This is the safety of the application and is achieved by eliminating vulnerabilities at *points of attack* on, among other things *the assets* of an application, e.g., protection against *interception* [208]. For a financial application, the asset is money. For a travel application, the actual transportation system (application) and the travelers are the asset. Hence the asset is people, while the *threat* is people too.

Biometrics distinguishes itself from traditional authentication protocols in various ways. Perhaps biometrics' most notorious liability is vulnerability to *impersonation*, which is also perceived as a security flaw of biometric technology. Impersonation covers the scenarios where biometric information is presented when the "owner" is not present. This encompasses things ranging from lifting latent fingerprints from objects to violent acts like cutting off fingers. Such scary stories about biometrics, along with the legacy law enforcement uses, influence the acceptance of different biometrics to different extents. These acceptance problems are in essence due to the fact that impersonating a biometric has a stigma associated with it because it amounts to violating somebody's identity. It should be noted, however, that lifting a latent fingerprint and using the print to impersonate its owner is functionally the same as covertly observing password input. There is a large difference in the difficulty of applying knowledge about a biometric as opposed to applying knowledge about a password to gain unauthorized access. This is even the case if a fingerprint scanner can be easily fooled by fake biometrics: impersonating a biometric is never as easy as impersonating an identity by using a stolen password.

The security of biometric authentication systems, like legacy authentication systems, depends on the strength of the linking of enrolled subjects to the universally better accepted "seed documents," such as passports. It also depends on the quality of the seed documents themselves.

In any authentication system, the weakest point is the most serious vulnerability since this is the easiest attack point. This is restated as the Principle of Easiest Access: "An intruder can be expected to use any available means to access the application and it can be expected that the least protected points of access will be attacked." The key here is that introducing biometrics should not create novel vulnerabilities and security loopholes. This means that for any biometric authentication system where security is important, the introduction of a biometric needs to be soundly justified. The biometric authentication should be an integrated aspect of the overall security of the application, which includes preventing security breaks of the biometric system itself. Note that the security of biometric systems is related to protection against career criminals. Where security issues are not a concern, biometrics can be used simply to improve convenience of an application (say setting a car radio to a user's favorite station, based on face recognition).

We believe that society as a whole sooner or later will figure out how to deal with biometrics. At this point in time, the technology is really not at a state where biometrics can be used the way the popular press sometimes describes it. Manufacturers of biometric equipment and systems also tend to exaggerate the state of the art in biometrics but in different ways.

2

Authentication and Biometrics

Biometrics is often used for establishing trusted communication between two parties to negotiate access to an application. This is also a topic that has been studied in the field of information security for quite a while now. Therefore, in the next sections we introduce some security terms and tools that can be found, for example, in [109,162,208]. Indeed we take the liberty of following Chapter 4 of [162] fairly closely.

But first let us give some definitions pertaining to authentication that can be found in [109]:

- *Access control:* A mechanism for limiting the use of some resource to authorized users.
- *Access control list:* A data structure associated with a resource that specifies the authorized users and the conditions for their access.
- *Authenticate:* To determine that something is genuine; to determine reliably the identity of a communicating party.
- *Authentication:* The process of reliably determining the identity of a communicating party.
- *Authorization:* Permission to access a resource.

2.1 Secure authentication protocols

Using conventions between two parties, cryptosystems can be used for purposes other than just secret communication, these conventions are called protocols. A *protocol* is an orderly sequence of steps two or more parties take to accomplish some task. The order of the steps is important in this activity, so the protocol regulates behavior of both parties. The parties agree to the protocol or at least the protocol is understood.

Let us use a telephone conversation as an example. Upon dialing, the person dialing hears a telephone ringing at the other end and, after a while, a click when the phone is answered by the second party. Protocol, or standard practice, is then that the receiver speaks first, saying "hello," or even gives some identifier like a name. The originator then identifies him or herself. It is only after this protocol that the intended communication is initiated.

By simply picking up the phone and not answering, the reader can easily verify the utter failure in communication if this established protocol is not followed. Even if the dialing party hears the click, without this confirmation of connection, the dialing party will often not initiate communication.

The initiation of a phone conversation is an example protocol. An authentication protocol should have the following desirable characteristics [162]:

- *Established in advance:* The protocol is completely defined and designed before it is used. The work flow of the protocol is defined, the rules that determine the work flow are defined, and it is defined what it means that two authentication credentials "match."
- *Mutually agreed:* All parties to the protocol agree to follow the steps, and agree to follow these steps in the prescribed order.
- *Unambiguous:* No party can fail to follow a step properly because the party has misunderstood the step.
- *Complete:* For every situation that can occur there is an a priori defined action to be taken. This means, for example, that the exception handling process is completely defined.

The modern, heavily networked and traveled world requires the use of computers and communications as tools for access to services, privileges, and applications. Operators of such systems are typically not acquainted with their users and access must increasingly be authorized without human intervention. Because of anonymity and distance, a user will not and should not trust the managers or other users of a system. Protocols need to be developed by which two suspicious parties can interact, these protocols in essence regulate behavior. Authentication then becomes the following of a behavioral protocol between user and application so the user can be authorized to use the application or to access the premises.

In itself, a protocol does not guarantee security. For example, the protocol that controls access to a business may specify the opening and closing hours and may even require a certain dress code but does not contribute to the security.

2.2 Access control security services

Two security services, at least, should be offered by any access control system [208]:

1. *Authentication:* This service is concerned with assuring that a communication is authentic. In the case of ongoing communication, two things are involved: (*i*) At the time of connection initiation, the service ensures that the two entities are authentic, i.e., that each entity is the entity it claims to be. (*ii*) The service must assure that the connection is not interfered with by a third party masquerading as one of the legitimate parties.

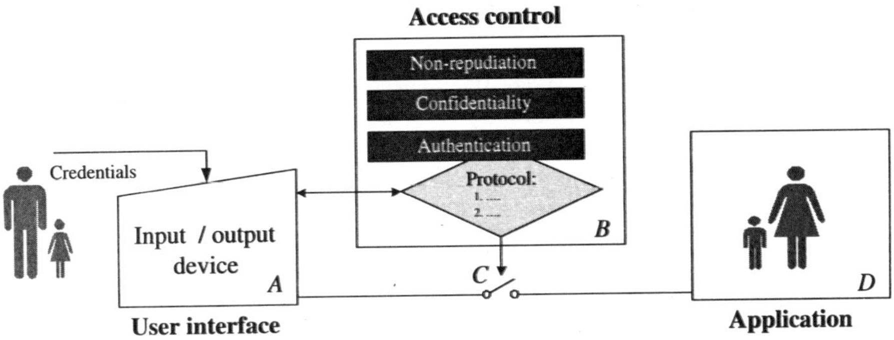

Figure 2.1: Security of access control involves authentication of user and authentication of the application.

2. *Non-repudiation:* This prevents either the sender or receiver from denying a transmitted message. Thus, when a message is sent, the receiver can prove that the alleged sender in fact sent the message. Similarly, when a message is received, the sender can prove that the alleged receiver in fact received the messages.

   This is perhaps best called reciprocal non-repudiation, which is generally not incorporated into authentication systems between a human subject and a physical or logical system.

So, from a security point of view, authentication is ***only one*** of the security services that access control should offer; there are a few more required services, like *confidentiality*; see [208]. An authentication system is then organized as outlined in Figure 2.1. A user interface A gathers credentials from a subject through input devices such as smartcard readers and fingerprint scanners; the user interface may also include output devices, possibly to give feedback on the quality of the biometric sample that has been acquired. The access control system B offers a number of security services, including authentication. The authentication is ensured through a prescribed protocol. Upon successful fulfillment of this protocol, the subject is given access to the application D, through the mechanical or logical operation of a switch C.

The access control system, as indicated in Figure 2.1, should also offer traditional security services such as confidentiality and non-repudiation.

In the following we look at some authentication protocols as they have evolved over the centuries and more recently have been established due to developments in photography and computing machinery.

2.3 Authentication methods

There are three basic *modes* for authorizing subjects, described by Miller in [136]. These have evolved with the technology of the printing press, photography, and automation; the methods were in use long before the widespread need for automated, electronic authentication:

P *Possession:* Anyone in the possession of a certain physical object, e.g., keys or magnetic stripe card, is able to receive the associated service or access (and is understood to be authorized to receive it). For example, anyone who possesses the car keys has the privilege to drive the car.

K *Knowledge:* Individuals with certain knowledge are eligible to access the service. Authentication here is based on secret knowledge, such as passwords, lock combinations, and answers to questions. The important word here is *secret;* the knowledge needs to be secret in order to be used for secure authentication.

However, we can also distinguish non-secret knowledge that is important in authentication. A computer user ID or a bank account number are often required for authentication and while not secret they are not universally known, preventing superficial impersonation attacks. Indeed, we can distinguish a continuum of secrecy — *mother's maiden name* is commonly used for authentication, because while not secret, it is not easy to find out. In Chapter 11 we will describe *conversational biometrics*, which combines several pieces of such semi-secret information (such as favorite color, birthplace) with biometrics to increase security.

B *Biometrics:* Personal traits of humans that can be somehow measured (sampled, acquired) from a person in the form of a biometric identifier and that uniquely distinguish a person from the rest of the world population. These are properties that are somehow intrinsically related to a human and are largely determined by either genetics or phenotypes, inheritable characteristics as opposed to characteristics formed during the foetus phase. They are difficult to share, steal, or forge, and unlike possessions and knowledge, they cannot be changed.

Possession (as used in this book) and knowledge, in the form of

$$(\textit{account number}, \textit{password}) = (\textit{Possession}, \textit{Knowledge}) = (P, K)$$

is probably the most widely used authentication method (protocol). This authentication method is used for computer accounts, Internet accounts, intranet accounts, email account, phone mail, etc. These authentication modes P and K only require exact matching but do not link a user (a real person) to some more or less established "identity." But rather they link some *well-defined* identity determined by possession P to the *anonymous knowledge of a password* K and *not* to the authentic enrolled person.

Authentication mode B (biometrics) offers additional security because it is hard, if not impossible, to share biometrics and therefore the one security aspect of access control, namely authenticity of the participants, is much more confidently and therefore more securely established.

Method	*Examples*	*Properties*
What you have (P)	User IDs, accounts Cards, badges Keys	Can be shared Can be duplicated May be Lost or stolen
What you know (K)	Password, PIN Mother's maiden name Personal knowledge	Many passwords are easy to guess Can be shared May be forgotten
What you have and what you know (P, K)	User ID + Password ATM card + PIN	Can be shared PIN is a weak link (Writing the PIN on the card)
Something unique about the user (B)	Fingerprint Face Iris Voice print	Not possible to share Repudiation unlikely Forging is difficult Cannot be lost or stolen

Table 2.1: Existing user authentication methods with some examples of positive and negative properties.

Table 2.1 shows four methods for user authentication that are widely used [136]. Because a biometric is an intrinsic property of some individual, it is difficult to duplicate surreptitiously and nearly impossible to share; additionally, a biometric of an individual can be lost only in the case of serious accidents, severe illnesses, and extreme wear and tear on the biometric. Therefore, biometric identifiers offer certain assurances about the real identity of the user in an authentication protocol; something that the use of other modes for authentication, possession and knowledge, does not guarantee. When combining the last row (B) of Table 2.1 with possession P and/or knowledge K, we get additional biometric methods like (P, B) (e.g., passports, smartcards with biometric template) and

$$(P, K, B) \Rightarrow \{P = \text{credit card}, K = \text{mother's maiden name}, B = \text{signature}\},$$

a much-used authentication method for credit cards.

The boundaries between possessions and knowledge can be indistinct. For example, the identifying parts of physical possessions can be digitized, often very concisely, such as recording the depths of cuts on a key. This in some sense converts a possession into knowledge. Nevertheless we consider these identification modes to be physical as it is a physical instantiation that is authenticated, not the information itself, even when such an instantiation can be created from the information. A credit card number (that can be used over the internet or telephone) is knowledge, but a credit card (that can be used in an ATM) is a possession. Moreover, remembered secret knowledge could even be argued to be a biometric as it is a measured, unique property of a person, but for most situations the categories as defined above are clear and useful distinctions.

One biometric, signature (and to a lesser extent voice), incorporates a significant knowledge component. This means that a signature can be changed at will but also means that signatures are more easily forgeable. This has motivated the automated signature research community to address active impersonation attacks (i.e., forgeries) very early on. There are other confusions between knowledge K and biometrics B as will be discussed later.

2.4 Authentication protocols

As pointed out in [18], any authentication protocol with multiple modes (*and* multiple biometrics) can be defined and can be executed against a set of presented credentials. An *authentication protocol* is the (automated) decision process and work flow to determine if a subject's credentials are sufficient proof of identity to authorize the subject for access based on credentials, or tokens.

The first column of Table 2.1 shows the necessary credentials that a user needs to have in order to be authenticated. Here the term *authenticate* is used in the strict security sense of the word, "the process of reliably determining the identity of a communicating party." In order of the rows of the table we have—

1. This can be the possession P like a physical key to gain access to a certain locale. This can also be the simple possession of a credit card at a gas station that enables the legitimate owner (or a illegitimate owner) to authenticate credit card transactions without much scrutiny (at least in the US).

2. In the case of a vault, the knowledge K is the combination to the lock that authorizes a person to enter the vault; this key knowledge implicitly authenticates removal of objects from the vault. Here the authentication protocol is simple: a user demonstrates his or her authorization to the vault through the secret knowledge of the pass code.

3. A well-known everyday authentication protocol requires an ATM card P and a PIN K for banking through an automated teller machine. Anyone that has these credentials *and* is aware of the rest of the authentication protocol is authorized to access a banking application and therefore can perform transactions through secure authentication.

4. The fourth row in Table 2.1 refers to the pure identification protocol, which is simply the presentation of a biometric *and nothing else*, no other tokens or forms of interaction with the user interface of an authentication system are involved.

   Without the concept of biometrics, i.e., time-invariant distinguishing attributes of a person, there is no hope of solving the automated human identification by machines processing sensory signals. It is biometric identifiers that truly distinguish one *person* from the next.

Hence for any secure authentication protocol there are one or more authentication credentials involved that we call *tokens;* hence, we have possession tokens P, knowledge tokens K, and biometric tokens B. The enrollment then is a communication between user and the access control system to exchange authentication tokens. The system supplies possession and knowledge tokens to the user, while the user may supply biometric tokens (samples) to the system. This process is defined by an *enrollment policy* (Chapter 9).

A set of tokens $T = \{x_1 \ldots x_n \,|\, x_i \in (P, K, B)\}$ is only part of an authentication protocol. What is needed further is a set of rules that define the authentication protocol $\mathcal{A}_P$ that uses T according to precisely defined orderly sequences of steps, or rules of behavior. An example of such an authentication protocol is

$$\mathcal{A}_P(T) = \mathcal{A}_P(P, K, B). \tag{2.1}$$

A simple example would be an authentication protocol that only used a possession like a house key $\mathcal{A}_P(P)$, or an ATM card used in conjunction with a PIN number $\mathcal{A}_P(P, K)$.

Combining multiple authentication methods, especially biometrics B, into an *authentication protocol* improves the certainty of authentication and therefore decreases chances of repudiation and fraud. For example, an authentication protocol,

$$\begin{aligned} \mathcal{A}_P(T) &= \mathcal{A}_P(\{P, K, B_1, B_2\}) \\ &= \mathcal{A}_P(\{credit\ card,\ PIN,\ photograph,\ signature\}) \end{aligned} \tag{2.2}$$

specifies what authentication modes are used in a protocol and specifies how these modes are to be used in the protocol, $\mathcal{A}_P$, a set of rules operating on $T = \{P, K, B_1, B_2\}$. The actual authentication protocol $\mathcal{A}_P$ in (2.2) may be described loosely as, Anyone in possession of a credit card P with a signature and a picture, who has the ability to produce a signature B_1 that appears similar to the signature on the credit card and has a likeness B_2 to the picture, and additionally has knowledge K of the associated PIN, has the privilege to use the credit card.

In fact, not surprisingly, authentication protocols are found throughout the biometrics literature. A few examples are—

- The authentication rule: "*Three strikes and you're out.*" A subject is given three chances to match the reference biometric but after the third failure, the subject is denied further access to the system.

- The Galton-Henry system [9] of manual fingerprint classification using ten-print cards, which was published in June 1900 and officially introduced at Scotland Yard in 1901 for its criminal-identification records, is an authentication protocol. Fingerprints are classified in a three-way process: by the shapes and contours of individual patterns, by noting the finger positions of the pattern types, and by relative size, determined by counting the ridges in loops and by tracing the ridges in whorls.

  Adaptation of the Henry system for computerized large-scale searches [139] is an automated authentication protocol.

- The fascinating science of latent fingerprint identification is described in very precise minutiae matching protocols [157], thereby defining as precisely as possible what in a latent print matches a reference print.

- Manual identification for law enforcement purposes follows a protocol. Here all the authentication modes and methods are used in an interrogative and investigative way, i.e., some long string of user tokens $P_1, B_1, B_2, K_1, P_2, B_3, \ldots, X_i$ with biometrics modes $B_1, B_2, B_3, \ldots$ is matched to tokens in a multitude of (government agency) databases.

> This of course is the optimal way of positively identifying a subject. Clearly the frequent use of multiple biometrics in law enforcement greatly enhances the accuracy in terms of error rates of the investigative process, thereby decreasing the false conviction rate. From the point of view of biometrics identification, the False Positive Rate is minimized.

Hence, in addition to a set of tokens $T = \{P, K, B\}$ and rules operating on these tokens $\mathcal{A}_P(T)$, an authentication protocol needs to define what it means that two tokens of any kind, P, K, and B, "match."

2.5 Matching biometric samples

Clearly, the execution of authentication protocol (2.1) requires matching capabilities. The credentials P and K can be checked simply by exact comparison. The credential B can only be compared through pattern recognition techniques, since two machine representations, obtained from two samplings of real world biometric $\mathcal{B}$, are never the same because of noise and distortions in the acquisition process.

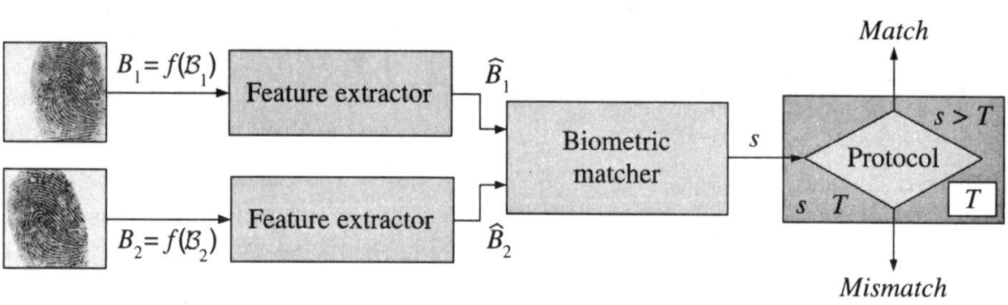

Figure 2.2: A matcher determines the similarity of two real-world biometrics $\mathcal{B}_1$ and $\mathcal{B}_2$, or rather the similarity between the two biometric templates $\hat{B}_1$ and $\hat{B}_2$, denoted $s(B_1, B_2)$.

A biometric template $\hat{B}$ is a machine representation of the biometric sample B in terms of features (e.g., fingerprint minutiae, distance between eyes, length of fingers). Templates often have domain specific names (e.g., voiceprint, iriscode, fingercode, eigenface, etc.). For simplicity, since both sample and the template are machine representation, we drop the hat from $\hat{B}$ and use them synonymously, i.e., $B \equiv \hat{B}$.

Figure 2.2 shows a generic biometric matcher that executes a simple protocol to determine if two real world biometrics $\mathcal{B}_1$ and $\mathcal{B}_2$ are the same, i.e., are from the same subject d. A biometric matcher computes a *score* s, which is the *similarity* $s(\hat{B}_1, \hat{B}_2)$ between the templates derived from the biometric sample $B_1 = f(\mathcal{B}_1)$ and $B_2 = f(\mathcal{B}_2)$:

$$s = s(\hat{B}_1, \hat{B}_2) = s(B_1, B_2) = s(f(\mathcal{B}_1), f(\mathcal{B}_2)). \tag{2.3}$$

The biometric part of the authentication protocol (2.1) uses this score s to arrive at a decision, based on a threshold T:

$$\begin{array}{ll} \text{if } s > T: & \text{decide } \mathcal{B}_1 \text{ and } \mathcal{B}_2 \textit{ match}, \\ \text{if } s \leq T: & \text{decide } \mathcal{B}_1 \text{ and } \mathcal{B}_2 \textit{ do not match}. \end{array} \quad (2.4)$$

From (2.3) and (2.4), three crucial design aspects of a biometric system become clear:

1. The biometric sampling or signal acquisition $B = f(\mathcal{B})$.
2. The similarity function $s = s(B_1, B_2)$ between two templates.
3. The decision, or similarity, threshold T that decides on a *match* or *mismatch*.

Next we define the various biometric authentication systems using the notation we have established.

2.5.1 Identification

Perhaps the most ambitious possible use of biometric technology is automated positive identification of an input subject d against some member database M of m subjects d_i. The authentication protocol demands *only* the presentation of the biometric to the authentication system and a subject is identified or not (and granted access or not) based *only* on the biometric.

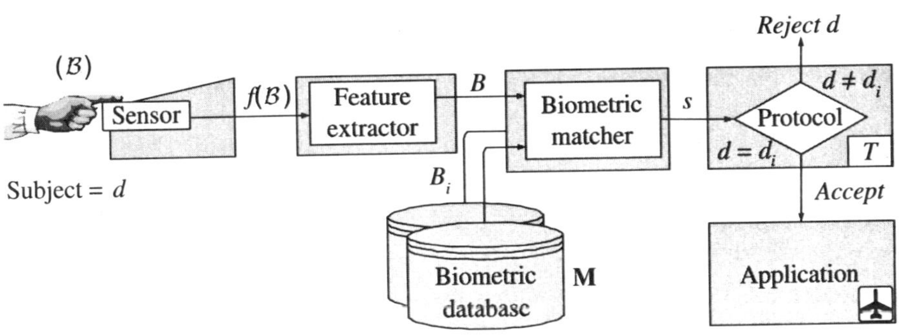

Figure 2.3: Positive identification is achieved by matching an input biometric $\mathcal{B}$ of subject d against databases M of biometric information.

This process is shown in Figure 2.3. Subject d presents real world biometric $\mathcal{B}$, and the biometric sensor acquires a sample $f(\mathcal{B})$ thereof. From this sample, a biometric template B is extracted by the feature extractor. Subsequently, the biometric matcher determines the similarity $s_i = s(B, B_i)$, between template B and template B_i corresponding to a database entry.

The authentication protocol then declares $d = d_i$ if $s_i > T$ and $d \neq d_i$ if $s_i \leq T$. Multiple $d_i \in$ M might satisfy that criteria and the output of the system in reality is a candidate list $C = \{d_a, d_b, ...\}$. For efficiency, a more precise, secondary matcher, or an additional biometric may be needed to reduce this candidate list to size 1.

2.5.2 Screening

Screening is *negative identification* (using Wayman's terminology [231]), establishing that a person is *not* on some watch list of "interesting" people. A research area that is developing is passive "screening" of crowds based on cataloging of the face images. Face cataloging is a concept that is being popularized by the Identix Inc. [220] and is the technology of building a repertoire of face images of people in a space, based on ordinary visual light cameras.

In general, screening is the authentication protocol that prescribes matching all tokens, or credentials, of a subject (passenger) to a database without any claim of the user's identity (or without trusting one that is given). The authentication protocol defines that a subject can be authenticated if a string of the subject's tokens (credentials) $P_1, B_1, B_2, K_1, P_2, B_3, ..., X_i$ does *not* match with "credentials" of, say, the list of most wanted criminals. Biometrics can play a role in this screening process; however, with the current state of the art none of the biometrics can be expected to narrowly pinpoint true subjects based on a biometric *alone* (see Section 7.2). Such biometric matches will create many False Positives, too many to handle if the list of the criminals in the database is large. Biometric identifiers can be used, on the other hand, to match against the return lists of parametric searches, based on name, date of birth, etc. Such lists may be sufficiently small to be able to support decisions based on biometric matches.

As already mentioned in Section 2.4, manual *positive identification* of an (arrested) subject for law enforcement purposes follows a protocol, which is similar to a screening protocol. Here all the available authentication modes and methods are used in an interrogative and investigative way, i.e., some long string of user tokens $P_1, B_1, B_2, K_1, P_2, B_3, ..., X_i$ with biometrics modes $B_1, B_2, B_3, ...$ is matched to tokens among a multitude on file. When properly matched, from the point of view of biometric identification, the False Positive chances are minimized.

2.5.3 Verification

For biometric verification, the credentials are (ID, B), optionally with knowledge included. Hence besides just the presentation of a biometric identifier, the authentication protocol includes the presentation of possessions and/or knowledge. These additional credentials uniquely define an enrolled identity in database **M** and hence an associated biometric machine representation (template) B_i.

A typical biometric verification system is shown in Figure 2.4. A subject d claims an ID number m and offers biometric $\mathcal{B}$ for sample acquisition $f(\mathcal{B})$. The feature extractor computes biometric template B and template B_i associated with i is retrieved from database M. The biometric matcher then computes a match score $s = s(B, B_i)$: if the score (similarity between B and B_i) is high enough $s > T$, decide $d = d_i$, and the subject is accepted to the application; otherwise, decide $d \neq d_i$, and the subject is rejected.

2.5.4 Continuity of identity

When a biometric is matched, we have some measure of confidence that the person being verified or identified was present at the time the biometric was acquired (an instant in the

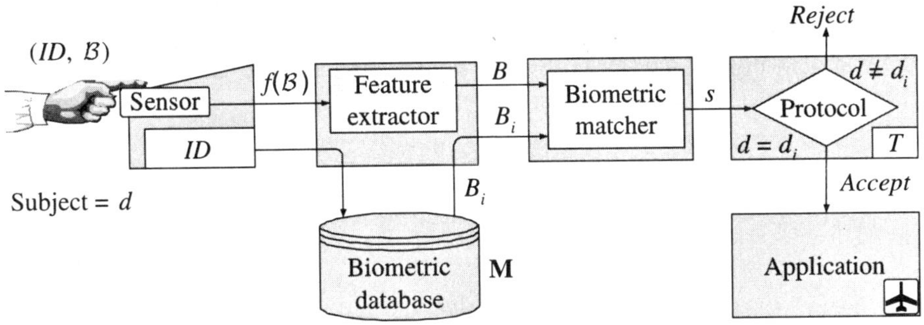

Figure 2.4: For verification, a subject claims identity d_i by supplying an identification number ID_i.

case of physical biometrics or some short period for behavioral biometrics). However, at some later time, what guarantees do we have about the identity of a person using the application? An ATM may make no assumptions and require the PIN to be reentered for every important transaction. Similarly a biometric authentication system may reestablish authentication credentials during a transaction. This is particularly important for those biometrics than can be passively acquired (particularly face and voice) since reevaluation of the biometrics does not require any interruption.

However, other systems can make inferences about "continuity of identity" to extend credentials over time. One example might be an ATM booth with a camera. With this we might be able to tell that the original authenticated user had not left and no other person had arrived, so we can infer that the user throughout some transaction is the same person. Continuity of identity becomes particularly important in the context of visual surveillance, where a system can track a person as they move through a space (see the proceedings of recent workshops in tracking [68,69]). So long as a system is tracking a person, it can infer that the identity for that person does not change. Thus any authentication of the individual applies for the whole track. This might be useful if there is a fingerprint reader at the entrance of a building — the authentication that takes place at the reader can be applied so long as the person is tracked. Or, as another individual moves through an airport lounge there may be no time when a clear view of the person's face is obtained, but multiple identification results based on partial views over an extended period of time can be integrated together if a tracking system provides us with an assurance of continuity of identity.

2.6 Verification by humans

When people were still living in small communities, authentication by biometrics was mostly limited to "databases" of personal acquaintances remembered by individuals (though people could be verified by very restricted "templates" transferred through verbal descriptions, and a very few by portraits). The advent of photography and the increase

Method	*Examples*	*Properties*
Something you have with something unique about you (P, B)	Passport with face image Credit card with signature	Forging is possible Easy to share
	smartcard with biometric	Hard to tamper with

Table 2.2: Some existing user authentication methods that use biometrics.

in literacy enabled the issue of credentials that contained a biometric (face photograph or signature) that could be verified by someone who had never seen the subject before. The rise of cities, bureaucracy, and personal mobility rapidly brought a demand for such credentials as the number of people being dealt with by civil government officials outstripped the ability of humans to remember everyone for whom they were responsible. Therefore in addition to the authentication methods in Table 2.1, authentication credential (P, B) is in widespread use.

Due to many convenience factors and its naturalness, "face" and "signature" have evolved as the biometrics of choice for manual authentication. These are the first two of Table 2.2, though signature is almost never visually verified or matched when using a credit card (in the US). The direct extension of this (P, B) authentication method is the smartcard with a stored biometric. In terms of acceptance, people have used possession plus a biometric as an authentication method for many years. Of course, it is very hard to estimate at what False Accept and False Reject Rates human beings can verify faces.

Clearly, there is large variation in verification accuracy from one person to the next and verification accuracy depends on the state of fatigue of the human inspector. There is, however, no reason to believe that a human can perform face verification very well; the error rates are on the order of 1 in 1,000.

The error rates for human verification of signatures are equally hard to establish. It is well known that handwriting experts can perform signature verification very accurately, but clearly the cursory inspection of signatures on credit card slips is not a very secure process.

2.7 Passwords versus biometrics

In Section 2.4, we pointed out that for any authentication protocol there is a set of tokens $T = \{P, K, B\}$ that are the required credentials for access to an application, though for a given application we may only require one or two of P, K, B.

For authentication, the credentials of a subject need to be matched, including the biometric templates. For biometric user verification, this amounts to an authentication protocol that is very much the same as for today's password systems (see Figure 2.5). A subject lays claim on an identity typically by presenting a possession (bank card) or non-secret knowledge (user ID), but instead of exact matching of a password, the authentication protocol of a biometric system requires matching one or more biometric samples, which can only be done probabilistically or in some fuzzy fashion.

In terms of the authentication system Figure 2.5, A is the input device, a keyboard or

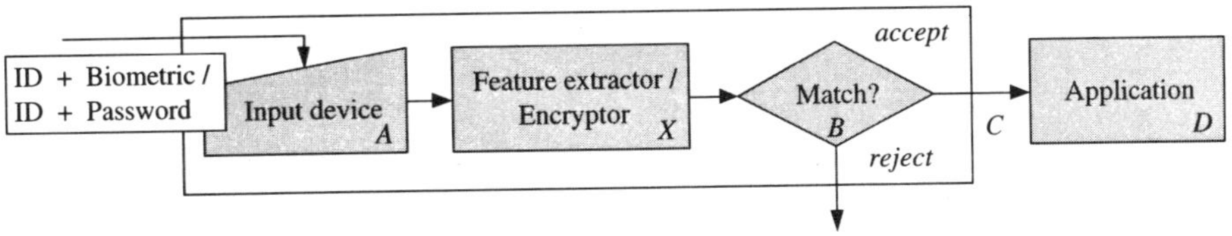

Figure 2.5: In essence, biometric user authentication is not *that* different from authenticating the password of a user.

some sensor; B is a matcher that matches either passwords or biometric templates according to some protocol $\mathcal{A}_P$; C is an access control mechanism; and D is the application protected by the access control system. Subsystem X is the password encryptor, or in case of biometric input it is the processing module that extracts biometric templates.

A fundamental difference between biometric identifiers (tokens) and other authentication methods (tokens) is the notion of the *degree of a match*, i.e., the underlying matching technology. A password authentication protocol always provides a crisp, binary result: if the passwords match, it grants access and otherwise it refuses access. Thus, there is *no concept of the probability of a match*. Consequently, there is no confusion about the *precise definition* of a match. By necessity, however, biometrics have to adopt probability theory and the use of statistical techniques to analyze the likelihood of a match. This has resulted in the concepts of *application error rates* (False Accept Rate and False Reject Rates) and *intrinsic error probabilities* (loosely the minimal achievable error rate for a given biometric) that are associated with biometric authentication systems and biometric identifiers. These are discussed in Chapter 5 and Chapter 14, respectively.

Poor definition of *what exactly* is measured when estimating the error rate for a particular biometric installation or biometric, and misleading, or unclear, testing methods contribute much to the hype surrounding biometrics as we will see later (Chapter 7).

2.8 Hybrid methods

The ability to match both parametric identifiers, such as passwords and knowledge, and biometric identifiers is at the heart of the problem of biometric authentication.

The decision to authenticate is based on one or more authentication methods, or tokens from $T = \{P, K, B\}$. For person authentication, each token supplied by a user needs to be matched with the token as it is recorded and stored during enrollment. The decision on whether these tokens match or not is to be made by integrating the output of the separate match engines that verify the tokens or by integrating the token matching more fundamentally. Comparing possession tokens and simple knowledge tokens such as passwords amounts to exact matching, though it is conceivable that for low security transactions, passwords that do not match exactly can be accepted. For example, a PIN code may be accepted if two numbers in the PIN that is entered are transposed.

Two issues here need attention:

1. *Integration of credentials:* It is best of course to combine two or more authentication methods. As noted before, associating possession P (an ID) or knowledge K with a biometrics B reduces the problem of biometric identification to biometric verification, i.e., reduces the problem from the 1:*many* matching to 1:1 matching of biometrics.

2. *Integration of biometrics:* The required credentials T may include multiple biometrics, e.g., $\{B_1, B_2\}$ with, say, B_1 finger and B_2 face. Integrating multiple biometrics is and has been a subject of considerable interest and will be discussed later (Chapter 11).

Integration of different biometrics is a problem that lies in the domain of pattern recognition. Integration of different non-biometric authentication modes, on the other hand, is very much a topic that is part of traditional security of authentication applications (e.g., [208]). Integrating traditional authentication methods with biometrics is only beginning to be studied together in the context of the overall security architecture of the application [18,181].

Again, in general, the use of any of the above modes P, K, or B means that one has to be able to match by verification of account numbers or pass phrases and comparison of biometric information, such as faces. Possession and knowledge tokens involve exact matching if done by machine. Biometric matching, on the other hand, involves some type of approximate matching, as we will explain in the rest of this book.

3

The Common Biometrics

In this chapter we provide a brief description of the six most widely used (or widely discussed) biometrics. These most commonly used automated biometric identifiers are (i) finger, (ii) face, (iii) voice (speaker recognition), (iv) hand geometry, (v) iris, and (vi) signature. Chapter 4 describes other biometrics that are not currently as common.

In this chapter, for these six biometrics, we describe the characteristic that is being measured, the devices used to measure the biometric, the features which are extracted to represent it, as well as a brief indication of the algorithms used to compare two samples of the biometric. In Chapter 8 we discuss criteria for selecting which biometric is more appropriate for a given purpose and Section 8.5 lists advantages and disadvantages of the main biometrics.

3.1 Fingerprint recognition

Fingerprinting has a long and interesting history as a biometric [120]. The inside surfaces of hands and feet of humans (and, in fact, all primates) contain minute ridges of skin, with furrows between each ridge, as shown in Figure 3.2. The purpose of this skin structure is to (i) facilitate exudation of perspiration, (ii) enhance sense of touch, and (iii) provide a gripping surface. Fingerprints are part of an individual's phenotype and hence are only weakly determined by genetics. In [158], it is shown, for instance, that identical twins have fingerprints that are quite different. Fingerprints indeed are distinctive to a person. Within the forensic community it is widely believed that no two people have identical ridge details. By this it is meant that no two fingerprints from different fingers have ever passed the test for identity as applied by an expert. There is evidence that the Chinese were aware of the individuality of fingerprints well over 5,000 years ago [9]. The belief in the uniqueness of fingerprints led to their widespread use in law-enforcement identification applications.

Early in the 20th century, an ingenious recognition system based on ten prints developed by Sir Edward Henry was brought into operation [9]. This system is now known as the "Henry system" and was adopted and refined by the FBI [66]. It allows for correct identification of offenders by manual indexing into databases of known criminals. It classifies the overall flow pattern of the fingerprint into a number of distinct patterns such as "arch," "left whorl," "tented arch," etc. These classes are not uniformly distributed over the population [9,66]. All ten fingers are classified in this way to yield a signature vector of the form $[Arch, Whorl, Arch_{Tented}, Loop_{Left}, Loop_{Right}, ...]$ (usually referred to as "ten-print cards"). While not unique to each person, this sequence can at least be used to rule out *some* suspects. Much research has been done on this sort of automated fingerprint classification, e.g., [34,100,198], for Automated Fingerprint Identification System (AFIS) applications.

In the early sixties, however, the number of searches that needed to be done on a daily basis simply became too large for manual indexing and AFIS started to be developed [139]. Just like the matching of *latent* prints that have been found at crime scenes [66], an AFIS still requires much manual labor because a search may result in many False Positives. In the 1980s, the development of cheap document scanning technology and personal computers enabled the use of fingerprint matching technology in everyday applications. Subsequently, the advent of several ink-less fingerprint scanning technologies coupled with the exponential increase in processor performance has taken fingerprint recognition beyond criminal identification applications. There are now a number of non-criminal, civilian applications such as access control, time and attendance tracking, and computer user login.

3.1.1 Acquisition devices

Fingerprinting for person identification had an advantage over most other biometrics in that fingerprint acquisition has been possible for centuries in the form of impressions of inked fingers on paper and direct impressions in materials like clay.

Over the last decade, many novel techniques have been developed to acquire fingerprints without the use of ink. The basic principle of these ink-less methods is to sense the ridges on a finger, which are in contact with the surface of the scanner. These scanners are known as "livescan" fingerprint scanners. The livescan image acquisition systems are based on four technologies:

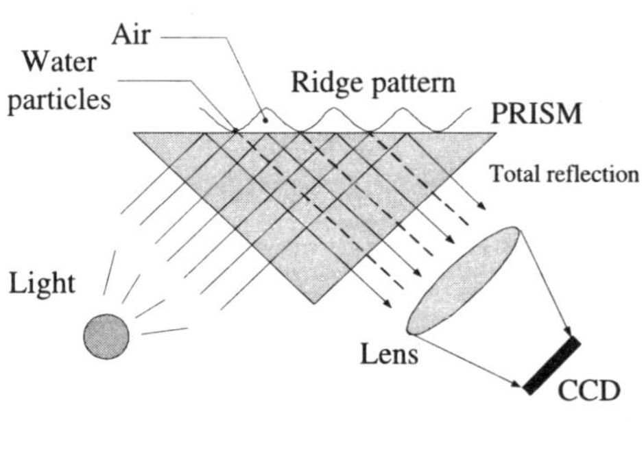

Figure 3.1: Optical fingerprint sensing by Frustrated Total Internal Reflection.

- *Frustrated total internal reflection (FTIR) and other optical methods* [71,82,83]: This technology is by far the oldest livescan method. As shown in Figure 3.1, a camera acquires the reflected signal from the underside of a prism as the subject touches the top of the prism. The typical image acquisition surface of 1" × 1" is converted to 500 dpi images using a CCD or CMOS camera. This technique for fingerprint image acquisition has been around for quite some time and was used, for example, in medical studies in 1966 [2]. Many variations of this principle, such as the use of holographic elements [135], are also available.

An issue with these reflection technologies is that the reflected light is a function of skin characteristics. If the skin is wet or dry, the fingerprint impression can be "saturated" or faint, respectively, and hard to process. These problems can be overcome to some extent by FTIR imaging using ultrasound instead of visible light, but the resulting system is bulkier.

- *CMOS capacitance* [104]: The ridges and valleys of a finger create different charge accumulations when the finger touches a CMOS chip grid. With suitable electronics, the charge is converted to an intensity value of a pixel. There are various competing techniques using AC, DC, and RF to do this. Unfortunately, these CMOS devices are sensitive to electrostatic discharge and mechanical breakage or scratching ("tap and zap").

  These devices normally image at 500 dpi and provide about 0.5" × 0.5" of fingerprint surface scan area. This can be a problem as two impressions of the same finger acquired at two different times may have little overlap (see fingerprint matching techniques described below). The images also tend to be affected by the skin dryness and wetness.

- *Thermal sensing* [128]: This sensor is fabricated using pyroelectric material, which measures temperature changes due to the ridge-valley structure as the finger is swiped over the scanner and produces an image. This works because skin is a better thermal conductor than air and thus contact with the ridges causes a noticeable temperature drop on a heated surface. The technology is claimed to overcome the dry and wet skin issues of optical scanners and can sustain higher static discharge. The resultant images, however, are not rich in gray values, i.e., dynamic range. The swipe-type of sensors are becoming more attractive because of their smaller footprint and lower cost. Swipe sensors based on optical and CMOS technology are also available commercially.

- *Ultrasound sensing* [10]: An ultrasonic beam is scanned across the finger surface to measure directly the depth of the valleys from the reflected signal. This can theoretically be implemented as a non-contact sensor. Skin conditions such as dry, wet, and oils do not affect the imaging and the images better reflect the actual ridge topography. However, these units still tend to be very bulky and require longer scanning time than the optical scanners.

However, there is a property of automatic fingerprint recognition systems that is not shared by many other pattern recognition systems, i.e., the process of sensing, or acquiring, the biometric involves *touching* some input device with the pattern *itself*. Because of this touch sensing, the actual pattern that is being sensed is distorted during the acquisition of the pattern. Figure 3.2 shows that this type of elastic distortion can be quite different for

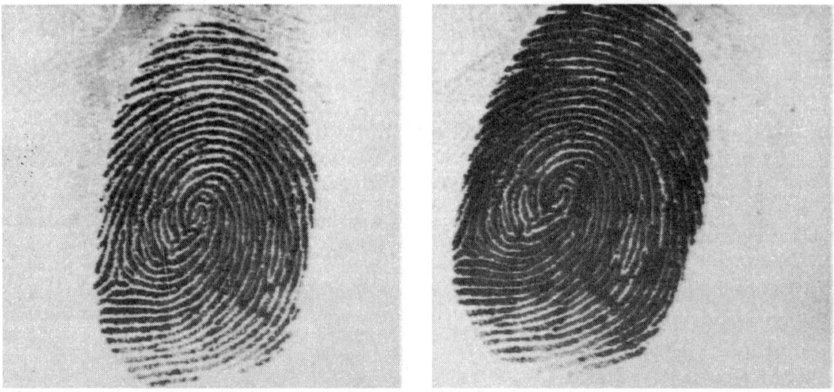

Figure 3.2: As these livescan fingerprint images show, impressions of the same finger can be quite different due to elastic distortion of the skin.

prints of the same finger—simply overlaying the images will not work. Recently, non-contact [50] fingerprint scanners have been announced that avoid problems related to touch sensing methods, including the elastic distortion of the skin pattern.

In general, a representation may be derived at the client end of the application or, alternatively, the raw image may be transmitted to the server for processing. Such transmission (and storage) of fingerprint images typically involves compression and decompression of the image. Standard compression techniques often remove the high frequency areas around the minutiae features. Therefore, a fingerprint compression scheme known as Wavelet Scalar Quantization (WSQ) is recommended by the FBI [26,67].

3.1.2 Matching approaches

A fingerprint authentication system reports some degree of similarity or some sort of "distance" (dissimilarity) between two fingerprint images and should report these measures accurately and reliably, irrespective of all the imaging problems discussed in the next section. Ideally, the similarity between two impressions (as in Figure 3.2) should be large, or equivalently, the distance between the images should be small. Hence, the similarity or distance between two impressions of the same finger should be invariant to (i) translation, (ii) rotation, (iii) the pressure applied, and (iv) elastic distortion between the impressions due to the elasticity of the finger skin.

Fingerprint matching has been studied over several decades by many researchers, see for example, [137,138,177,230]. Two broad classes of matching techniques can be distinguished:

- *Image techniques* (e.g., [38,78,103,122,176,189,210]): This class includes both optical as well as numerical image correlation techniques. Several image transform techniques have been also explored. These matching techniques will become important when the area of the finger that is sensed is small (e.g., as with CMOS sensors).

- *Feature techniques:* This class of techniques extracts interesting landmarks (features) and develops different machine representations of a fingerprint from these features. This is the most widely used approach to fingerprint matching, which we will discuss in more detail.

There exists a third class of algorithms for matching fingerprint images that combine the above approaches:

- *Hybrid techniques:* A third class of matching techniques [43,101,221,234] combines both image and feature techniques or uses neural networks in interesting ways to improve accuracy. For example, Hamamoto [79] describes an identification method based on Gabor filters. Jain et al. [101] present a matching algorithm that uses features such as responses to Gabor filters of different frequency and orientation.

3.1.3 Minutiae

Human experts use many details of the ridge flow pattern to determine if two impressions are from the same finger. Figure 3.3 shows a piece of thinned fingerprint structure with a few examples of these features: (a) ridge endings, (b) ridge bifurcations, and (c) an independent ridge ([9,66] give more complete listings, such as, lake, spur, crossover). Automated fingerprint matching algorithms attempt to match fingerprint impressions in a similar fashion. However, the most commonly used fingerprint features are only *ridge bifurcations* and *ridge endings*, collectively known as *minutiae*, which are extracted from the digitized print. Many matching algorithms do not even distinguish between bifurcations and endings because during acquisition and fingerprint image processing, depending on the amount of pressure exerted by a subject, a ridge ending may change into a bifurcation and *vice versa*. For example, for feature (d) in Figure 3.3 it is unclear if it is an ending or a bifurcation.

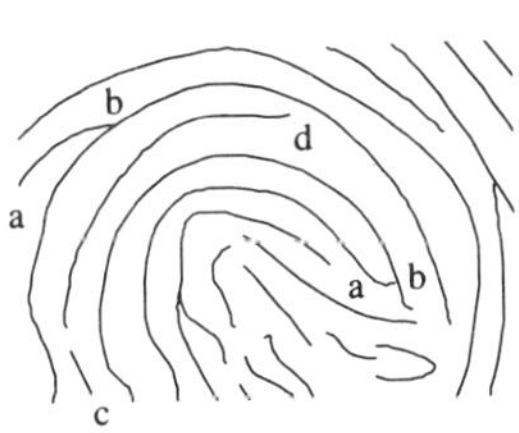

a: ridge ending
b: bifurcation
c: independent ridge
d: ambiguous ridge ending / bifurcation

Figure 3.3: Ridge patterns of individual fingers have minute details, known as minutiae, that distinguish one print from another.

The process of fingerprint feature extraction typically starts by examining the quality of the input image as discussed in Section 9.3 (see also, e.g., [239]). Virtually every published method of feature extraction (e.g., [129,180]) then proceeds by computing orientation of the flow of the ridges, which reflects the local ridge direction at each pixel. The local ridge orientation is then used to tune filter parameters for image enhancement and ridge segmentation. From the segmented ridges, a thinned image (Figure 3.3) is computed to locate the minutia features. Usually, a minutia post-processing stage cleans up several spurious minutiae resulting from either fingerprint imperfections (dirt, cuts), enhancement, ridge segmentation, or thinning artifacts.

X	Y	Θ
x_1	y_1	θ_1
x_2	y_2	θ_2
$\vdots$	$\vdots$	$\vdots$

Table 3.1: A minimal machine representation of a fingerprint image in terms of minutiae positions and orientations.

The machine representation of a fingerprint is critical to the success of the matching algorithm. A minimal representation of a processed fingerprint is a set $\{(x_i, y_i, \theta_i)\}$ of minutiae, i.e., a set of points (x_i, y_i), expressed in some coordinate system with a ridge direction at this point θ_i as in Table 3.1. Such a representation and point set matching is used in [183]. The representation used by Jain et al. [98] is a string and matching is performed through string matching algorithms.

Neither of these techniques takes into account the local topological information available in the fingerprint image. Graphs have been used in fingerprint analysis, primarily for fingerprint classification and matching. Isenor and Zaky in [93] use a graph representation, where nodes correspond to a ridge and edges to neighboring ridges or intersecting ridges. A graph matcher where nodes correspond to minutiae is presented in [184]. Other feature matching methods can be found in [205,212].

3.2 Face recognition

Face appearance is a particularly compelling biometric because it is one used every day by nearly everyone as the primary means for recognizing other humans. Because of its naturalness, face recognition is more acceptable than most biometrics. Since the advent of photography faces have been institutionalized as a guarantor of identity in passports and identity cards. Because conventional optical imaging devices easily capture faces, there are large legacy databases (police mug-shots and television footage, for instance) that can be automatically searched. The fact that cameras can acquire the biometric passively means that it can be very easy to use. Indeed, surveillance systems rely on capturing the face image without the cooperation of the person being imaged—either overtly or covertly—with consequent privacy concerns.

3.2.1 Imaging

Face recognition systems are often required to deal with a wide variety of image acquisition modes. NIST (National Institute of Standards and Technology) has proposed a recommended set guideline for face image acquisition [149].

- Single image: Optical methods include digitizing hardcopy documents using optical scanners. This is important because legacy data is mostly available in the form of still photographs, either black-and-white or color. Analog and digital cameras may also be used for live face image acquisition. Generally images are taken cooperatively (as in the case of driver's licenses) and under well-controlled lighting conditions in order to normalize the appearance of samples in the database.

- Video sequence: Surveillance cameras acquire video sequences, often including face images. Regular camera footage has proved to be not been very useful for face recognition because the spatial resolution is too low. Even using techniques such as hyper-resolution, where detail is built up by integrating successive frames, has not borne much fruit. This is because the frame rates for many surveillance systems are quite low (1–4 frames per second) and hence very few good images of a face are acquired from a moving target. Tracking techniques in conjunction with a pan-tilt-zoom (PTZ) camera might be used to improve the resolution by physically zooming in on suspected faces (at the cost of diminishing the overall field of view).

- 3D image: Many newer face recognition techniques are based on skin or skull geometry and require 3D images of the face instead of just a 2-D image. There are a number of techniques for acquiring such images including stereo, structured light, and phase-based ranging.

- Near infrared: One of the problems in face acquisition is related to obtaining robust images under poor lighting conditions. Low-power infrared illumination (invisible to the human eye) can be used to supplement the face detection process.

Facial thermograms [173] have also been investigated. While they have some attractions, they lack the compelling advantages of visible-light face recognition. A particular disadvantage is the relatively high cost of such non-standard sensors.

3.2.2 Local and global models

In general, face recognition systems proceed by detecting the face in the scene, thus estimating and normalizing for translation, scale, and in-plane rotation. Many approaches to finding faces in images and video have been developed; all based on weak models of the human face that model face shape in terms of facial texture.

Once a prospective face has been localized, the approaches then divide [28] into two categories:

- *Face appearance:* The underlying idea behind these approaches is to reduce a facial image containing thousands of pixels to a handful of numbers. The trick is to capture the distinctiveness of the face without being overly sensitive to "noise" such as lighting variations. To do this, a face image is transformed into a space that is spanned by basis image functions, just like a Fourier transform projects an image onto basis images of the fundamental frequencies. In its simplest form, the basis functions, known as *eigenfaces*, are the eigenvectors of the covariance matrix of a set of training images (see Figure 3.4).

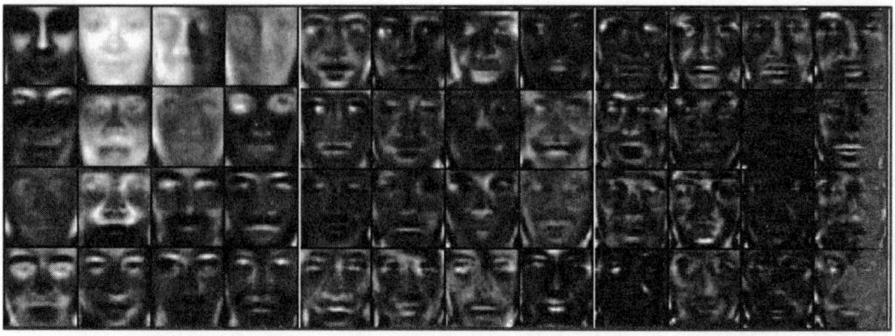

Figure 3.4: A face image decomposed as a sum of weighted eigenfaces; some consider the 1st eigenface (top left) "beauty" and the other eigenface deviations from beauty—caricatures.

- *Face geometry:* Here the idea is to model a human face in terms of particular face features, such as eyes, mouth, etc., and the geometry of the layout of these features. Face recognition is then a matter of matching feature constellations (see Figure 3.5). Automatic face recognition has a 30-year history; one of the first attempts was a feature-based system designed by Kanade [106].

Turk and Pentland [216] popularized the eigenface approach. Similar face image transforms which were introduced by Kirby and Sirovich [113,203] for representing and compressing face images. Kirby and Sirovich also developed a computationally efficient matrix computation of the transform. This development, and the publication of [216], resulted in a flurry of activities in face recognition research [37,194]. The underlying transform, originally known as the Karhunen-Loève transform [108,124], is now known by names like Principle Component Analysis (PCA).

Figure 3.5 shows an approach to face recognition based on face features. Panels a–e show the local face feature detection processes. The automatically detected geometric features are marked at the corresponding image positions on the face below. Features like the rim of the nose and the cheeks of the subject are detected and their geometric relationships are used for recognition of the face. Local face feature appearance models often are similar to the eigenface models for complete faces like shown in Figure 3.4. These are then "eigen-eyes," "eigen-noses," etc.

Indeed, there are many hybrid approaches. While appearance methods can be global [7,58,113,216] in that the whole face is considered as a single entity, they can also be local, where many representations of separate areas of the face are created. [199,213,235]. The work of von der Malsburg and colleagues (e.g., [235]) is one of the more sophisticated geometrical approaches, where a face images is represented by a graph of local appearance models. In many such face recognition systems geometry and appearance are combined. That is, to apply appearance-based methods in the presence of facial expression changes requires first generating an expressionless "shape-free" face by image warping.

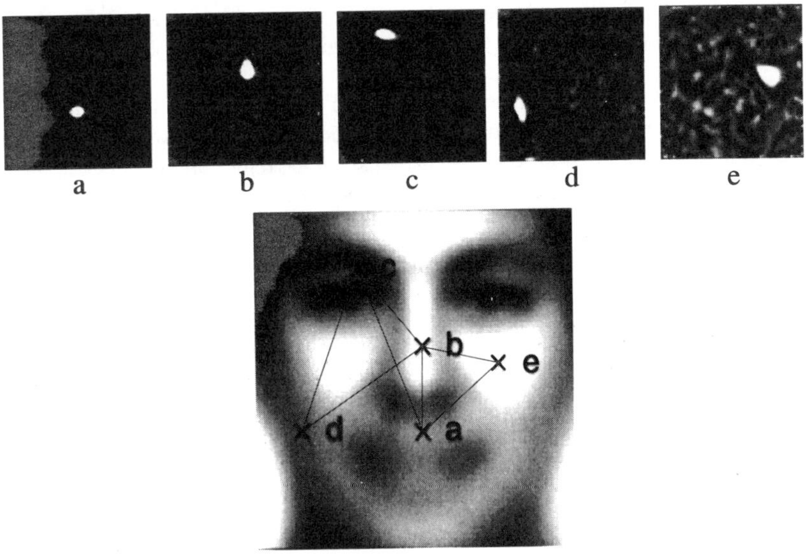

Figure 3.5: Feature-based face recognition. a–e: Local feature detection and localization. The image below shows the local features and geometric relations (courtesy J.J. Atick (Identix Incorporated) & P.S. Penev [159]).

3.2.3 Challenges

Considerable progress has been made in recent years, with much commercialization of face recognition. Yet, despite all this work, face recognition is not yet sufficiently accurate to accomplish the large-population identification tasks tackled with fingerprint or iris recognition. Much remains to be done towards the "general" face recognition problem. To date, only a limited amount of variation in head pose can be tolerated by 2D face recognition systems, and little progress has been made toward modeling and matching faces in 3D.

Determining the identity of two photographs of the same person is hindered by a number of problems. One particularly clear obstacle is the similarity of appearance of identical twins. In general the challenges for face recognition can be broken into four broad categories of variations that must be handled:

- *Physical appearance:* These include quick changes in facial expression change such as blinking, speech, emotional, and communicative facial expression. Also of importance are slow appearance changes because of aging and personal appearance changes. The latter may be due to the application of make-up, wearing of glasses, facial hair changes, changes in hairstyle, and intentional disguises.

- *Acquisition geometry:* In general the face in the image is in an unknown location, has an unknown in-plane rotation, and is of unknown size (scale). These are just the geometry changes when a person is looking straight in the camera. Rotations of a face in depth, i.e., facing the camera in profile or obliquely, introduce a host of differences of appearance of a face from one image to the next.

- *Imaging conditions:* The lighting of a human's face can have large effects on the appearance of the face in an image. Just consider, for example, frontal lighting compared to side illumination. Intrinsic camera characteristics and parameters may further change the appearance of a face in an image, quite independently of the ambient light. These camera characteristics include things like automatic white balancing, gain control, and noise reduction.

- *Compression artifacts:* These are often quite unexpected image degradations because of image compression (for transmission and/or storage) and subsequent decompression. The commonly used compression standards like JPEG and MPEG are based on the compression of image blocks, and are not particularly designed to preserve the appearance of the human face. This can significantly impact the performance of face recognition algorithms on archived data, such as legacy mug-shot databases and broadcast video.

No current face recognition system can claim to handle all of these problems well. In particular, there has been little research on making face recognition invariant with respect to aging effects. In general, to avoid the above problems, constraints on the problem definition and image capture situation are used to limit the amount of invariance that needs to be built into the algorithms. Even under the best imaging conditions, however, recognition error rates need to be improved. It has been reported that automated face recognition systems can be used for matching mug-shot face images to face appearances in a crowd. In practice, however, the performance is not yet high enough for a reasonable degree of automation without severely constraining the system. Such systems will require extensive human intervention for some time to come.

3.3 Speaker recognition

Speaker recognition [31,72] (sometimes referred to as"voiceprint recognition" or *voice recognition*) attempts to identify individuals by how they sound when speaking. Note that, although often sharing the same front-end processing, speaker recognition should not be confused with *speech recognition*, where it is the words, not the speaker, that must be determined. Like face appearance, speaker recognition is attractive because of its prevalence in human communication and human day-to-day use. We expect to pick up the phone and be able to recognize someone by his or her voice after only a few words, although clearly the human brain is also very good at exploiting context to narrow down the possibilities. That is, the scenario of the conversation provides sometimes very significant cues (Who is likely to call me ... at the office ... at this time ... from a cell phone?) about the identity of the speaker.

Voice is a behavioral biometric but is dependent on underlying physical traits, which govern the type of speech signals we are able and likely to utter. Properties like the fundamental frequency (a function of the vocal tract length which, in some systems, is explicitly estimated), nasal tone, cadence, inflection, etc., all depend in the identity of the speaker. A segment of speech is shown Figure 3.6.

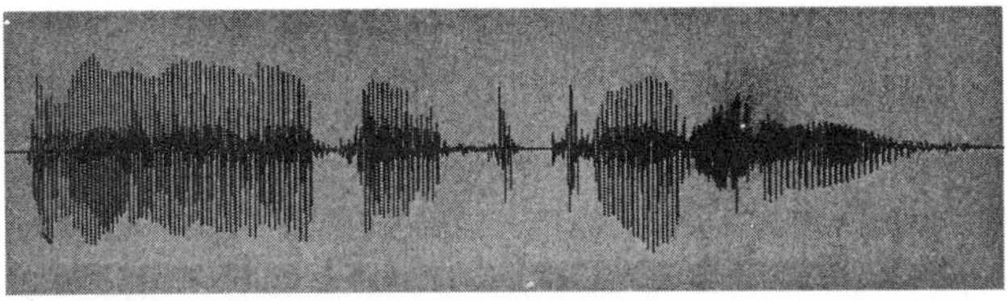

Figure 3.6: A segment of a voice amplitude signal (e.g., voltage measured across a microphone) as a function of time.

3.3.1 Application categories

We can categorize speaker authentication systems depending on requirements for what is spoken; this taxonomy, based on increasingly complex tasks, also corresponds to the sophistication of algorithms used and the progress in the art over time. In the list below we also highlight some of the vulnerabilities of each protocol. Speaker verification is particularly vulnerable to replay attacks because of the ubiquity of sound recording and playback devices (see Chapter 12).

- *Fixed text:* The speaker says a predetermined word or phrase, which was recorded at enrollment. The word may be secret, so it acts as a password, but once recorded a replay attack is easy, and re-enrollment is necessary to change the password.

- *Text dependent:* The speaker is prompted by the authentication system to say a specific thing. The machine aligns the utterance with the known text to determine the user. For this, enrollment is usually longer, but the prompted text can be changed at will. Limited systems (e.g., just using digit strings) are vulnerable to splicing-based replay attacks.

- *Text independent:* The speaker authentication system processes any utterance of the speaker. Here the speech can be task-oriented, so it is hard to record and replay speech that also accomplishes the imposter's goal. Monitoring can be continuous, and the more that is said, the greater the system's confidence in the identity of the user. Such systems can even authenticate a person when they switch language. The advent of trainable speech synthesis [53,206] might enable attacks on this approach.

- *Conversational:* During authentication, the speech is recognized to verify identity by inquiring about knowledge that is secret, or at least is unlikely to be known or guessed by an imposter. False Acceptance Rates below 10^{-12} are claimed to be possible by this combination of biometric and knowledge, making conversational biometrics very attractive for high-security applications [127,175].

Telephony is the main target of speaker recognition, since it is a domain with ubiquitous existing hardware where no other biometric can be used in a practical way at the moment. Increased security for applications such as telephone banking and "m-commerce" (commerce over the mobile phone) means the potential for deployment is very large. Speaking solely in order to be identified (an "active" authentication protocol) can be somewhat unnatural, but in situations where the user is speaking anyway (e.g., a voice-controlled computer system, or when ordering something by phone) the biometric authentication protocol becomes "passive" and unobtrusive.

Physical and computer security by speaker authentication has received some attention, but here it is less natural and poorer-performing than other biometrics. Speaker authentication is necessary for audio and video indexing. Where a video signal is available, lip-motion identification has also been used [55,116,125].

While traditionally used for verification, more recent technologies have started to address identification protocols, one particular domain being in audio and video indexing [6]. As noted above, a recent, interesting development [127,175], combining voiceprint recognition with the exchange of knowledge in an interactive authentication protocol (called *conversational biometrics*), can provide higher accuracy and is discussed further in Section 11.2.3 on dynamic protocols.

The advantages of speaker recognition are nicely summarized in [17] when used in text-independent mode (i.e., no constraints on the words to be spoken):

- Remote authentication over legacy phone lines is possible.
- Users do not have to remember passwords or pass phrases.
- Users do not have to go through a separate process for verification, since anything they say as part of the transaction dialog can be used to verify their identities, resulting in a truly integrated and non-intrusive verification process.

3.3.2 Acoustic features

One reason for the attractiveness of speaker recognition is the ubiquity and low cost of the sensor required to acquire the speech signal. Microphones are already present in many devices: telephones, cell phones, laptops, and may desktop computers. All of these can serve as the sensor for acquiring the speech signal.

Speaker authentication, however, suffers considerably from any variations in the microphone [84,187] and transmission channel. Performance typically deteriorates badly when enrollment and use conditions are mismatched (for example, enrollment over a land line and authentication over a cell phone). This, of course, inevitably happens when a central server carries out speaker authentication based on telephone signals. Background noise can also be a considerable problem in some circumstances, and variations in voice due to illness, emotion, or aging are further problems that have received little study.

To process the speech signal, the output of the microphone is first digitized. The next step in feature extraction is to separate the speech from the non-speech portions, such as silence, in the signal. After this, most speaker recognition systems extract some form of frequency-based features similar to those used in speech recognition systems. For instance, the use of short-term spectral analysis with 20-ms windows placed every 10 ms to compute

Fourier coefficients is typical. These magnitude spectra are then converted to cepstral features (a method for extracting the spectral envelope independent of the voicing signal). The cepstral features are further processed to compensate for channel mismatch before being used to generate or match models of individual speakers.

Matching techniques in speaker recognition vary significantly, as many of the features used in the representation are algorithm specific. Reynolds [185] classifies the matchers to four categories. In template matching, a fixed text utterance is used to generate a stored reference which is then compared to the newly acquired feature vector to generate a matching score. An optimization technique called *dynamic time warping* is used to obtain the best alignment between the two signals. In a variation, called nearest-neighbor matching, the match score is computed as the sum of distances between the query vector and the k nearest neighbors (reference templates) corresponding to the speaker's purported identity. Neural network based matchers essentially develop more precise and statistically accurate decisions boundaries but require extensive data driven training to discriminate the speakers. Finally, Hidden Markov Models (HMMs), a common technique in speech recognition, encode not only the feature vectors themselves but also the evolution of the features over the course of an utterance. They can also compensate for statistical variation of the features but, again, require large amounts of training data.

Misspoken or misread prompted phrases.
Extreme emotional states.
Time varying (intra- or inter-session) microphone placement.
Poor or inconsistent room acoustics (e.g., multipath and noise).
Channel mismatch (e.g., different microphones for enrollment and verification).
Sickness (e.g., head colds can alter the vocal tract).
Aging (the speech characteristics can drift away from models with age).

Table 3.2: A voice biometric can be "corrupted" by the physical and emotional state of the subject, as well as by environmental issues.

Of course speaker recognition is not suitable for all applications and, like face recognition, there are a number of known areas of difficulty. Campbell [31] enumerates a number of problems with voice recognition that we repeat in Table 3.2.

3.4 Iris recognition

The colored part of the eye bounded by the pupil and sclera is the iris, which is extremely rich in texture (see Figure 3.7). This has been purported to be a universal biometric identifier with good discriminating properties. Like fingerprints, the appearance of the iris is the result of the developmental process and is not dictated by genetics. Quoting Daugman [46]: "Just as the striking visual similarity of identical twins reveals the genetic penetrance of overall facial appearance, a comparison of genetically identical irises reveals that iris is a phenotypic feature, not a genotypic feature." Although iris [232,233] is a relatively new biometric, it has been shown to be very accurate and stable. Unlike fingerprints, there is no elastic distortion from one sample to the next (except pupil dilation). So far, in the literature, there have been only a few iris recognition systems described. Perhaps the most well-known iris recognition system is the one designed by Daugman [46,47].

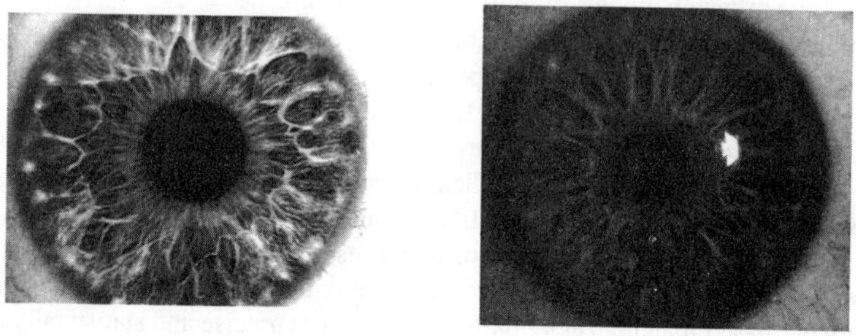

Figure 3.7: Iris images acquired under ideal circumstances (courtesy of J. Daugman, University of Cambridge, UK).

The design of an iris image capture device that is convenient and unobtrusive is a real challenge. Ideally it should be user-friendly and yet capture the iris image with minimal variation from one ambient lighting situation to the next. The iris image capture device should further be able to deal with specular reflection off the eyeballs, and with glasses and contact lenses (hard contact lenses create the most problems). To capture the rich texture present in iris patterns, Daugman recommends an imaging system with a minimum resolution of 70 pixels in iris radius. Most of the commercial systems use an iris radius of 100 to 140 pixels. CCD cameras (640 x 480 pixels) are then used to acquire the image in monochrome, since present feature extraction methods do not use the iris color.

Two different types of acquisition systems have been reviewed in [232]. One approach is first to find the human face in the image using, for example, stereo techniques as described in [232]. Then a high-quality image of the iris is acquired, for instance, by steering a PTZ camera to the detected eye position within the face. The majority of current commercial systems, however, require the user to position his or her own eyes within the field of view of a single narrow-angle camera. This is done by observing visual feedback via a mirror, which can be quite tedious for the user (and sometimes impossible for the non-dominant eye).

The first step in the feature extraction process is to locate the iris in the acquired image. This is typically done by estimating the center of the pupil, as well as the center of the iris, over an image pyramid of different scales until the estimates converge to single pixel precision. The isolated iris pattern is then demodulated to extract its phase information using quadrature 2D Gabor wavelets [47]. There are two advantages of using the phase: (i) phase is more discriminating than amplitude as amplitude depends on imaging contrast and illumination; and (ii) the phase angles are assigned irrespective of image contrast.

For each iris contrast independent 2K bits (256 bytes) are computed (see Fig 3.8). Another 2K bits representing a mask for the noisy areas of the image is also computed to improve matcher accuracy. The iris matcher then computes a normalized Hamming distance measure, a simple count of bit differences, between two iris templates. The normalization

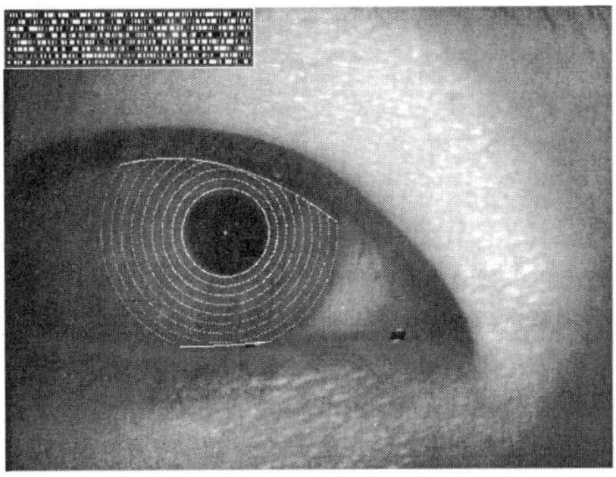

Figure 3.8: The iriscodes are extracted from concentric circular strips around the pupil (courtesy of J. Daugman, University of Cambridge, UK).

factor uses the mask to discount the areas where the images have been observed to be noisy. Since Hamming distance can be computed very fast, it is claimed that the same algorithm can be used for both verification and authentication.

3.5 Hand geometry

Hand geometry, as the name suggests, refers to the geometric structure, or ideally geometric invariants, of the human hand. Typical features include length and width of the fingers, aspect ratio of the palm or fingers, width of the palm, thickness of the palm, etc. [12]. To our knowledge, the existing commercial systems do not take advantage of any non-geometric attributes of the hand, e.g., color of the skin. Likewise, hand geometry does not involve extraction of detailed features of the hand (for example, wrinkles on the skin) which fall into the domain of palm print verification methods [240] or fingerprint recognition. Although the metrics in Table 3.3 do not vary significantly across the population, they can be used to authenticate the identity of an individual.

It is well known that the individual hand features themselves are not very descriptive and that hand geometry authentication has relatively high False Accept (FA) and False Reject (FR) rates. Devising methods to combine these non-salient individual features to attain robust positive identification is a challenging pattern recognition problem in its own right. Yet despite these error rates, hand recognition systems are surprisingly widespread, a testament to their user-friendliness.

Hand geometry authentication is attractive for a number of reasons. Almost all of the working population has hands and exception processing for people with disabilities could be easily engineered [241]. Hand geometry measurements are easily collectible and non-intrusive compared to, say, iris and retina. This is due to both the dexterity of the hand and a relatively simple method of sensing, which does not impose undue requirements on the imaging optics. As the computations are also fairly simple, a standalone system is easy to build. Further, hand geometry is ideally suited for integration with other biometrics,

Feature	Description	Feature	Description
F1	Width of thumb at second phalanx	*F2*	Width of index finger at third phalanx
F3	Width of index finger at second phalanx	*F4*	Width of middle finger at third phalanx
F5	Width of middle finger at second phalanx	*F6*	Width of ring finger at third phalanx
F7	Width of ring finger at second phalanx	*F8*	Width of little finger at third phalanx
F9	Length of index finger	*F10*	Length of middle finger
F11	Length of ring finger	*F12*	Length of little finger
F13	Width of palm at the base four fingers	*F14*	Width of palm at base of thumb
F15	Thickness of fingers at second phalanx	*F16*	Thickness of fingers at third phalanx

Table 3.3: An example feature set for hand geometry [102].

in particular, fingerprints. For instance, an authentication system may use fingerprints for more precise authentication and use hand geometry for less stringent authentication. It is easy to conceive a sensing system which can simultaneously capture both fingerprints and hand geometry.

Hand-geometry-based verification systems have been available since the early seventies. However, there is not much open literature addressing the research issues underlying hand geometry authentication; instead much of the available information is in the form of patents [61,94] or application-oriented description [136]. Notable exceptions are the prototype system described by Jain et al. [102], and the 3D hand profile identification apparatus used by Sidlauskas [201].

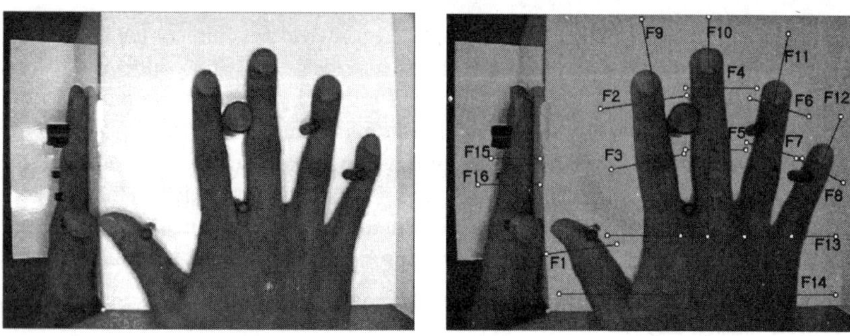

(a) (b)

Figure 3.9: (a) Hand geometry acquisition devices sense the top and the side of the hand, (b) Sixteen features used for hand geometry..

While it is conceivable that images of the hand can be more reliably captured using other modalities of sensing (e.g., thermal), the existing hand measurement acquisition systems typically rely on visual image(s) of the hand [102,201,241]. For example, Figure 3.9a shows a prototype of the hand geometry imaging system [102] capturing top and side views of the hand placed on a platen. The top view can be used for measuring lengths and widths of parts of hand while the side view can be used for assessing the thickness of the hand/fingers. The two views of the hand can be captured using a single camera by judicious placement of a single 45-degree mirror. Figure 3.9b shows a hand image from this device along with the positions of detected points (P_s and P_e) along each of 16 axes and the corresponding feature vector. In order to achieve consistent placement of the hand across different presentations of a hand, the hand placement platen may be mounted with hand/finger guiding pegs. The device may be further instrumented with additional sensors to verify whether the user has indeed placed the hand erroneously on the platen.

To enroll a person in the database, typically two snapshots of the hand are taken and the "average" of the resulting feature vectors is computed and stored. To match hands, a newly sensed feature vector is then compared with the feature vector stored in the database associated with the claimed identity. Jain et al. [102] explore four different distance metrics, (absolute, weighted absolute, Euclidean, and weighted Euclidean), corresponding to the following four equations:

$$\sum_{j=1}^{d} \mid q_j - r_j \mid \quad < \quad \epsilon_a \tag{3.1}$$

$$\sum_{j=1}^{d} \frac{\mid q_j - r_j \mid}{\sigma_j} \quad < \quad \epsilon_{wa} \tag{3.2}$$

$$\sqrt{\sum_{j=1}^{d} (q_j - r_j)^2} \quad < \quad \epsilon_e \tag{3.3}$$

$$\sqrt{\sum_{j=1}^{d} \frac{(q_j - r_j)^2}{\sigma_j^2}} \quad < \quad \epsilon_{we} \tag{3.4}$$

Here $Q = (q_1, q_2, ..., q_d)$ represents the query feature vector of the hand whose identity is to be verified, where the r_j's are features such as these listed in Table 3.3. Similarly, $R = (r_1, r_2, ..., r_d)$ represents the d-dimensional reference feature vector in the database associated with the claimed identity. The value σ_j^2 is the variance of the jth feature computed over all enrolled templates (a measure of the feature's importance). The verification is declared positive if the distance between R and Q is less than a threshold value: ϵ_a, ϵ_{wa}, ϵ_e, or ϵ_{we} for each respective distance metric. These sorts of scoring functions are common for many different biometrics. For the hand geometry application [102], the Euclidean weighted metric performed better than the other metrics.

3.6 Signature verification

Signature verification (e.g., [142]) is another biometric that has a long pedigree. It was in service before the advent of computers, and has seen wide usage in document authentication and transaction authorization in the form of checks and credit card receipts. Signature recognition is an instance of writer recognition, which has been accepted as irrefutable evidence in courts of law. Signatures also come in many forms (see Figure 3.10), thereby giving the signatory the ability to choose the "distinctiveness" and "uniqueness" of the signature, which will influence his or her FA and FR rate (see Chapter 9 on sheep and wolves).

The signature verification community also has an interesting way of looking on False Accepts in terms of defining the sophistication level of the forger (attacker), in categories

Figure 3.10: Signatures come in a many forms.

like *zero-effort forgery, home-improved forgery, over-the-shoulder forger,* and *professional forgery*. There is no such clear definition of level of sophistication of impersonating for any of the other biometrics.

However, as mentioned earlier, prerequisite characteristics for a biometrics are (i) universality; (ii) uniqueness; (iii) *permanence*, meaning invariance over time; and (iv) collectability. The characteristic of permanence of signature is questionable since a person can change his or her signature at any time. A person could even insist on a different signature for every day of the week. In a sense, the genetically and environmentally determined muscle dexterity of the hand is translated into a visual and machine-readable token. Hence, this biometric is (like face and voice) affected by illness, emotion, or aging—factors that have received little study. It is also not clear if any of the parameters estimated in signature verification correspond to underlying inherent physical attributes of the writer (which could not be forged).

In terms of technology, the natural division of automated signature verification is by the sensing modality:

1. **Off-line** or "static" signatures are scanned from paper documents, where they were written in the conventional way [168]. Off-line signature analysis can be carried out with a scanned image of the signature using a standard camera or scanner.

2. **On-line** or "dynamic" signatures are written with an electronically instrumented device and the dynamic information (pen tip location through time) is usually available at high resolution, even when the pen is not in contact with the paper.

The first comprehensive approach to static signature verification can be traced back to [141]. The problem of *writer authentication* (e.g., [88]) most frequently presents itself in the form of analyzing scanned paper documents and falls in this class. Approaches to writer authentication (and signature verification) are typically based on features such as *number of interior contours* and *number of vertical slope components* [207]. However, the lack of further information about the signature generation process makes static techniques very vulnerable to forgery.

The development of dynamic signature capture devices (see, e.g., [57]) resulted in much activity because a notion of time beyond two-dimensional space (paper) was introduced, for instance, $x(t)$ and $y(t)$, the location of the pen while signing. Sometimes even more sophisticated sensors are used. These devices record a stream of five-dimensional vectors $(x, y, p, \theta_x, \theta_y)$ sampled at equidistant time points. Here p is the axial pen force while θ_x

and θ_y describe the angles of the pen with the X-Y plane [52]). This additional information, beyond just position, is particularly useful in preventing forgery. Approaches to dynamic signature verification include measuring Euclidean distances between pen trajectories, regional correlation matching measures, and probabilistic temporal recognizers such as Hidden Markov Models (e.g., ([52]).

The strength of dynamic signature recognition, the richness of its representation, is also its weakness since special hardware is required to capture this information. Yet signatures are often captured electronically already, as simple bitmaps merely to reduce paper storage and transport. Moreover, the volume of signature authorized transactions today is huge, making automation very important. Recently, approaches for pen location and orientation estimation using visible light have been developed [140]. This may eventually lower the cost of signature acquisition and may even lead to three-dimensional signatures.

4

Additional Biometrics

Advances in sensor technology and an increasing demand for biometrics is driving a burgeoning biometric industry to develop new technology. In the preceding chapter we have described the principal biometrics that have received most attention and which have been most widely deployed. However, as commercial incentives increase, many new technologies for person identification are being developed, each with its own strengths and weaknesses and a potential niche market. A recent book [96] includes papers describing most of the less well-known biometrics. We here review these other biometrics, though we do not claim that this list is exhaustive, and we do not attempt to cover several techniques that are unlikely to leave the domain of forensic investigation, such as footprint, bite mark, lip print, and hand print. (The latter can include both hand-geometric aspects and analyses of minutiae and creases as in fingerprints. Many police authorities keep databases of hand prints as well as fingerprints.) For completeness, however, we begin by briefly describing the Bertillon system that began the modern science of biometrics.

This biometric system was developed by Alphonse Bertillon (1853–1914) for use in the Paris prison service. It consisted of a number of body measurements, such as a person's height and arm length and the length and breadth of the skull measured with calipers. The system proved useful [214] but ultimately fell into disrepute and was superseded by fingerprints. One critic was Galton, who greatly advanced the study of fingerprints:

> There was...a want of fullness in the published accounts of it, while the principle upon which extraordinary large statistical claims to its quasi-certainty had been founded were manifestly incorrect, so further information was desirable. The incorrectness lay in treating the measures of different dimensions of the same person as if they were independent variables, which they are not. For example, a tall man is much more likely to have a long arm, foot, or finger, than a short one. The chances against mistake have been overrated enormously owing to this error; still, the system was most ingenious and very interesting. [73]

The discovery in 1903 of two William Wests in Leavenworth Prison, with almost identical Bertillon measurements was final proof of the system's fallibility [111].

The Bertillon system included many other attributes (such as scars and moles) and interestingly included classification of ear shape (Section 4.6 below) and irises (Section 3.4).

4.1 DNA

DNA identification [144,145] is often cited as the ultimate biometric, since DNA codes identity information in a digital form that is available from every cell in the body. Its biggest theoretical drawback is the fact that identical twins will have the same DNA, and in practical terms it is still a slow (taking days or weeks), expensive, and complex operation to compare identities using two DNA samples (see Figure 4.1). Thus far its use as a biometric for identification is limited to forensic science, which DNA identification has revolutionized, though of course the application of DNA to determine whether people are related (for identification, paternity determination, or genetic studies) is very similar, and there are many other uses for DNA testing.

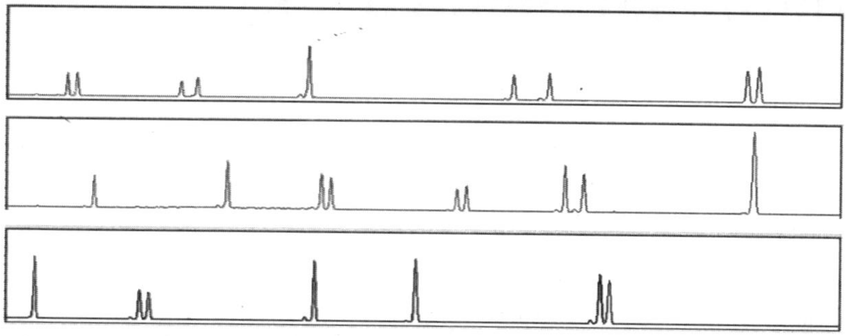

Figure 4.1: Electropherogram results of human DNA amplified using the AmpliType® Profiler Plus loci (D3S1358, vWA, FGA, Amelogenin, D8S1179, D21S11, D18, S51, D5S818, D13S317, D7S820) and detected using the Applied Biosystems Prism® 310 Genetic Analyzer. DNA matching is not completely automatic and involves hard to automate manual labor.

The basis of DNA identification is the comparison of alternate forms (alleles) of DNA sequences found at identifiable points (loci) in nuclear genetic material. Although around 99.5 percent of human genetic material is common between all people, this still leaves many loci that can be used for identification, and a useful number of loci with large numbers of alleles have been identified. For identification, a set of loci is examined to determine which alleles are present at each (usually there will be two different alleles for the locus — one on each chromosome). Any difference between enrolled and test samples indicates a difference of identity, while consistent identity between the samples indicates identity, or a coincidence whose probability can be determined. Typical tests give estimates of the order of one in a trillion of such a coincidence, with the proviso that close relatives (particularly identical twins) can be excluded for other reasons. (For example, consider a locus with 50 different alleles; a person can have 1,275 possible pairs of such alleles. If four independent loci are examined there are 2.6 trillion different possible combinations.) Of course strict procedures must be carried out to guarantee the purity of the sample and make sure that it does not get relabelled during the complex operations involved in sending the DNA for sequencing.

Privacy is an important issue in DNA identification, since DNA encodes information that can be put to other purposes besides identification, such as information about medical conditions and predispositions to certain diseases as well as race and paternity. The information about which allele is at each locus is of much more limited use and is less sensitive than the whole genetic sequence present in the original sample.

4.2 Retina recognition

Retina recognition seeks to identify a person by comparing images of the blood vessels in the back of the eye, the choroidal vasculature. Strictly speaking, when the image acquisition is done *only* in the near infrared, to which the retina is transparent, the retina itself is not imaged, though commercially available systems seem to also actively illuminate in the visual spectrum.

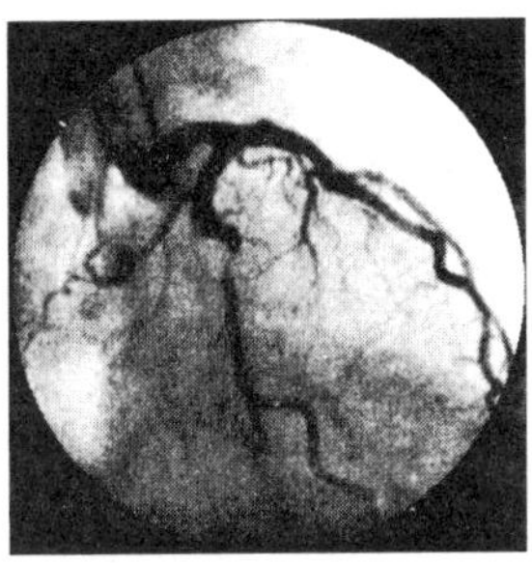

Figure 4.2: Image of the vein pattern of the human retina (courtesy of Buddy Boyet, EyeDentify Corporation).

Retina recognition can be traced back to the paper by Simon and Goldstein [202] which noted the individuality of the patterns of blood vessels (Figure 4.2) and has been developed commercially since the mid-1970s. Hill [85] provides an excellent overview of the subject and the technical developments of the EyeDentify system. EyeDentify [64] markets a standalone system for retina identification of up to 3,000 individuals. The template size is 96 bytes and identification among 1,500 people can be carried out in under 5 seconds. Hill notes that the system cannot reliably form images for people whose eyes suffer from strong astigmatism or very poor eyesight. The system requires the eye to be placed close to the sensor (0.75" in the production systems, up to 12" away in laboratory prototypes) and the sensors remain expensive compared to those for other biometrics. A great advantage of the retina for identification is the fact that it is a permanent structure, unaffected by anything but the most traumatic accident, and can be considered impossible to alter. Hill also claims that construction of a fake retina is extremely difficult because of the optical properties that must be simulated, and describes possible ways (such as a specialist challenge-response) to make such forgery even harder.

The EyeDentify system has been improved since Hill's overview [85]. According to the manufacturer [64], the current system uses a camera with an electromechanical sensor which measures from a short distance (< 3 cm) the natural reflective and absorption properties of the retina. The subject looks into the device with one eye. A 7-mW bulb illuminates the retina and the vein pattern is recorded with *visible* light and near infrared (NIR, 890 nm — see Figure 4.3).

4.3 Thermograms

A thermogram measures images in the various bands of the infrared spectrum of Figure 4.3, sometimes supplemented by visible spectrum imagery.

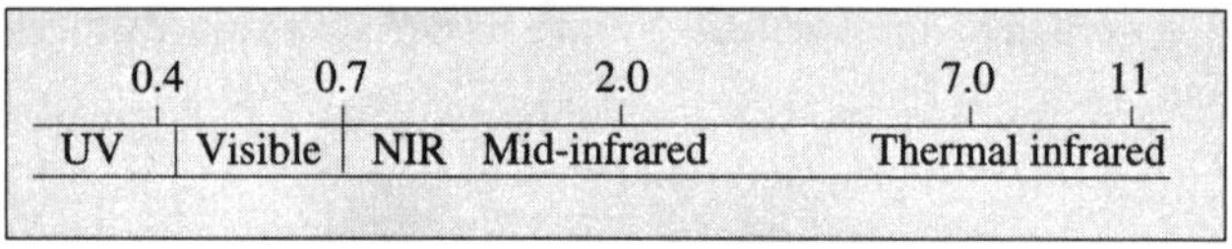

Figure 4.3: Spectral resolution is measured by the number of bands, their width, and their range within the electromagnetic spectrum (*in microns*).

That is, thermograms, for the purposes of biometrics, are images of parts of the body in the short (0.9–1.7 μm) mid (3–5 μm) or long (8–12 μm) infrared wavelengths (as in Figure 4.4). Early studies have been made, particularly of thermal images of the face and hands [173]. The great advantage of thermograms as opposed to visible light images for face recognition is their independence of ambient illumination. A facial thermogram can be captured when there is no illumination and the images are not subject to the appearance changes that lighting direction and shadowing cause and which plague unconstrained face recognition. Thermograms also offer robustness to disguise, or at least the ability to detect certain kinds of disguises.

Figure 4.4: Coregistered face images: a visible light image on the left, infrared images center and right (courtesy Larry Wolff [60]).

Equinox [60], a company specializing in novel imaging system design and sensor technology, is marketing a unique combined broadband visual and long wave infrared camera that produces images as in Figure 4.4. Equinox also is making available a thermal face image database (Chapter 13).

As with retina recognition (previous section), since the structures that are being (indirectly in this case) imaged are beneath the skin, it would be next to impossible to forge or modify them, and they are robust to aging and are unaffected except by traumatic accidents. A continuing problem, however, is the high cost of the sensors.

4.4 Gait

Gait recognition is a behavioral biometric that is still in its infancy. The strength of gait recognition is in its applicability to recognition of people at a distance in video images. Initial work has been carried out on identifying people by gait using motion capture equipment such as moving light displays or special markers. More recent research has extended to tracking people in less constrained video, though, for instance Nixon et al. [152] still use special clothing.

The recent DARPA "Human Identification at a Distance" program has spurred more research in the last two years [44] and researchers are beginning to understand how to recognize gaits from video images as shown in Figure 4.5. Many studies so far have been on small populations, and show susceptibility to variations such as ground surface, viewpoint, footwear, objects carried, as well as the differences in the person's speed and of course physical well being. Variations in clothing (particularly skirts, see Figure 4.5) also make the determination of gait more difficult.

A comprehensive study of recognizing people by their gate using a larger database (452 sequences from 74 subjects) can be found in [167]. A baseline algorithm is described that allows researchers to study this problem further. This algorithm first semi-automatically defines the bounding boxes of the walking subjects and then extracts silhouettes as shown in Figure 4.6. A third step is the scaling of the bounding boxes to 128×88 pixels to enable quick comparisons by "correlating" the silhouettes. Recognition performance depends very much on walking conditions; for example, it has been found to be difficult to recognize a person walking on one surface (say, concrete) when the system is trained with the person

Figure 4.5: Sample frames of a person walking in an indoor environment.

Figure 4.6: Sample silhouettes using the baseline algorithm (courtesy Kevin Bowyer and Sudeep Sarkar).

walking on another surface (say, grass). Much further study of this topic is needed; the authors of [167] made their database of videos available to the research community; see [217] and Chapter 13.

4.5 Keystroke

Keystroke identification is the identification of a person by their personal typing style. Each individual will have a characteristic typing ability and consequently the times between keystrokes and the hold time of keys in certain contexts should show less variation for a given typist than across the population in general. Obaidat and colleagues [153]

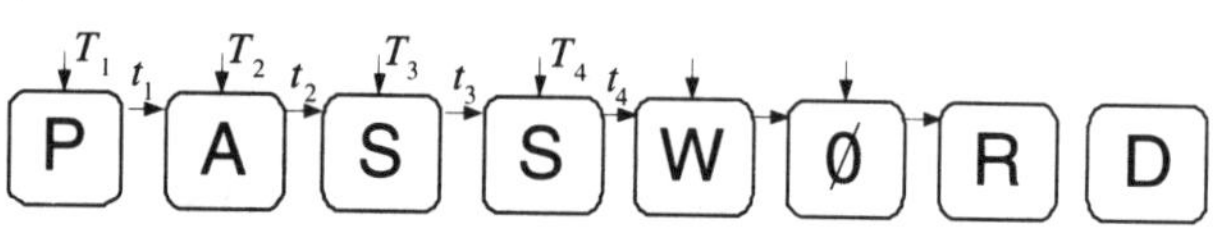

Figure 4.7: The sequences $T_1, t_1, T_2, t_2, ...$ vary with typing style.

have described a system using neural networks to discriminate between fifteen users. They have found that the discrimination is increased by using both the time intervals between keystrokes and the time that each key is held down (Figure 4.7). Such a system uses a fixed password, but could presumably be extended to be text independent in a way similar to speaker identification. Bergadano et al. [8] recently surveyed keystroke authentication. There is also a commercial product based on keystroke timing [147].

4.6 Ear recognition

Ears were originally a part of the Bertillon system of human measurement, being classified by trained observers into one of a few different types for use as one index in a larger biometric system. There is anecdotal evidence of latent ear prints being used to identify criminals, but in recent years some attention has also been paid to using ear shape as an automated biometric. One group of researchers [29,30] uses edge-finding techniques to extract from images the principal structures of the ear for comparison. These authors have also investigated using thermograms (i.e., infra-red images) for invariance to illumination and robustness to occlusion by hair.

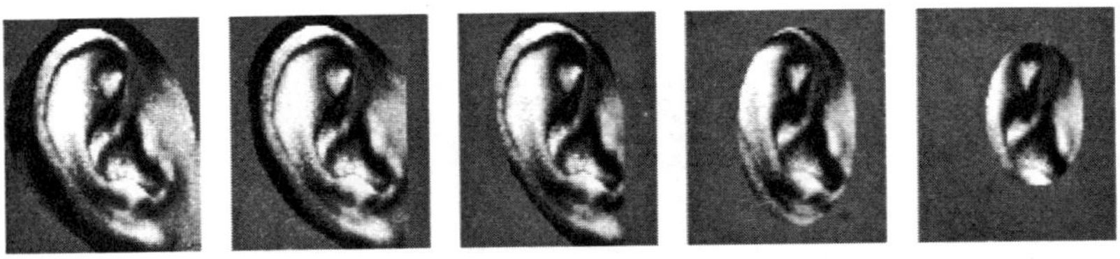

Figure 4.8: Five different ear mask sizes considered in tuning results (courtesy Kevin Bowyer and Sudeep Sarkar [219]).

There has been additional interest in ear shape recognition as a biometric that might be recognized at a distance in the Human Identification at a Distance (HID) project [44]. Resulting work can be found in, e.g., [219]; principal component analysis of ear images similar to the eigenface approach for face image recognition (see Section 3.2) is applied. The recognition results using the ear alone are not as good as those using the face alone, using combined face-plus-ear images offers improved performance. Figure 4.8 shows the five different levels of masking of ear images that were tried.

4.7 Skin reflectance

One example of a new biometric enabled by new sensors is skin reflectance. This uses a small chip developed by Lumidigm, Inc. [126] to measure the near-infrared light reflected by the skin in a range of wavelengths over a 6mm patch. While it can be used by itself, one potential application of this is in combination with a fingerprint sensor to make forgery harder. Sampling in a small area means that the chip can be small both physically and in terms of memory and processing power, and there is no registration problem as must be surmounted in fingerprint recognition.

4.8 Lip motion

Lip motion is a behavioral biometric that compares the characteristic lip motions of people as they speak. It can be seen as a visual counterpart to speaker identification, and could similarly be classified as fixed text, text-dependent, or text independent. Research has been spurred in recent years by the availability of the M2VTS (Multi Modal Verification for Teleservices and Security applications) databases [63], and a biometric product using lip motion has been marketed by BioId [89]. One of the most desirable characteristics of lip motion is the ease with which it can be combined with speaker identification and face recognition to make an accurate system which is extremely difficult to fake. This triple biometric system is typically intended for physical access control, with a user speaking in front of a microphone and video camera. The video image is used for both face recognition and lip motion recognition and the answers integrated with the speaker identification result. Any inconsistency in the identities or in the synchronization of the speech with the video can be used to detect replay-style attacks.

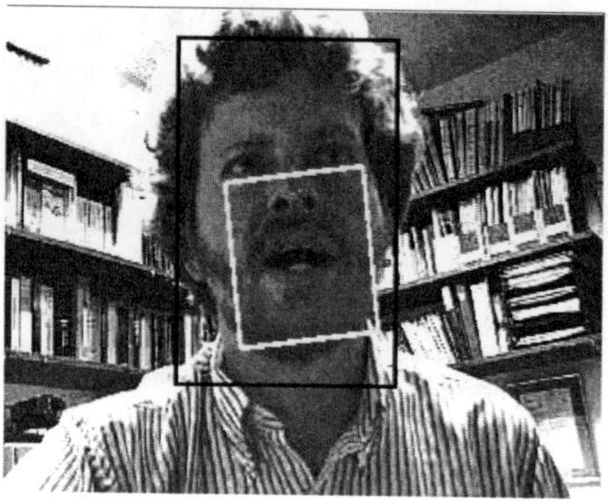

Figure 4.9: Under "reasonable" lighting conditions, the mouth and its expressions can be detected.

With a certain amount of control of the imaging and ambient light conditions, lip motion can be fairly well extracted (Figure 4.9). However, when the imaging conditions are less favorable, finding the mouth in a face video is a difficult task in the visual light. The non-visible bands of Figure 4.3, can also be used for lip-motion imaging:

1. *Infrared:* If security is of high concern, it may become cost effective to add thermal images as in Section 4.3.

   The same is true for gait and ear.

2. *Near infrared:* When the solution needs to be cheap, it may be worthwhile to use active, controlled NIR illumination as in [16].

   This particular type of active sensing, as with almost all attempts at controlling the imaging, make the act of sensing *detectable* and so less useful in covert $1{:}N$ screening.

4.9 Body odor

It has long been known that people can be identified from their individual odors, and this fact has been exploited in the use of dogs to track people. Recent advances in semiconductor-based chemical analyses have lead to the development of "electronic noses" that can measure the concentrations of a spectrum (as many as 32 [160]) of different chemicals. Such sensors do not yet have the range nor the sensitivity of the human nose, and suffer from a host of problems, such as the need for calibration and the risk of permanent drift or "poisoning." It is also clear that personal odor is susceptible to all manner of influences, from diet and state of health to usage of soaps, perfumes, and deodorants, and it is not yet clear whether these factors can be normalized away well enough to allow reliable identification of individuals.

Part II

Performance and Selection

5

Basic System Errors

The specifications for a biometric authentication system will usually include requirements like maximum allowable error rates. There are a number of biometric error types, expressed as error rates or error percentages, that need to be understood before a solution is designed and before a particular biometric is selected. Some of these errors are inherent with biometric authentication being just a kind of pattern recognition application; other errors are more specifically related to biometric authentication systems. What is clear is that any biometric authentication system will make mistakes, and that the *true* value of the various error rates for a matcher cannot be computed or theoretically established; it is only possible to obtain statistical estimates of the errors using test databases of biometric samples.

In this chapter, we present both intuitive and theoretical meanings of the various error types that are found in the biometrics literature, concentrating mainly on the errors made in a verification engine. A verification engine is a simple kind of biometric matcher that makes a 1:1 match decision based on a score s. In Chapter 6 we expand our analysis to identification systems, biometric search engines that make $1\colon m$ match decisions. In this chapter, we define both the problem of matching biometric samples and the problem of checking the credentials of a subject for biometric authentication in terms of hypothesis testing. An error is made when the wrong hypothesis is accepted as the true one. The definition of the errors therefore depends on the formulation of the hypotheses.

5.1 Matching

A matcher is a system that takes two samples of biometric data and returns a score that indicates their similarity. In a verification system we use this score as a means of determining whether the two biometric samples are from the same original "real-world" biometric.

Before going further, we must define what we mean by a matcher. For this we introduce some notation. Two real-world biometrics (e.g., two fingers or two faces) are denoted by $\mathcal{B}$ and $\mathcal{B}'$ and associated machine representations of these biometrics are denoted by $B = f(\mathcal{B})$ and $B' = f(\mathcal{B}')$, where f represents the process of sampling the data with a sensor and, perhaps, applying some processing to extract features B and B'. Unfortunately, the real-world biometrics $\mathcal{B}$ and $\mathcal{B}'$—the *actual* subjects—are functions of time, and

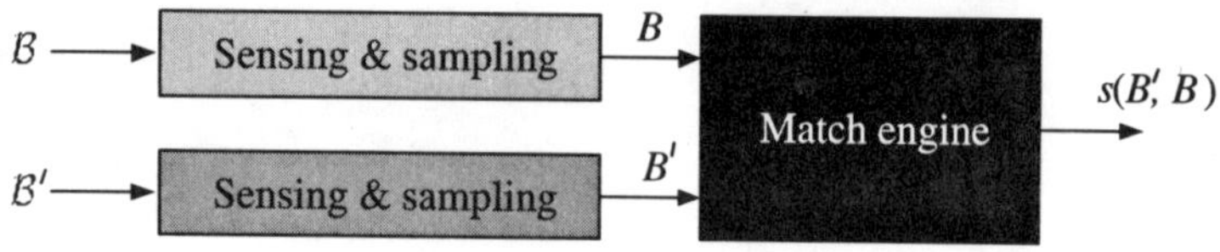

Figure 5.1: A match engine determines the similarity score s between a sample B of $\mathcal{B}$ and sample B' of $\mathcal{B}'$.

perhaps more importantly, the sensing function f is also a function of the time the sensing takes place, and other environmental factors; so we indicate this variability by writing f_t. Therefore biometric representations B and B' are functions of time, i.e.,

$$B = B(t) = f_t(\mathcal{B}(t)) \quad \text{and} \quad B' = B'(t') = f_{t'}(\mathcal{B}'(t')).$$

Thus we see that the much touted property of *uniqueness* of a biometric $\mathcal{B}$ and the "uniqueness" of samples or representations B of this biometric is in practice not as valid as it is often claimed to be. This degree of uniqueness over time varies a lot from one biometric identifier to the next (e.g., from finger to face) and therefore the *scalability* (Section 8.1.3) properties of the various biometrics are very different.

Biometric match engines make a decision by computing a measure of the likelihood that the two input samples from two persons (*Subject* 1 and *Subject* 2) are the "same" and hence that the subjects are the *same* real-world identity. This measure is typically an algorithmically defined similarity measure. This similarity measure is highly dependent on the precision of the acquisition device and highly dependent on the precision of the machine representation of the biometric samples. However, as we will see, if the similarity measure is able to capture nuances in biometrics that differentiate one person from the next, this similarity should then be related to the match probability $Prob\,(\text{subject}_1 \equiv \text{subject}_2)$, in some monotonic fashion.

Nevertheless, quite independent of the type of biometric identifier, biometrics in its most general form proceeds by computing a score, $s(B', B)$, as shown in Figure 5.1, with $s(B', B)$ the similarity of samples B' and B and hence the similarity of $\mathcal{B}'$ and $\mathcal{B}$. Put in the simplest way, a match engine takes $\mathcal{B}'$ and $\mathcal{B}$ as input and computes a score:

$$s(B', B) = s\left(B'(t'), B(t)\right) = s\left(f_{t'}\left(\mathcal{B}'(t')\right), f_t\left(\mathcal{B}(t)\right)\right). \tag{5.1}$$

Typically B is the enrolled sample (at time t)—it could be a biometric template—and is rarely changed unless desired for specific reasons, and B' is the live query sample. This score $s(B', B)$ only expresses some sort of likelihood that the true biometrics $\mathcal{B}'$ and $\mathcal{B}$ are the same or "match." Of course, if we have a probabilistic matcher, $s(B', B)$ may be derived from an estimate of $Prob\,(\mathcal{B}' \equiv \mathcal{B})$ itself, which gives desirable properties, but in general we assume little about the scores, except that they are monotonically increasing with probability, i.e., a higher match score $s(B', B)$ means that two biometrics are more likely to come from the same $\mathcal{B}$.

An alternative way to compute match scores is to determine distances, or dissimilarities, $d(B', B)$ between the samples B' and B. Such distances might be determined as "edit distances" (e.g., [98]) from one biometric pattern to the next, or distances between exemplars in some vector space of biometric features. These could be translated to a similarity score $s(B', B)$ by taking, for example,

$$s(B', B) \sim \exp\{-d(B', B)\},$$

which maps the $[0, \infty)$ distance range to a $[0, 1]$ similarity score range.

5.1.1 Two kinds of errors

Most error rates of a biometric authentication application are determined by the accuracy with which the internal biometric match engine can determine which one of the following two hypotheses is true. Given two biometric samples, we can construct two possible hypotheses:

$$\begin{array}{lll} \text{The } null \text{ hypothesis :} & H_o \Rightarrow & \text{the two samples match;} \\ \text{The } alternate \text{ hypothesis :} & H_a \Rightarrow & \text{the two samples do } not \text{ match.} \end{array} \quad (5.2)$$

There are different biometric applications that have different definitions of the hypotheses H_o and H_a, the decisions that the biometric application can make, and therefore different applications can have different definitions of errors. Consequently, there is much terminology around that expresses the accuracy of an application, such as *False Match Rate, False Accept Rate, False Positive Rate,* etc. In this chapter and the next we define various terms for the error rates found in the literature and describe the underlying applications that give rise to the definitions.

We define the error rates of a biometric matcher according to its correctness in deciding between *two* outcomes as in (5.2). The match engine either decides H_o is true or decides H_a is true. Consequently, there are two types of errors that a matcher can make. Here we use terminology introduced by Wayman [227], since this terminology nicely allows for the differentiation between a *biometric matcher* and a *biometric application:*

I. *False Match* (FM): Deciding that two biometrics are from the same identity, while in reality they are from different identities; the frequency with which this occurs is called the *False Match Rate* (FMR).

II. *False Non-Match* (FNM): Deciding that two biometrics are *not* from the same identity, while in reality they *are* from the same identity; the frequency with which this occurs is called the *False Non-Match Rate* (FNMR).

These are also called *Type I* and *Type II* errors, respectively; a Type I error is erroneously deciding H_o (i.e., when H_a is true), and a Type II error is erroneously deciding H_a (when H_o is true). These errors and this terminology, the False Match (FM) and the False Non-Match (FNM), are defined solely in terms of the hypotheses of (5.2).[1]

Next to the erroneous decisions that can be made, we also have the correct decisions: a *Correct Match* and a *Correct Non-Match* corresponding to correctly deciding that two biometric samples match and correctly deciding that the samples *do not* match, respectively.

[1] We later introduce the well-accepted terms *False Accept* and *False Reject* (and *False Positive* and *False Negative*) for biometric applications or *scenarios*.

5.1.2 Score distributions

The hypotheses as formulated in Expression (5.2) simply translate to

$$\begin{array}{ll} H_o: & \mathcal{B}' \equiv \mathcal{B} \\ H_a: & \mathcal{B}' \neq \mathcal{B}, \end{array} \tag{5.3}$$

and to decide between the two hypotheses, the score $s(R', R)$ is computed. The subsequent *"YES/NO"* decision is then often based on some fixed threshold T:

$$\begin{array}{ll} \text{Decide } H_o \text{ is true}: & \text{if } s > T, \\ \text{Decide } H_a \text{ is true}: & \text{if } s \leq T. \end{array} \tag{5.4}$$

The decisions of (5.4) are so-called "hard" decisions, in that the application is not allowed to make a *"don't know"* or *"too close to call"* decision. This is called decision making without *exception handling*.

The reliability of the score in comparing two biometric samples is influenced by many factors. There is variability in the live real-world biometric input signal $\mathcal{B}$, there is variation from sensor to sensor, and there is much variability in the sampling process $R = f(\mathcal{B})$. The latter is mainly due to variability in the way a subject presents the real-world biometric $\mathcal{B}$ for sampling. This variability is much higher from one sampling to the next as compared to (say) the variability for authentication methods like password entry, where the error is generally only due to input (typing) error. Rather, when we have two samples from an identical biometric (i.e., $\mathcal{B}' = \mathcal{B}$), the similarity score $s(R', R)$ is rarely 1 (or whatever score represents a perfect match) except when the two samples are copies of each other. Similarly, when we have two biometrics from two different individuals ($\mathcal{B}' \neq \mathcal{B}$), the similarity score $s(R', R) \neq 0$ (or whatever is the lowest possible score). What we can say is that when $\mathcal{B}' = \mathcal{B}$, the match score is usually high, while when $\mathcal{B}' \neq \mathcal{B}$, the score is usually low. This is shown in Figure 5.2, where the probability density $p_n(s)$ of the non-match scores (left) and the $p_m(s)$ of the match scores (right) are depicted. The non-match or *imposter scores* corresponding to $\mathcal{B}' \neq \mathcal{B}$ tend to be low and the match or *genuine scores,* for $\mathcal{B}' = \mathcal{B}$, tend to be high.

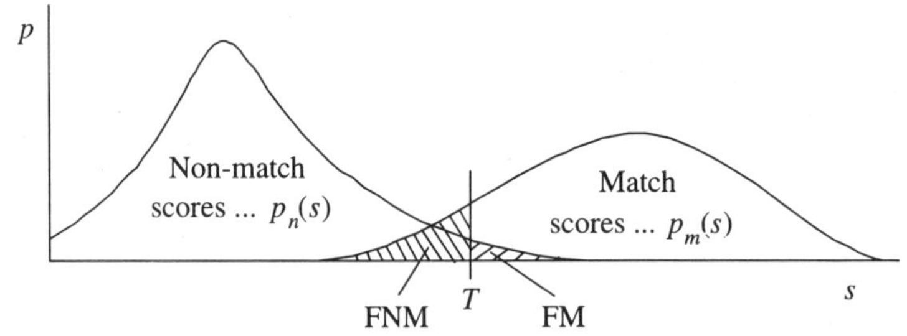

Figure 5.2: The non-match scores are on average lower than the match scores; in this case, the threshold T is set high to minimize False Accept.

We can now describe the two errors that can be made more precisely:

I. *False Match* (FM): Decide $\mathcal{B}' = \mathcal{B}$ because $s(R', R) > T$, when in fact $\mathcal{B}' \neq \mathcal{B}$ is true in reality. Accept H_o while in reality H_a is true.

An imposter has somehow generated a high ($s > T$) non-match score, i.e., a subject $\mathcal{B}'$ has impersonated a subject $\mathcal{B}$.

II. *False Non-Match* (FNM): Decide $\mathcal{B}' \neq \mathcal{B}$ because $s(R', R) \leq T$, when in reality we have $\mathcal{B}' = \mathcal{B}$. Reject H_o and accept H_a when H_o is true in reality.

A genuine subject has caused a low match score s less than T, something that easily happens if biometric $\mathcal{B}$ is poorly presented and hence poorly recorded and digitized.

The *False Match Rate* (FMR) and the *False Non-Match Rate* (FNMR) are then the frequencies at which FM and FNM occur, respectively:

1. **FMR:** This is the proportion of the time that a biometric sample R' matches R, when $\mathcal{B}' \neq \mathcal{B}$.

   From a probabilistic point of view, a random score s is drawn from the non-match distribution $p_n(s) = p(s|H_a)$ of Figure 5.2 and $s > T$. The FMR is the area of the hatched region on the right under the non-match (mismatch) score density curve $p_n(s)$, the proportion of the time $s > T$ when $\mathcal{B}' \neq \mathcal{B}$. We get the FMR as a function of T:
$$\text{FMR}(T) = 1 - \int_{s=T}^{\infty} p_n(s)\,\mathrm{d}s. \tag{5.5}$$

2. **FNMR:** The proportion of the time that a biometric sample R' does not match R, when $\mathcal{B}' \equiv \mathcal{B}$.

   A random sample s is drawn from the match distribution $p_m(s) = p(s|H_o)$ of Figure 5.2 and $s \leq T$, which means that the FNMR is the area of the hatched region under the genuine score density. This area is the proportion of the time $s \leq T$ when $\mathcal{B}' \equiv \mathcal{B}$
$$\text{FNMR}(T) = \int_{s=-\infty}^{T} p_m(s|H_a)\,\mathrm{d}s. \tag{5.6}$$

Unfortunately, for biometric applications, the non-match score distribution $p_n(s)$ and the match score distribution $p_m(s)$ in Figure 5.2 always overlap, so it is not possible to choose a threshold for which both $FMR = 0$ and $FNMR = 0$. Therefore the threshold T in decision rule (5.4) needs to be selected in such a way that the biometric system operates in an "optimal" fashion. Choosing the threshold involves assessing the consequences of the two types of errors. In a first approximation, a biometric matcher needs to be "tuned" to operate at acceptable False Match and False Non-Match rates for a given population of subjects. This threshold can only be determined through a process of training and testing where some more-or-less representative sample of the user population is available. In Figure 5.2, an operating point T is selected so that the FM rate is less than the FNM rate.

The threshold T controls the trade-off between $\text{FMR}(T)$ and $\text{FNMR}(T)$ as expressed by the Receiver Operating Characteristic (ROC) curve (see, e.g., [161]) discussed in Section 5.2. A matcher operating at a high threshold T has a low FMR but high FNMR; a low threshold, conversely, means high FMR and low FNMR.

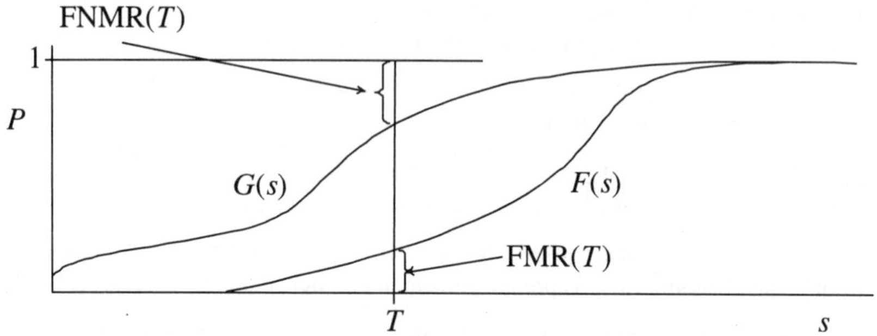

Figure 5.3: Cumulative probability distributions of non-match and match scores, $G(s)$ and $F(s)$; the $\mathrm{FNMR}(T)$ and $\mathrm{FMR}(T)$ at T are indicated.

The cumulative probability distributions $G(y)$ and $F(x)$ of the non-match scores and match scores respectively are defined as follows:

$$\begin{array}{lcl} G(y) & = & \int_{-\infty}^{y} p(s|H_a)\ \mathrm{d}s = 1 - \mathrm{FNMR}(y) \\ F(x) & = & \int_{-\infty}^{x} p(s|H_o)\ \mathrm{d}s = \mathrm{FMR}(x). \end{array} \tag{5.7}$$

These can be used interchangeably with FMR and FNMR because there is 1:1 correspondence, as shown in Figure 5.3. The notation "FMR" and "FNMR" is probably more accepted than probabilistic notation in terms of cumulative distributions (as $G(y)$ and $F(x)$). Note, however, that the function $\mathrm{FNMR}(y)$ is *not a distribution*.

5.1.3 Estimating errors from data

For the matcher of Figure 5.1, the output is a score $s(R', R)$, which is somehow related to the probability $Prob\,(\mathcal{B}' \equiv \mathcal{B})$. To estimate matcher $s(R', R)$ performance we need access to data and estimates of $G(y)$ and $F(x)$.

More precisely we, need a set of M match (genuine, suspect) scores **X** and a set **Y** of N non-match (imposter, legitimate) scores. We further need estimates of the true underlying match score distribution $F(s)$ and non-match score distribution $G(s)$. Estimates of the match distribution $F(x)$ is given by

$$\hat{F}(s) = \frac{1}{M} \sum_{i=1}^{M} \mathbf{1}\ (X_i \leq s) = \frac{1}{M}(\#\ X_i \leq s), \tag{5.8}$$

derived from the match scores $\mathbf{X} = \{X_1, ..., X_M\}$. The mismatch scores $\mathbf{Y} = \{Y_1, ..., Y_N\}$, on the other hand, give an estimate $\hat{G}(s)$ of the non-match score distribution:

$$\hat{G}(s) = \frac{1}{N} \sum_{j=1}^{N} \mathbf{1}\ (Y_j \leq s) = \frac{1}{N}(\#\ Y_j \leq s). \tag{5.9}$$

Again, these probability distribution estimates are called *empirical distributions*.[2]

Given the score densities as in Figure 5.2 one would like to compute the probability of an FM and the probability of a FNM, for some given score s. These probabilities cannot be computed exactly, but can only be estimated using test databases that represent the users and ideally, of course, test databases of the "enemies" of an application who try to gain access without being authorized. The true value of the False Match Rate of a matcher at some score s is hard to estimate, since even though there may be a lot of test data available, there is usually little or no data available about these forged or impersonated biometrics. Deliberate imposters and forgeries is an area of biometrics which needs much more study. Signature verification is the only area of biometrics where an imposter's forgery capabilities are taken into account, but even there the bulk of testing is against "zero-effort" forgeries. The FMR and FNMR for a given decision threshold T can only be estimated if the training data represents the target population well and there is enough of it. Generally training data will not accurately reflect the data seen in practical usage, so we can never really talk about *the* FMR and FNMR of a matcher, only estimates thereof made with a given set of data.

Along with estimates of the False Non-Match Rate and False Match Rate, ideally measures of confidence in these estimates of these errors should also be computed. This allows for the application to make an informed decision (using decision theory, risk analysis, etc.). If standard practices are followed, a biometric system is trained by using all $\mathcal{B}' \neq \mathcal{B}$ as imposters for $\mathcal{B}$. Consequently test data usually results in significantly fewer genuine scores than imposter scores, and the confidence in the FNMR is typically much lower than that in the FMR. However, the type of False Accept that is implicitly meant here are False Accept Rates associated with the zero-effort "attacks." The False Match Rates when a biometric system is attacked by intentional and professional forgers, imposters, and impersonators *will be much higher.* It is therefore important to note that all error trade-off studies implicitly assume some weak "*average*" type of imposters that do not try to impersonate true users.

Because the practical False Match Rate for any biometric will not be zero, repudiation and accountability are not perfectly guaranteed by using a biometric matcher. Here repudiation is the denial of a transaction by a subject. Since a password or a possession can be shared, a person can deny that they personally carried out a transaction authenticated by these means. A biometric "proves" that a particular person was present, but with some probability of error.

5.1.4 Error rates of match engines

Up to this point we have phrased the errors in terms of matching or not matching biometric samples, as in expressions (5.3), the two hypotheses a biometric matcher can decide between when using the decision rule of (5.4). For a core biometric matcher, we have labeled the errors associated with deciding the wrong hypothesis a False Match (FM) and a False Non-Match (FNM), following the terminology from Wayman [229].

On the other hand, one could use the more conventional pattern recognition terminology of *False Accept* (FA) and *False Reject* (FR); where "accepting" and "rejecting" refer to

[2] The empirical cumulative distributions $\hat{G}(y)$ and $\hat{F}(x)$ are sometimes used interchangeably with FMR and FNMR, which are also used to refer to $G(y)$ and $F(x)$, respectively.

the accepting and rejecting of the null hypothesis, *not* to the accepting and rejecting of a subject for accessing a location or an application. Hence in biometrics there are two *different* notions of accepting and rejecting:

1. *Accepting/rejecting* hypotheses H_o and H_a from a pattern recognition point of view as we have discussed up to this point.

2. *Accepting/rejecting* a subject for access to an application based on biometric authentication, which we discuss in this section.

A biometric application distinguishes itself from the simple exercise of testing match hypotheses using a matcher as in Figure 5.1, because for an application there is the notion of *enrollment* in some member database M. In this chapter we deal with verification applications which claims of the form: "I am X" and thus only consider a single member of the database at a given time. In the next chapter we look at positive and negative identification where the database will typically have more than one person.

5.1.5 Definitions of FAR and FRR, positive authentication

Such positive verification ("*I am who I claim I am*") and positive identification ("*I claim I am uniquely enrolled in the database*") systems can make two kinds of errors, a *False Accept* or a *False Reject:*

I. *False Accept* (FA): Deciding that a (claimed) identity is a legitimate one while in reality it is an imposter; deciding H_o when H_a is true. The frequency at which False Accept errors are made is called the *False Accept Rate* (FAR).

II. *False Reject* (FR): Deciding that a (claimed) identity is *not* legitimate when in reality the person is genuine; deciding H_a when H_o is true. The frequency at which False Rejects occur is called the *False Reject Rate* (FRR).

When biometrics are used for securing a logical or physical site the errors have certain consequences. A FA results in security breaches, with an unauthorized person being admitted. A FR results in convenience problems, since genuinely enrolled identities are denied access to the application, or at least will have to undergo some further check to be admitted.

5.2 The Receiver Operating Characteristic (ROC)

Suppose for the moment that the integrals in (5.5) and (5.6) can be evaluated for any threshold T. Then the functions $\text{FMR}(T)$ and $\text{FNMR}(T)$ give the error rates when the match decision is made at some threshold T. The error rates can be plotted against each other as a two-dimensional curve:

$$\text{ROC}(T) \doteq (\text{FMR}(T), \text{FNMR}(T))\,. \tag{5.10}$$

That is, the FMR and FNMR behavior is expressed in terms of a Receiver Operating Characteristic (ROC) curve [75,161]. An example of a ROC curve is shown in Figure 5.4. The

False Match Rate and the False Non-Match Rate, as functions of T, are mapped as

$$\mathrm{ROC}(T) = (\mathrm{FMR}(T), \mathrm{FNMR}(T)) \rightarrow \begin{cases} (1,0) & \text{as} \quad T \rightarrow \ -\infty \\ (0,1) & \text{as} \quad T \rightarrow \ \infty. \end{cases} \tag{5.11}$$

That is, when the threshold is set low, the FMR is high and the FNMR is low; conversely when T is high, the FNMR is high and the FMR is low.

We can operate the matcher using any threshold T, which then defines a point on the ROC. This is the *operating point* of the matcher and it can be specified by choosing any one of T, FMR or FNMR, with the other two then being implicitly defined. For a given matcher, operating points are often given by specifying the threshold since this is something that can be chosen in the matcher, and FMR and FNMR can only be estimated given a threshold. On the other hand, when specifying a biometric application or a performance target, or when comparing two matchers, the operating point is specified by choosing FMR or FNMR since the threshold is a number that is only meaningful for a particular implementation of a particular matcher.

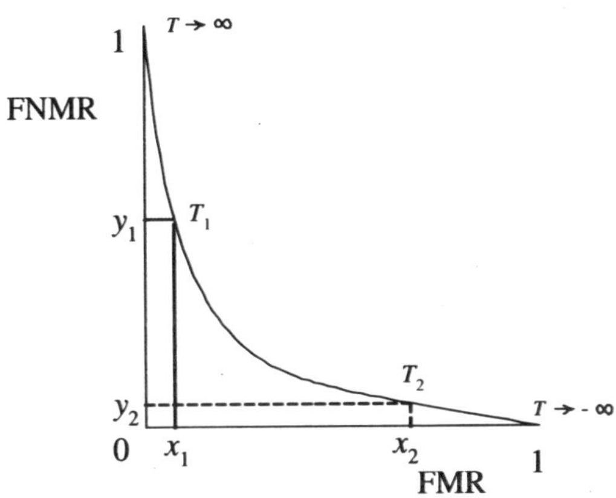

Figure 5.4: The ROC curve expresses the trade-off between FMR and FNMR.

Most biometric identifiers cannot guarantee error rates that are low enough for applications that have to be both secure (low FMR) and convenient (low FNMR). Suppose for instance that we still have *the exact* ROC curve of a matcher. Figure 5.4 shows that we can choose—

- The probability of a False Match can be fixed at some (low) FMR $= x_1$, and the probability of a False Non-Match (FNMR) is consequently FNMR $= y_1$,
- The probability of a False Non-Match can be set at some (low) fixed value FNMR $=$ y_2, and the probability of a False Match is consequently FMR $= x_2$.

Given a representative test data and the matcher error estimators, one needs to choose an operating point implied by a matcher threshold, T, to reach an "optimal" trade-off between

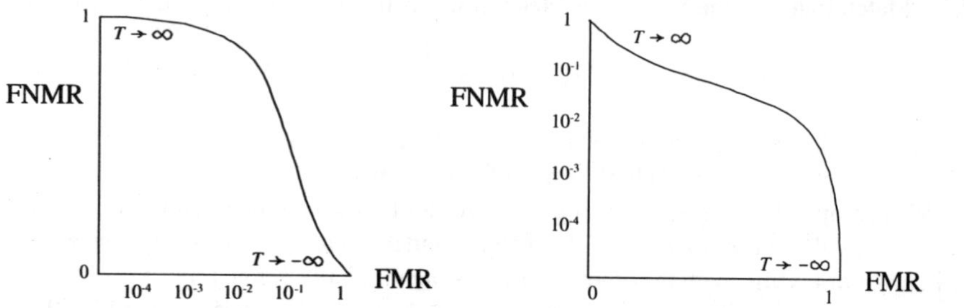

Figure 5.5: The ROC with one probability scale in logarithmic form; on the left the FMR is expressed in logarithmic form, on the right the FNMR is in logarithmic form.

the probabilities of an FM and FNM. This tuning of matcher performance can be considered as training on ROC.

5.2.1 Variations on ROCs

There are a number of variations of ROCs that are used for expressing the same information. Naturally, the FAR and FRR (Section 5.1.4) can equally well be plotted in ROCs, as can the FNR and FPR we will see in Section 5.4, or the axes can be inverted by plotting correct rates instead of error rates. In addition, quite often one or both of the probabilities (FMR and FNMR) in the ROC is plotted on a logarithmic scale. Semi-log plots of ROCs are shown in Figure 5.5. More commonly, a log-log plot of ROC is used as it magnifies the typical region of interest (operating area). A sample ROC in log-log scale is shown in Figure 5.6.

Sometimes researchers plot the Correct Match Rate (i.e., $1 - \text{FNMR}$) versus the FMR. This form of curve comes from detection theory, and is called the *Detection Error Trade-off*

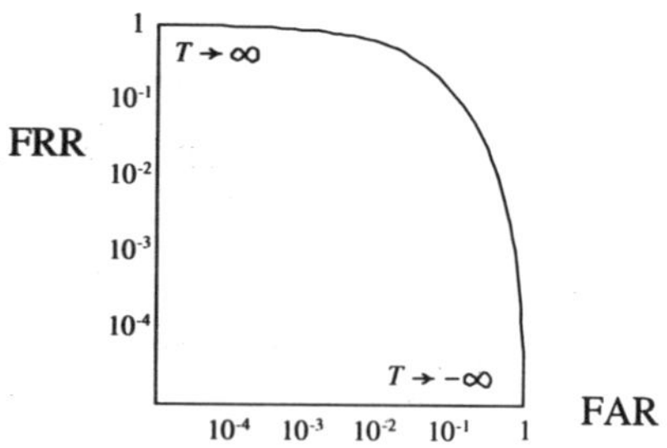

Figure 5.6: The ROC with both probability scales in logarithmic form.

(DET) curve. Along the y-axis, we have $(1 - \text{FNMR})$, which is the correct detection rate, or simply the detection rate; along the x-axis, we still have the FMR, here more usually termed the False Accept Rate or the False Alarm Rate. Figure 5.7 shows an example DET curve. The detection rate goes to 1 when the False Alarm Rate goes to 1.

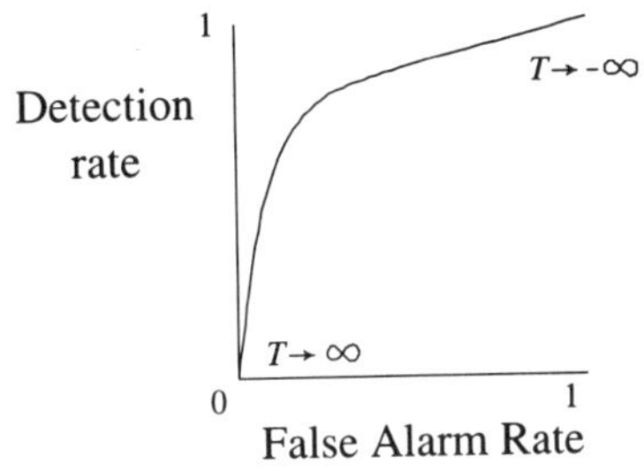

Figure 5.7: A detection error curve with the detection rate $(1 - \text{FNMR})$, plotted against the False Alarm Rate.

One has to be careful when reading a ROC to interpret the axes properly, especially if the families of ROC curves are not plotted in the same graph. There does not appear to be a particular convention in the expression of the error trade-off as function of T in biometrics; there are many variations but all boil down to the same thing.

5.2.2 Using the ROC curve

While the ROC is a precise, complete specification of a single biometric matcher's performance, its real usefulness comes when we wish to compare two matchers. In particular one of the most often asked questions in biometric evaluation is, Given two matchers a and b, which is more accurate?

It is rare that such a question can be answered unequivocally, since it usually depends on the operating points of the matchers. The ROC shows the trade-off between FMR and FNMR over a wide range of thresholds, but any operational matcher will be making decisions with a particular threshold T and thus have a particular FMR(T) and FNMR(T). Figure 5.8 shows the ROCs for two matchers and their operating points for some specified target FNMR. In this case it is clear that matcher b is always better than matcher a since for every FMR that might be specified it has a lower FNMR, and similarly for every possible FNMR, its FMR is lower. This is a rare case when we can unequivocally say one matcher, b, is better than another, a (always assuming the estimates of FMR and FNMR are sufficiently accurate—confidence intervals on these estimates are dealt with in Chapter 15). It should be noted that even in this case, it is easy to choose widely differing operating points for the two matchers and make either a's FMR or FNMR lower than b's, but not at the same time. Hence the importance of always ensuring that the quoted FMR and FNMR were measured at the same operating point.

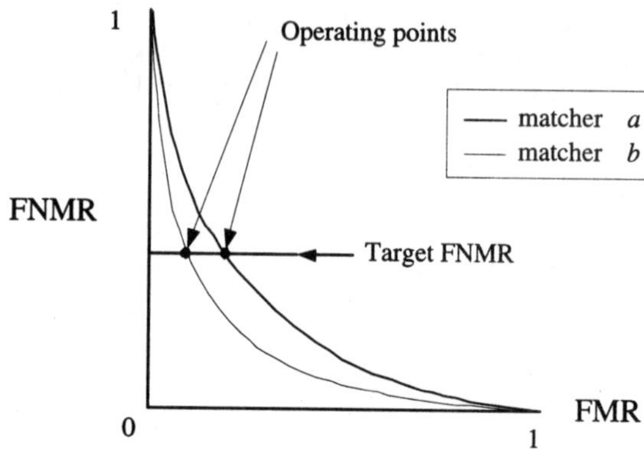

Figure 5.8: Two distinct ROC curves and a sample operating point specified with a target FMR. Matcher b is more accurate than a for all T.

Now consider the two ROC curves of Figure 5.9. It can be seen in Figure 5.9 that at some operating point $(\mathrm{FMR}_c, \mathrm{FNMR}_c)$, both matchers are equally accurate (though estimates of the crossover point will be unstable).

At operating point FMR_1, matcher a is more accurate, while at operating point FMR_2 the reverse is true. We can conclude from Figure 5.9 that if a low FMR (i.e. lower than FMR_c) is important, it is best to use matcher a at (say) operating point FMR_1. If, on the other hand, a low FNMR (lower than FNMR_c) is desired, matcher b might be the better choice, e.g., at operating point FMR_2.

Comparing two matchers by "eye-balling" the associated ROC curves, ROC_a and ROC_b, as Figure 5.9 tries to depict, is useful, but the conclusions can only be tentative without associated confidence intervals to show that perceived differences in performance are statistically significant.

5.2.3 Expressing the "quality" of a ROC curve

Comparing the FMRs or FNMRs of two matchers for some specified FNMR or FMR respectively, is the easiest way to compare matcher performances, but it presupposes having a known operating point. When comparing the performance of two or more matchers, for instance in independent evaluations (Chapter 16), a single number representing the quality of a matcher independent of operating point is often desired. This requirement has led to various expressions which attempt to reduce the information in the ROC to a single number. If one had complete trust in such a number as a representation of a matcher performance, choosing which matcher is the best would simply be a matter of choosing the one with the best performance figure. Here we present some performance summary figures that attempt this: the Equal Error Rate, d-prime, expected overall error and expected cost. These measures have severe limitations but in some circumstances but can be useful for summarizing

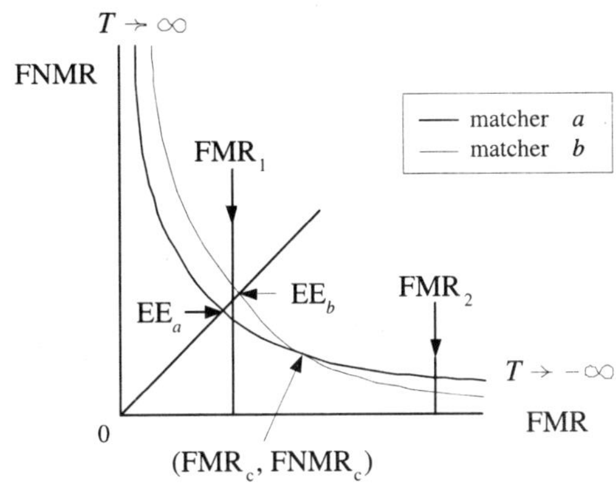

Figure 5.9: Which matcher is best can depend on the operating point chosen.

matcher performance and comparing matchers when the ROCs do not cross.

The Equal Error Rate

The Equal Error (EE) operating point, as shown in Figure 5.9, leads us to the Equal Error Rate (EER), which is an obvious, simple choice to judge the quality of a matcher. The Equal Error point of a biometric matcher is the operating point at the intersection of the line FMR $=$ FNMR with the ROC of the matcher. The Equal Error Rate is the value of the error rates at this point $\mathrm{EER} = \mathrm{FMR}_{\mathrm{EE}} = \mathrm{FNMR}_{\mathrm{EE}}$. In the figure, the Equal Error Rate EER_a of matcher a is clearly less than the Equal Error Rate EER_b.

The EER (subject to statistical significance which we deal with in Section 16.4) can certainly tell us that one matcher is better than another, but it only does so for a narrow range of operating points: $\mathrm{FMR} \in [\mathrm{EER}_a, \mathrm{EER}_b]$ or $\mathrm{FNMR} \in [\mathrm{EER}_a, \mathrm{EER}_b]$. Beyond these ranges, the ROCs may cross over and a decision based on the EER would be invalid. Very often matchers operate with highly unequal FMR and FNMR, making the EER an unreliable summary of system accuracy.

d-prime

Another way of judging the quality of a matcher is to measure how well the non-match score probability density $p_n(s)$ and the match score probability density $p_m(s)$ are separated, (see Figure 5.2). A measure of this separation for a matcher is d':

$$d' = \frac{\mu_m - \mu_n}{\sqrt{(\sigma_m^2 + \sigma_n^2)}}, \tag{5.12}$$

as suggested by Daugman [48]. Here μ_m and σ_m are the mean and variance of the match scores of genuine users; μ_n and σ_n are the mean and variance of the (non-)match scores of mismatching finger prints (or estimates thereof).

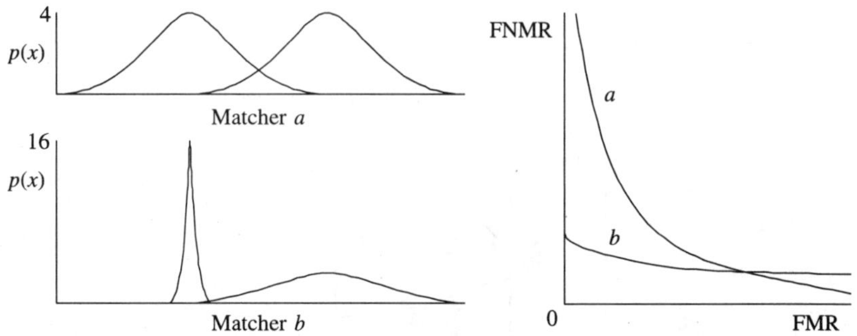

Figure 5.10: Different ROCs for two hypothetical matchers a and b with identical d'. Here Gaussian score distributions with identical means and different variances lead to the same d' but different ROCs.

The measure d' does provide a way of comparing matchers but can only be relied on to differentiate matchers when there is a significant difference in performance. Figure 5.10 shows that the relative performance of matchers with identical d' depends on the operating point chosen. Both matchers have the same μ_n, μ_m and matcher a has $\sigma_m = \sigma_n$, but for matcher b the variances are adjusted to be unequal but give the same d'. Clearly the ROCs for the two matchers are different, and cross, so our choice of matcher depends on the operating point. Small perturbations in the parameters would give differing d', whose ranking may or may not correspond to the performance of the matchers, depending on the operating point chosen: a matcher with a higher d' may perform worse over a wide range of operating points, even at the Equal Error point or the point of minimum expected error which we now define.

Expected overall error

The Equal Error Rate and d' implicitly treat the False Match and a False Non-Match errors as equally likely and of equal importance or cost, which is typically not the case for a biometric authentication application. For instance, a nuclear installation may hardly ever see an imposter, but the consequences of a False Accept would be tremendous. On the other hand, employees are being authenticated hundreds of times a day with minor delays and annoyance being perhaps the only costs of False Rejections.

A measure which takes into account the likelihood, but not the cost, of different errors is the expected overall error rate, $E(T)$, which can be calculated from the ROC for any threshold T, based upon the prior probability $P_{\mathcal{I}}$ of a random user being an imposter and the prior probability $P_{\mathcal{G}}$ of a user being genuine:

$$\mathrm{E}(T) = \mathrm{FMR}(T) \times P_{\mathcal{I}} + \mathrm{FNMR}(T) \times P_{\mathcal{G}}, \tag{5.13}$$

with $P_{\mathcal{I}} + P_{\mathcal{G}} = 1$. This is the expected probability that a random trial will result in an error. The minimum overall error of a matcher is defined as

$$\mathrm{E}_{\min} = \min_{T} \mathrm{E}(T) \tag{5.14}$$

This can be visualized (Figure 5.11) as the point where the ROC intersects a diagonal line (marked L_d in Figure 5.11) from the family FMR $\times P_{\mathcal{I}}$ + FNMR $\times P_{\mathcal{G}} = k$, with k the lowest value for which such an intersection takes place. Changing the prior weights changes the gradient of the lines for which we seek an intersection. Given the priors, this is another matcher-independent way of specifying an operating point, as well as giving a performance measure for comparing matchers.

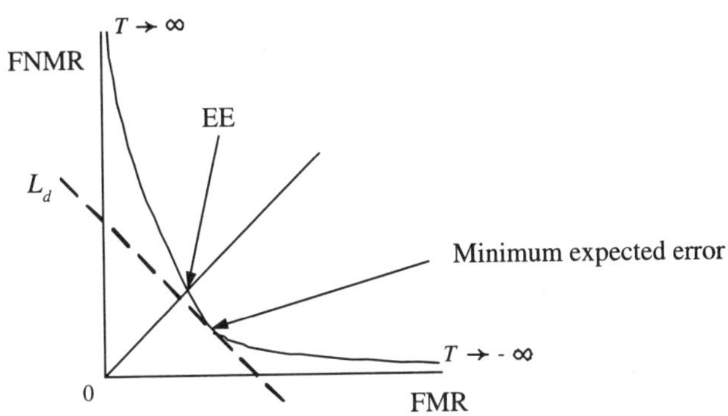

Figure 5.11: The minimum expected error will not generally be found at the same operating point as the Equal Error Rate.

For simplicity, and in the absence of knowledge about the values of $P_{\mathcal{I}}$ and $P_{\mathcal{G}}$, a single (but not necessarily meaningful) measure of accuracy can be derived by setting $P_{\mathcal{I}} = P_{\mathcal{G}} = 0.5$. This gives a measure akin to, but (as shown in Figure 5.11) not in general the same as, the Equal Error Rate discussed above.

Cost functions

A refinement of the overall expected error is to associate a cost with each of the errors and to calculate the expected cost of each match decision:

$$Cost = C_{\mathrm{FNM}} \times \mathrm{FNMR} \times P_{\mathcal{G}} + C_{\mathrm{FA}} \times \mathrm{FM} \times (1 - P_{\mathcal{G}}).$$

The minimum-cost operating point can be visualized in the same way as the minimum expected error point, with the relative costs affecting the gradient of the family of straight lines. If $C_{FM} = C_{FNM}$, the minimum-cost operating point will be the same as the minimum expected overall error point, but by putting real costs into the equation (e.g., dollar amounts at risk) the cost function becomes a meaningful number: the expected cost per use of the system. Cost functions are used in the NIST speaker identification evaluations [150] and this usage is discussed in more detail in Section 16.3.

5.3 Error conditions "specific" to biometrics

In addition to the fundamental type I and type II misclassification errors, and their multiple aliases, there are error rates associated with error conditions (listed in Table 5.1) that are more specific to biometrics.

1. The *Failure to Acquire* (FTA) rate is the percentage of the target population that does not possess a particular biometric, i.e., does not deliver a usable biometric sample.

   Here what the exact definition of a FTA is subtle. It can be that a subject does not possess the biometric that is needed for enrollment, i.e., the subject is missing an eye; or it can be that a subject's biometric cannot be measured, say, the fingerprint of a brick layer (the ridges have been worn away). Technology may well be improved so that this latter subject can be enrolled at some future point.

2. The *Failure to Enroll* (FTE) rate: Another variable is the FTE rate, which is the proportion of the population that somehow cannot be enrolled because of limitations of the technology or procedural problems [14].

These two *failure errors* are specific to biometrics and therefore are very basic error rates that are encountered in any biometric scenario, and are errors that are encountered both in verification and identification systems. Both FTA and FTE are partially due to intrinsic biometric properties and limitations in state of the art of the biometrics.

We introduce another application variable, mainly for voluntary applications, the "Failure to Use" (FTU) rate (also a random variable, at least during the design stage of the application). This is FTE rate plus the proportion of the population that for some reason does not enroll, or enrolls and fails to continue using thc biometrics system.

For voluntary biometric authentication applications, the difference between the FTU and FTE rates will be due to convenience (usability) problems with the voluntary applications. For involuntary applications, these rates are in theory the same (if it is affordable). There exists some intrinsic lower bound on the FTA for each biometric because some portion of the population, cannot show, or does not possess the particular biometric. This is sometimes referred to as a biometric's *universality* [97] properties. The FTE rate, on the other hand, can be used as a system parameter to design and build an installation within budget. The FTE can also be controlled by biometric sample quality control, i.e., raising quality control will improve error rates because poor biometric samples will not be used.

For a given application, a particular (FMR, FNMR) operating point needs to be determined. This operating point, obviously, is hard to establish beforehand. Even when an installation is in place, the genuine occurrence of a False Accept, or a False Negative in a negative identification scenario, may never be detected. For a voluntary application, the

FTA	Failure to Acquire causing a Failure to Enroll.
FTE	Failure to Enroll causing Failure to Use.
FTU	Failure to Use—a significant cost factor any biometric application.

Table 5.1: Quantitative variables of a biometrics application.

FTU rate, of course, can be determined when the installation is in place. The technical reasons for a high FTU rate will most probably be convenience problems (although there may also be non-technical reasons).

For an involuntary application, on the other hand, the FTE rate can be measured, which is usually done in laboratory-type (e.g., more supervised, controlled) situations. Tests of biometric authentication generally use subjects on a voluntary basis and it is in general unclear how the FTU (which could be interpreted as not volunteering) influences the FTE. On the other hand, given a database of test samples, by increasing the FTE the corresponding ROC curve can be made better and better.

The design parameter FTE can be artificially increased for a given installation, which will improve the overall quality of the enrolled population at the expense of increased exception handling and inconvenience. The system variable FTE enables a trade-off between *manual* versus *automated* authentication, which in turn is related to maintenance cost *versus* upfront cost.

What the ROC curve of an operational biometric installation is, and at what point the system is operating, will depend entirely on the enrolled population and the desired security—which should be defined with respect to the real enemies of the system. For positive identification systems, False Accepts may never be detected; for negative identification systems may never be detected. In such situations, it is possible to use statistics that only use scores from genuine users, like the Cumulative Match Curve (CMC, Section 16.5) and Rank Probability Mass (RPM) functions that are used for large system performance characterization.

5.4 Negative authentication

Up to this point, the null hypothesis H_o has been a "positive statement." However, we can turn the hypotheses of (5.2) around and have a negative authentication scenario. Consider two biometric samples, we again have two hypotheses:

$$\begin{array}{lll} \text{The } \mathit{null} \text{ hypothesis :} & H_o \mapsto & \text{the two samples do } \mathit{not} \text{ match,} \\ \text{The } \mathit{alternate} \text{ hypothesis :} & H_a \mapsto & \text{the two samples match.} \end{array} \tag{5.15}$$

Again, by computing a match score $s(B, B')$, the biometric match engine determines which *one* of the *two* hypotheses is true.

This biometric application amounts to deciding that a presented biometric sample is *not* some specific subject d_n. This could, for instance, be a criminal on a "watch list" with real world biometric $\mathcal{B}$ and biometric sample B. The hypothesis test of (5.15) verifies in an overt or covert (surveillance) fashion that input biometric $\mathcal{B}'$ is not the most-wanted biometric $\mathcal{B}$. The hypotheses are formulated as

$$\begin{array}{ll} H_o : & \mathcal{B}' \neq \mathcal{B}; \\ H_a : & \mathcal{B}' \equiv \mathcal{B}. \end{array} \tag{5.16}$$

This type of biometric capability is often associated only with biometric identification, which is discussed in Chapter 6. However, when phrased as hypothesis testing (5.15) or (5.16) it is just 1:1 matching with errors that are defined differently. The hypothesis test of (5.16) amounts to screening for the appearance of biometric $\mathcal{B}$.

For the negative authentication scenario we have two error conditions; two types of errors can be made when deciding (5.16):

1. Falsely missing $\mathcal{B}$,
2. Incorrectly matching a subject $\mathcal{B}'$ to biometric $\mathcal{B}$.

Using terminology from detection theory, we call these a *False Negative* and a *False Positive*, respectively:

I. *False Negative* (FN): Deciding that a subject is legitimate while in reality the subject is "most wanted" subject with biometric $\mathcal{B}$; that is, accepting H_o while H_a is true. The frequency with which this occurs is called the *False Negative Rate* (FNR).

II. *False Positive* (FP): Deciding that a subject is most wanted biometric $\mathcal{B}$ while in reality the person is legitimate; accepting H_a while H_o is true. (This is also called a *False Alarm*.) The frequency with which this occurs is called the *False Positive Rate* (FPR).

Here, in contrast to positive authentication, a mismatch or False Match is called an FN (False Negative) and results in possible security breaches; an undesirable person is somehow accepted to the application. A FP (False Positive) results in convenience problems since legitimate subjects are denied access and need to be biometrically checked further or otherwise examined to get access.

Figure 5.12 shows the distribution $p_m(s)$ of legitimate scores and the distribution $p_n(s)$ of scores of undesirable subject d_n. Of course, these distributions are similar to those in Figure 5.2. The difference between a negative authentication application and a positive authentication application is that a mismatch error in negative authentication results in a False Negative, while for positive authentication a mismatch error merely results in a False Reject. Therefore, if security is an issue, the operating point T for negative authentication will in general be selected lower (than for positive authentication) so as to minimize False Negatives.

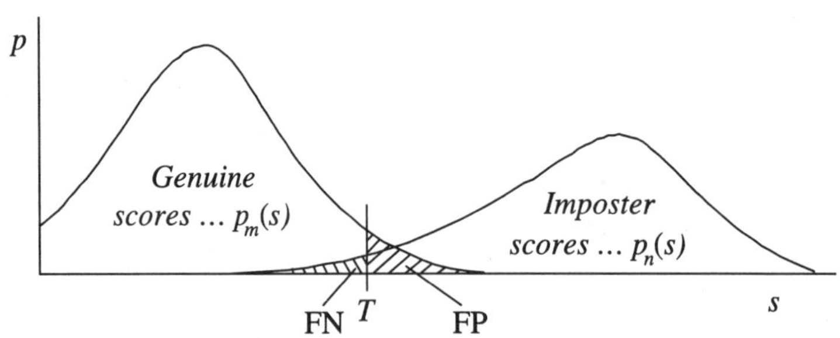

Figure 5.12: The scores are higher on average when input biometric matches an undesirable subject d_n; decision threshold T is set low to minimize the False Negative Rate.

We will use the terms *False Positive* (FP) and *False Negative* (FN) exclusively for "screening" (negative identification) applications. We use the terms *False Accept* (FA) and *False Reject* (FR) for positive authentication (verification or identification) applications.

5.5 Trade-offs

The two types of errors a verification system can make have different outcomes, affecting different people. A False Accept into a secure system means an unauthorized person has gained access—a breach in security. A False Reject means an authorized user has been denied access—which does not impact security but inconveniences the user and can have other repercussions by preventing them from going about their business.

Hence, the trade-off between FAR and FRR translates into a trade-off between security and convenience. Taking this to extremes, by setting $\text{FMR} = \text{FAR} = 1$, we have $\text{FRR} = 0$ and a very convenient but insecure access control system which permits access by everyone. Conversely, setting $\text{FNMR} = \text{FRRR} = 1$, we have $\text{FAR} = 0$, and we have a completely secure biometric system that denies everyone access.

5.5.1 Convenience versus security

When talking about "convenience," the biometric literature can be somewhat confusing, in that really two ideas are described by convenience:

- *The convenience of a biometric:* This is a somewhat nebulous concept of an intrinsic user-friendliness of a biometric. This is expressed by properties of biometric identifiers as described in Section 8.5.

  The problem is that the biometrics which are more convenient in this sense (since they can be acquired without intentional action by the user), like face and voice are also relatively weak biometrics (Chapter 14), meaning that they have inherently higher FMR and FNMR than more cumbersome biometrics such as fingerprint and iris.

- *The convenience of a particular implementation:* This is the ease with which a correctly enrolled person is authenticated on access of the application. This includes things like availability, the work flow of the authentication process, the exception handling process, and the False Non-Matches (FNM) of the authentication (i.e., the False Rejects of the application).

Following the latter usage, it is clear that a high False Reject Rate will inconvenience the legitimate users and we might wish to define a measure of convenience of an authentication system as follows:

$$Convenience = 1 - \text{FRR}. \tag{5.17}$$

The higher the FRR, the less convenient an application is because more subjects are incorrectly *identified* and therefore subject to denial of service or the exception handling process.

Similarly, the FAR is often used as some measure of security of a verification system:

$$Security = 1 - \text{FAR}. \tag{5.18}$$

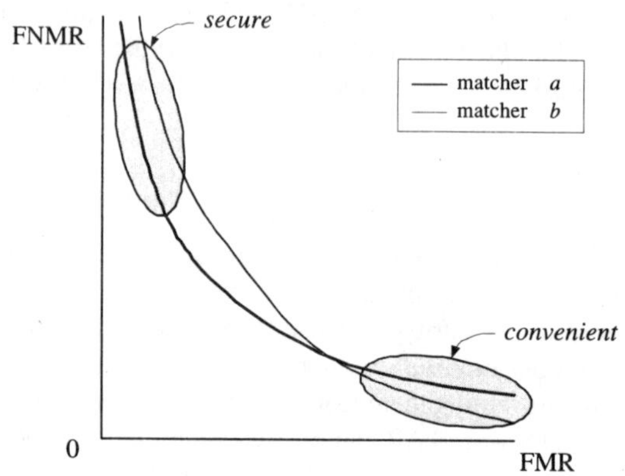

Figure 5.13: ROCs for Matcher a and b. Matcher b may be preferred for convenience and Matcher a for security..

Hence, security is a trade-off with convenience in any biometric authentication and using these numeric definitions, the trade-off can be shown in an ROC.

5.5.2 Cost versus security of positive authentication

What is perhaps more important is the trade-off between security and cost of the biometric authentication system. This is also related to the trade-off errors. Obviously, by setting $\text{FRR} = 0$ and $\text{FAR} = 1$, we have a very cheap but totally insecure system. By setting $\text{FAR} = 0$ and $\text{FRR} = 1$, the biometric system is not accepting anybody and one has to rely on costly human and manual labor.

Hence the FRR can be used as some measure of cost of an authentication system:

$$\mathit{Cost} = \text{FRR}.$$

The higher the FRR, the more expensive an application is because more subjects are incorrectly *identified* and therefore subject to denial of service or the exception handling process. As in (5.18) above, the FAR is some measure of security.

Hence, security is a trade-off with cost in any biometric authentication, and using these numeric definitions, the trade-off can be shown in an ROC.

This means that using the same biometric matcher for *both* a secure application and a related convenient application may *not necessarily* be the optimal solution. This is, for example, the case for the ROC curves in Figure 5.13. Here the ROC curve for matcher a corresponds to a more secure matcher since, for low FMR (FAR)—close to the y-axis—the associated FNMR (FRR) is lower than that of matcher b. If convenience is an issue, matcher b may be selected operating in the area of low FNMR (ROC approaching the x-axis) or high convenience. In this area of low FNMR matcher b has lower FMR characteristics.

Typically, a system is designed by selecting an operating point by specifying a FMR (for security) or a FNMR (for convenience) and hence only a portion of ROC curve of the core matcher is important. A particular biometric system is designed to operate at one point on the curve, another system may be designed for another point on the curve.

Moreover, the operating point does not necessarily need to be static. For example, screening processes like those at airports can be multi-modal, in the sense that demographics, text, sound, image, and video can all be used. Depending on demographic match, biometric authentication can operate at different operating points on the ROC curve. For example, it may be the case that wanted person X is from state Y. One can then set the FMR and FNMR differently when matching with a subject Z from state Y. That is, for all subjects Z from state Y the matcher will be operated at a lower FNMR when matching with wanted person X. This is an example of a dynamic authentication protocol, where the match errors of the application are somehow optimized by selecting different FMR and FNMR for different scores s.

The trade-off between the FMR and FNMR rates largely expresses itself as a security versus convenience trade-off. Therefore when designing a biometric authentication system, the first question that should perhaps be asked is: "For this application, is security of prime concern, or is convenience the real issue in this application?" The latter would, for example, be the case in voluntary applications because there the convenience may be the *deciding* factor in the success of a particular installation. As shown in Figure 5.4, when security is most important, security can be fixed as FMR $= x_1$, which implies some FNMR $\leq y_1$, which could be quite high. Alternatively, convenience could be selected at some low FNMR $= y_2$ and the corresponding (lack of) security is expressed as FMR $\leq x_2$.

An important thing to remember is that the FNMR (FRR) of biometric authentication systems is often given without taking the FTE (Failure to Enroll) or the FTA (Failure to Acquire) into account. Using $(1 - \text{FRR})$ as a measure of "convenience" then is overly optimistic and misleading because the inconvenience for users who are unable to use the biometric system is not taken into account.

5.5.3 Cost of negative authentication

For screening (or negative authentication) systems,

$$\textit{Security} = (1 - \text{FNR});$$

i.e., the lower the chances of missing undesirable subjects d_n, the higher the security. The convenience of such a negative authentication system is related as

$$\textit{Cost} = \text{FPR},$$

where falsely matched innocent subjects *d* are inconvenienced because they will be subject to further interrogation. For the negative identification scenario, when the FPR and FNR rates are too high, the authentication system becomes inconvenient to all users, since the access control point will no doubt get congested. In Section 7.2, we quote some dramatic numbers from [156] related to high FNMR/FMR of biometric identifiers in a well-defined authentication applications.

In this chapter we have dealt with the errors in biometric systems, and have concentrated on positive authentication, described in terms of False Accept and False Reject errors.

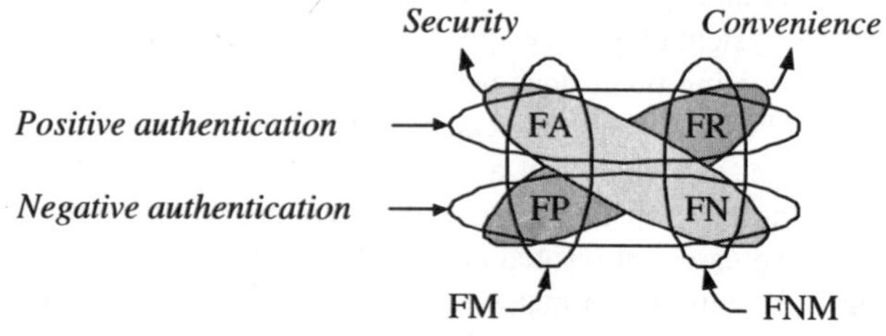

Figure 5.14: How the different errors imply loss of security or convenience.

However, in the case of negative authentication of Section 5.4, we adopted a different terminology. Here subjects are not being "accepted" by the application, instead they are being screened not to be an undesirable subject d_n. We term a match against d_n a "Positive" and a determination that someone is not d_n a "Negative". Thus the errors such a system can make are False Positive—believing there is a match of input subject with d_n—or a False Negative—failing to determine that the input subject is subject d_n.

We list these errors in Table 5.2. and summarize their trade-offs in Figure 5.14.

FM	Falsely matching two biometrics.
FNM	Falsely *not* matching two mated biometrics.
FA	False Accept of an intruder causing security problems.
FR	False Reject of an authorized user in **M** causing inconvenience.
FN	Erroneously missing a match in **N** causing security problems.
FP	Detecting a False Match in a screening database **N** causing inconvenience; this is also called a *False Alarm.*

Table 5.2: Quantitative error rate pairs for biometric 1:1 core match engines, authentication applications, and screening applications.

Table 5.3 lists the two authentication scenarios with the associated errors and the consequences of these errors to the application. For positive authentication, a FA (caused by FM) is a security problem and a FR (caused by FNM) is a convenience problem. For negative authentication, on the other hand, an FN (caused by FNM) is a security problem and a FP (caused by FM) is a convenience problem. When a subject d is denied access, either through a False Reject, or by erroneously matching undesirable subject d_n, this certainly amounts to inconvenience, both to the subjects and to the staff operating the biometric system.

Note that the causes of a FA and a FP are the same, i.e., the occurrence of a False Match. The consequences of such a False Match are quite different. It is a security problem in case of a FA and a convenience problem in the case of a FP. Similarly, the cause for a FN and a FR are the same, a False Non-Match, but here a FN is a security problem and a FR

Positive authentication	Negative authentication
FA = FM ⇒ *security problem*	FN = FNM ⇒ *security problem*
FR = FNM ⇒ *convenience problem*	FP = FM ⇒ *convenience problem*

Table 5.3: Error parameters of our two biometric applications..

is a convenience problem (see Figure 5.14). The definitions of the errors depend on the application and hence on the definition of the hypotheses H_o and H_a.

6

Identification System Errors

In this chapter we extend the statistical error analyses of the previous chapter to biometric identification systems. Identification systems have biometric representations of many (m) users enrolled and, when a biometric is presented to them, determine which of the enrolled biometric templates, if any, matches. Identification systems in effect try to answer the question, Who is this? Such systems base these answers on biometric data *only*.

There are two scenarios where we wish to apply identification: positive (Section 5.1.5) and negative (Section 5.4), distinguished principally by whether the *enrolled* subjects are cooperative users who wish to be identified to receive some benefit or access, or users who are to be denied access or benefit and consequently may not want to be recognized. There are other identification scenarios (e.g., the watch-list scenario) and other possible distinctions (e.g., *overt versus covert*). The important differences are in the *enrolled subjects*. Identification systems by definition contain a database (see Figure 6.1). For positive identification we denote this database as $\mathbf{M}$ (positive database); for negative identification we denote it as $\mathbf{N}$ (negative database), but for the moment we assume that we have a positive identification system with database $\mathbf{M}$ as in Figure 6.1.

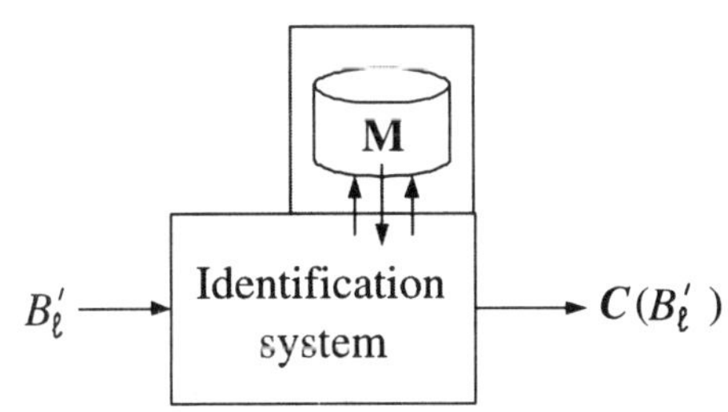

Figure 6.1: An identification system contains by definition an enrollment database $\mathbf{M}$, and identification involves searching for any matches to input B'_ℓ. Note that $\mathbf{M}$ is an *integral* part of the identification system.

Mostly, this chapter is a description of the biometric identification protocol or search problem. It describes the terminology, system implementation issues, and test practices related to this *biometric search* problem. Section 6.4 introduces a rank-based solution to identification, because the notion of *biometric rank* is often used in describing $1{:}\,m$ systems. A more detailed, and perhaps more controversial, description of rank-based identification can be found in Chapter 16 (Sections 16.5 and on).

6.1 Problem overview

The basic problem of identification is to generate an output identifier in response to some input biometric signal. Note that the desired output here is typically a multi-valued ID, not a binary "Yes/No" decision as in a verification system. Note also that no other credentials or claimed identity are submitted to the system—the mapping of biometric to ID is performed on the basis of the biometric information only.

Given some input B_i', an identification system searches a database, M, of biometric templates B_i each with corresponding identity ID_i, to determine if $B_i' \in \mathbf{M}$. Of course, if $B_i' \notin \mathbf{M}$, ideally a $1{:}\,m$ biometric search engine detects this error condition. The database $\mathbf{M}$ is just a set of biometric identifiers B_i, each associated with a true identity ID_i

$$\mathbf{M} \;=\; \{(B_i, \text{ID}_i);\; i = 1, ..., m\},$$

or a looser and less precise description of $\mathbf{M}$ is a vector

$$\mathbf{M} \;=\; \{B_i;\; i = 1, ..., m\} \;=\; (B_1, B_2, ..., B_m)^T. \tag{6.1}$$

The enrollment database $\mathbf{M}$ is constructed through an enrollment process and policy (the subject of Chapter 9), but $\mathbf{M}$ in its simplest form is just a (column) vector as in (6.1). However, it cannot be stressed enough that any $1{:}\,m$ search engine needs a database $\mathbf{M}$ in order to exhibit any meaningful behavior at all; without $\mathbf{M}$, we have no $1{:}\,m$ search system. Consequently, in any testing scenario of a system as in Figure 6.1, this enrollment process, or construction of $\mathbf{M}$, needs to be simulated or mimicked. This, of course, has implications that we try to address in this chapter.

Ideally, an identification system returns a candidate set $\mathbf{C}(B_\ell')$ of K matching identities

$$\mathbf{C}_K(B_\ell') \;=\; \{\text{ID}_{(1)}, \text{ID}_{(2)}, \ldots, \text{ID}_{(K)}\}, \tag{6.2}$$

where the (k) are a relabeling of $\mathbf{M}$. This indicates which biometric templates, $B_{(k)} \in \mathbf{M}$, were found to match input query B_ℓ' according to some criterion. Often the candidate list will be ordered using some matcher with the most similar identities first, so can rewrite it as a vector:

$$\mathbf{C}(B_\ell') \;=\; (\text{ID}_{(1)}, \text{ID}_{(2)}, \ldots, \text{ID}_{(K)})^T. \tag{6.3}$$

This vector then ideally consists only of the rank of the true identity of the user B_ℓ', or no answer when the user is not enrolled in the database

$$\mathbf{C}(B_\ell') \;=\; \begin{cases} (\text{ID}_i) & \text{when} \quad \mathcal{B}_\ell' = \mathcal{B}_i \in \mathbf{M}. \\ \emptyset & \text{when} \quad \mathcal{B}_\ell' \notin \mathbf{M}. \end{cases} \tag{6.4}$$

This is the *expected behavior* of the biometric search engine of Figure 6.1. Essentially, the search engine containing vector **M** "relates" a biometric B'_ℓ with a $B_i \in \mathbf{M}$. Or it associates a single index i (number) with input biometric B'_ℓ. A search engine should behave like a biometric database: the input is a query (B'_ℓ); the output is an entry i of **M** (which would be the identity ($\mathrm{ID_i}$) associated with B'_ℓ). The fuzzy concept of a match is a problem here.

In Expressions (6.2) and (6.3) and in the rest of this book, we use as notation for a relabeled set of biometric objects, $\{\mathrm{ID}_{(k)}, k = 1, ..., m\}$. More strictly, this should perhaps be denoted as $\{\mathrm{ID}_{(k)}, (k) = 1, ..., m\}$ to indicate that, in general, $k \neq (k)$. We refrain from this because it does not create confusion.

6.1.1 Winnowing

Now that we understand this goal, we can start describing the terminology, system implementation issues, and test practises related to this *biometric search* problem. We do need to introduce the notions of "filtering" and "binning" first, however. The goal of identification is to determine a unique identifier for each input biometric, or return "not known" if they are not enrolled (in many cases a biometric may be too weak, or a database too large, to be able to reliably generate such clear answers). In other words, **M** may become too problematic. In these cases some applications may simply try to "create" a smaller portion of database entries (6.1) that the test subject *might* be.

Thus identification can be viewed as (1) the "winnowing" down of a vector of m possible identities to vector **C** as in (6.3); (2) followed by a second system relying on some other agency to make the final decision. This second system, or "external agency" could be

- a human supervisor;
- additional circumstantial meta-data such as sex (see "filtering" in Chapter 11);
- another biometric system, perhaps even operating on the same biometric (see "binning" in Chapter 11).

In the latter case, the shortlist effectively redefines the database (i.e., a subset of **M**) used by the second) identification system. It is hoped that this shortlist of possibilities includes the true identity and that, by returning a list instead of a single candidate, the probability of the correct identity appearing somewhere in the output list will be increased. Basically, such approaches make m as small as possible before attempting a $1\!:m$ search; here, fuzzy biometric input and/or exact other information may be used. The effects on the search accuracy of such fusion or integration of information are hard to establish [228]. This problem is discussed in Chapter 11. The computational burden of the biometric search is certainly less but the problem *itself* is still the same.

6.1.2 Approaches and implementations

Before further considering the errors that an identification system can make, we describe how a practical identification system operates to see the types of answers it returns. Typically an identification system will run in one of three modes of operation, depending on the application for which it is being used. In each mode, some subject presents a biometric to the system and that biometric is compared to the biometric samples enrolled in the database $\mathbf{M}$. In some cases every enrolled sample in the database will be compared, but in others only some subset of $\mathbf{M}$ is compared. For simplicity we consider the former case, with a database $\mathbf{M}$ with m enrolled subjects, i.e., the $1\!:\!m$ search problem in its most general case. The three modes of operation are related to the three primary criteria for choosing the subset of $\mathbf{M}$:

- **Threshold-based:** This approach is effectively the same as repeating the operation of 1:1 verification for each person in the database. The query biometric template B'_ℓ is compared with each of the enrolled biometrics to obtain a match/non-match decision. This is typically done by computing a score $s(B'_\ell, B_i)$, $i = 1, \ldots, m$ for each enrolled template $B_i \in \mathbf{M}$ and considering as matches all those candidates with scores exceeding some threshold t_o. The complete list of all matching identities is returned. If no candidate matches (e.g., no score exceeds the threshold), the person is presumed not to be in the database.

- **Rank-based:** The system always returns some vector of fixed size, K, of the enrolled identities that best match the presented biometric. This requires an ability to rank (sort) the items in the database. The ranking might be based on scores. With $K = 1$ the system returns a single candidate corresponding to the person in the database most likely to be the same as the input query biometric B'_ℓ. Note that it is usually not necessary to rank *all* the identities in $\mathbf{M}$. Producing a ranked short-list of the best K items can be accomplished more efficiently than sorting all m enrollees. However, in the most general (and most computational complex case) the output vector is just a permutation or re-ordering or ranking of database vector $\mathbf{M}$. It is a vector whose meaning depends upon the relation between the input biometric and the enrolled biometrics.

- **Hybrid:** Here the K highest-scoring candidates are returned, provided that their scores are above a threshold. When many candidates are above a threshold this acts like a rank-based system, but when a single (or fewer) candidates are above the threshold it acts like a threshold-based system. Thus, this approach can be viewed as a combination of the previous two approaches. It is the one most commonly used. We can also put in this category systems that dynamically change the criterion for results being in the candidate list—by varying the list length or threshold. Here the length K of output candidate vector is determined by some operating threshold t'_o on a sorted list of candidates and we have back a threshold based trade-off system.

We will see that each of these approaches has limitations, and that their performances must be considered differently because of the different types of answers they return.

6.1.3 Open and closed worlds

An important issue in identification is the *closed-world assumption*. Systems are often designed as closed-world systems, but this impacts how well they operate under both open- and closed-world assumptions. In a *closed world* it is assumed that the only people who will ever try to use the system are people that have previously been enrolled in the system. The "world" of possible users consists solely of the known database $\mathbf{M}$, and no imposter ever tries to "break in."

By contrast, in an *open world*, a completely unknown user may attempt to access the system. This has ramifications for the kind of answers identification systems should return. Just to list a few possible ramifications:

- In an open-world situation, a rank-based system, which always returns K candidates, will necessarily be making an error when presented with an imposter.
- Conversely, a threshold-based system can return no answer, which would always be an error in a closed-world situation.
- To overcome the former problem, operating in an open-world, a rank-based system needs an additional *rejection* mechanism—in practice this generally means turning it into a hybrid system.
- One rejection mechanism is a generic imposter model—an extra template that is ranked or scored in the same way as the enrolled templates, i.e., "*an anti-user*" so to speak. If the impostor model is similar enough to the test template, it will appear on the candidate list and indicate that the input biometric is an impostor (or at least a low confidence in the reliability of the identification).

One might ask if there really are any situations that correspond to closed worlds. Imagine that there is some governmental agency where primary access to the facility is through the use of a badge. Further suppose that, at the time of badge issuance (during enrollment), a certain biometric (such as fingerprint) is collected from every employee, i.e., we have $\mathbf{M}$. Now it may be that, for some especially sensitive area within the facility, access is further restricted by a biometric identification system. The job of the biometric system is then to determine the identity of an individual and then look up in some table whether they are to be granted permission to enter this area or not. In this case, the complete set of people who might conceivably access this system are known, as are their biometric identifiers. On a smaller scale, imagine tapping a telephone conversation that we believe only involves two parties. Having identified the participants, working out who is speaking at any particular time later is simply a matter of choosing one of the two known parties (the closed world of possibilities). In either of these situations, though, it is possible to imagine our closed-world assumption being violated (an intruder bypasses the installation's first round of security, an operator cuts into the phone conversation) so any system should be designed bearing in mind the likelihood of an imposter and how the system will behave if one tries to use the system, which could raise some serious doubts about the closed-world assumption.

6.1.4 Ranking versus scoring

As noted above, a common (but not the only) way to generate an ordered list is to rank the candidates by sorting matcher similarity scores. That is, the best scoring match becomes the first entry in the ranked list, the next best scorer becomes the second entry, etc. Ranking and scoring are different ways of looking at the same problem, but we can convert between the two. Consider the set of scores depicted in Figure 6.2. Given a set of similarity scores, as on the bottom, it is easy to see how to produce a list of ranked candidates, as on the top. Similarly, given a list of ranked candidates, one could produce artificial "scores" for them, where the value of the score is simply inversely related to the rank of the candidate.

Strictly speaking, scores contain more information than is actually needed to perform ranking. Notice that from the score values (bottom) we can see that there is a set of top

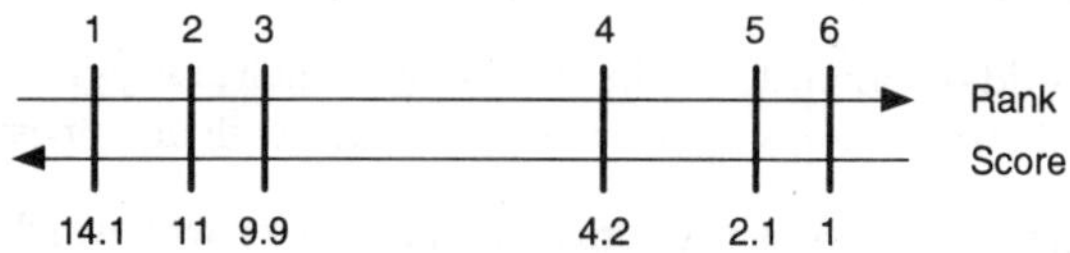

Figure 6.2: The ranking of candidates (top) in an output list is often based on matcher similarity score values (bottom).

candidates which are fairly similar (values 14.1, 11, and 9.9), and then a big gap until the next set of candidates is encountered (values 4.2, 2.1, and 1). We could exploit such separations or some other statistics of the search results dynamically to change the threshold (or change K). This gives another approach to making short lists that could again be termed "hybrid."

However, let us consider the situation of Figure 6.2 a little further. Given a database $\mathbf{M}$ of m reference biometric templates and some biometric matcher engine, one could generate a set of scores $\{s_1, ..., s_m\}$ and therefore a ranking $(\text{ID}_1, \text{ID}_2, ...)$ based on these scores. There is a subtle difference between scoring (matching followed by sorting) and ranking:

- A threshold-based system assigns a score s to every reference template $B_i \in \mathbf{M}$, and returns a variable length candidate set, $\mathbf{C}(B'_\ell)$:

$$\begin{aligned}\mathbf{C}(B'_\ell) &= (ID_{(1)}, ID_{(2)}, ...) \text{ such that } s(B'_\ell, B_{(k)}) > t'_o \text{ for } k = 1, ..., K. \\ &\quad \text{and } s(B'_\ell, B_{(1)}) > s(B'_\ell, B_{(2)}) > s(B'_\ell, B_{(3)}) \ldots > t'_o \end{aligned} \tag{6.5}$$

- A rank-based system assigns a rank $r(B'_\ell, B_i), 1 \leq r \leq m$, to each reference template $B_i \in \mathbf{M}$, indicating its position in a list sorted by similarity to the test subject, with 1 being the most similar. A rank-based system thus returns a candidate vector of fixed length K:

$$\mathbf{C}_K(B'_\ell) = (\text{ID}_{(1)}, \text{ID}_{(2)}, \ldots, \text{ID}_{(K)}), \tag{6.6}$$

  with $r(B'_\ell, B_{(k)}) = k$ for each $k = 1, ..., K$.

The score s is a continuous random variable, the rank r, on the other hand, is a discrete random variable.

It should be noted that one could devise ranking methods that do not rely on scores. That is, we might have a function that merely tells us which of two enrolled templates,

$B_i, B_j \in \mathbf{M}$, is more similar to the query biometric. This *rank engine* would allow us to rank the database $\mathbf{M}$ without using scores $s(B'_\ell, B_i)$ and $s(B'_\ell, B_j)$.

But what is perhaps more important is that a rank engine allows for ranking two pairs (B_i, B_j) and (B_m, B_n) in *different* ways. That is, a rank engine has the ability to compare the representations of enrolled biometrics in a pairwise fashion. For some pairs one might use a detailed ranking function as in Figure 6.3, for other pairs direct scoring might be used to rank. For example, working in the fingerprint domain, assume that each individual is

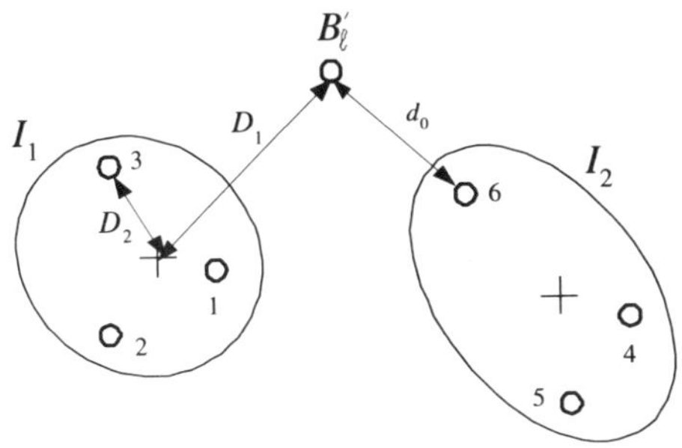

Figure 6.3: Ranking can sometimes be accomplished without scoring. Here the rank of individual samples (circles) is not directly related to their distance from the query B'_ℓ.

required to submit three separate impressions of the same finger (to account for placement and imaging variations). These are represented in the figure as labeled circles. It is not unreasonable to group these into distinct sets (the ellipses I_1 and I_2) and represent an individual by the average of this set of measurements (the cross within each ellipse). When a query print B'_ℓ is presented to the system, the ranking algorithm might then operate in two phases:

- Find the closest identity to B'_ℓ. This is identity 1 represented by ellipse I_1, which is determined by distance D_1.
- Then find the closest impression to B'_ℓ, impression 1 in I_1 determined by distance D_2.

Using such an algorithm imposes an overall ranking on the individual impressions as shown. All impressions belonging to individual I_1 are considered first and sorted by their distances to the centroid. Next, all impressions belonging to individual I_2 are likewise sorted and added to the tail of the output list. Notice that, by imposing this hierarchy, impression number 6 comes out *last* in the list despite the fact that its direct distance d_0 (the basis for a more conventional rank) is closer to the query print B'_ℓ than any of the other impressions in the database. In other words, this system generates a ranking of identities, which is *not* based on a similarity score between templates $s(B'_\ell, B_j)$, with B_j, $j = 1, ..., 6$. That is, individual I_1 is represented by ranked list (B_1, B_2, B_3) and individual I_2 by (B_4, B_5, B_6) and the overall ranking is based on these clusters.

6.1.5 Identities versus templates

Thus far we have been assuming that there is only one biometric template enrolled for each person. However, we may well enroll several templates for a given person to capture random variations, variations over time, or explicit variations (such as face images with/without glasses). In this case, we need to be careful to specify whether our identification engine is to return a list of the most similar enrolled templates, or the most similar enrolled identities. This is of particular concern when a human operator is to sort through the candidate list of the biometric engine. One might not want to waste labor by returning, for instance, mug-shots of the same person both smiling and frowning. The available human resource could be better spent checking up on different identities.

Moreover, the number of records in a database for each individual may not be the same. For example, the system might have one sample each from Dick Smith and Fred Jones, but ten samples from Tom Hardy. Suppose all these look like the current person being identified and the system is allowed to list its top four choices. By basing these choices on templates instead of individuals, it might generate the list $(Smith, Hardy_1, Hardy_2, Hardy_3)$, completely omitting the Jones identity. It may thus be best to return the shortlist identities, not templates, to a human operator, showing for each the most highly ranked template, but making available all highly ranked templates for closer inspection.

Still, there may be valid reasons to list samples instead of identities. For instance, if the top ten reported matches are all of the same individual, this may lend additional confidence to the identification. One could imagine incorporating this same heuristic into a system that instead returned only identities. To add a specific identity to the output list the system might, for example, require that some majority of the templates registered for an individual show up near the beginning of the ranked sample list (but not necessarily contiguous). As an example, only if five out of the ten samples for Tom Hardy were matched with good scores would he be added to the list. This is similar to the "k-nearest neighbor" classifier approach and would generate an output consisting of only high-confidence candidates. However, in applications such as screening, it may be important to list *all* identities for which there is even a shred of evidence (e.g., list Tom Hardy even if only one of his samples matches well). In Chapter 16, though, we devote some attention to analyzing this problem.

6.2 Generic evaluation criteria

At this point, we feel that it is warranted that we discuss how these $1:m$ systems are characterized and can be characterized in terms of their accuracy, i.e., the accuracy characterization of AFIS. The measures below can be applied to threshold-based, rank-based, or hybrid systems.

6.2.1 Reliability and selectivity

Reliability and *Selectivity* are accuracy measures used to characterize identification systems, particularly when the database is so large or the biometric so weak that the system cannot reliably pick a unique identity, and it is being used for "winnowing"—determining a candidate list of people who *might be* the subject. This is particularly the case in forensic applications, or when we have a *watch list* of "wanted" people, i.e., for negative authentication (Section 5.4) applications. These performance measures can be applied equally to all three of the approaches to forming candidate lists listed in Section 6.1.2, though we will see later that rank-based systems use slightly different measures in the *Cumulative Match Curve*.

The reliability, Rel, is the correct detect rate for tests when the true identity is actually in the database. That is, as defined in detection theory [170], it is the proportion of times that the proper enrolled identity is a member of the output candidate list. Thus, for a threshold-based system:

$$Rel = 1 - \text{FRR}. \tag{6.7}$$

Selectivity, Sel, is the average number of incorrect matches (FA) that the identification system reports per match against the entire database. This is a somewhat counter-intuitive use of the term *selectivity* because *lower* selectivity (few False Accepts on average) is a desirable system characteristic, while high selectivity is not. The selectivity can also be thought of as corresponding to the amount of work it will take a subsequent expert to select the correct answer from the list presented. Selectivity is easier to define assuming closed-world test conditions but will be slightly higher if estimated with imposter test data. For a rank-based system, returning K candidates, there will be $K - 1$ incorrect matches in the list except when the correct candidate is in the list, so

$$Sel \quad = \quad K - Rel. \tag{6.8}$$

For a threshold-based system, the selectivity Sel can be written as a function of m (the size of the identification database $\mathbf{M}$) and the False Accept Rate of the underlying matcher at the chosen operating point:

$$Sel \quad = \quad (m - 1) \times \text{FAR}. \tag{6.9}$$

By varying the criterion for an enrolled identity being returned in the candidate list, the identification system can be tested at a variety of operating points. In a threshold-based system the threshold is varied, in a rank-based system, K can be varied and in a hybrid system the list length, threshold, or other list cut-off criteria can be changed. Selectivity and Reliability can be measured at each of these operating points and plotted in a graph. As can be seen in Figure 6.4, there is a trade-off between Sel and Rel as the operating point is varied. If Rel is raised to give a higher chance of retrieving a matching biometric (e.g., by lowering t'_o or increasing K), a *higher* selectivity Sel will result, i.e., more False Matches per database query. Reliability increases toward one as selectivity increases toward the database size m. An ideal matcher would have selectivity of zero with reliability of one. An identification system can only work effectively to identify individuals (as opposed to winnowing down to a short-list) if False Alarms rarely occur for any search, that is, if the selectivity is significantly less than one. The Selectivity/Reliability curve can help in picking an operating point where this is true. As with the ROC for verification systems, the Reliability-Selectivity curve is a tool to allow us to choose an operating point for an identification system. A system specification might prescribe a desired reliability or selectivity, or the selectivity might be constrained by the amount of (machine or human) labor

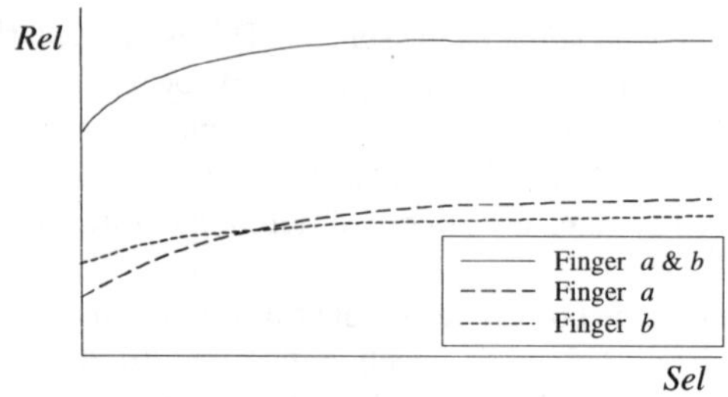

Figure 6.4: Characterization of a system in terms of the traditional large-scale system parameters selectivity and reliability.

available to further process the candidate lists. In each case, the graph tells us the performance achievable at the necessary operating point and how the selection criterion must be set to achieve this performance.

6.2.2 The recall precision curve

The field of document retrieval is attempting to solve a similar task to biometric identification (finding all matching entries in a database using an inexact matching function). For completeness and comparison we present here the performance measures used in document retrieval. The crucial difference now is that in response to some query, many documents may be considered as correct matches, and a document retrieval system must return as many of these as possible, while returning as few of the unrelated documents as possible. In contrast, a biometric identification system may have several biometric templates enrolled for a single identity, generally it is sufficient for any one of these to be returned, not as many as possible. It should be noted that recall-precision trade-off curves can be used to evaluate identification algorithms using any of the candidate selection criteria of Section 6.1.2.

Following the definitions of Witten, Moffat, and Bell [236], the precision, Pre, of a document retrieval system is the fraction of the returned documents that are relevant to the query

$$Pre = \frac{\text{number retrieved that are relevant}}{\text{total number retrieved}}. \tag{6.10}$$

This can be calculated for a particular query, or averaged over many queries to estimate expected performance. The recall, Rec, is the proportion of relevant documents in the database that are found by the system

$$Rec = \frac{\text{number of relevant that are retrieved}}{\text{total number relevant}}. \tag{6.11}$$

Recall and precision are often plotted against each other as a *recall-precision* curve. Figure 6.5 shows an example, loosely based on [236].

The *recall*, *precision*, and the associated *recall-precision* trade-off curve, do not apply very well to the $1\!:m$ biometric search problem. After all, we do not necessarily need the

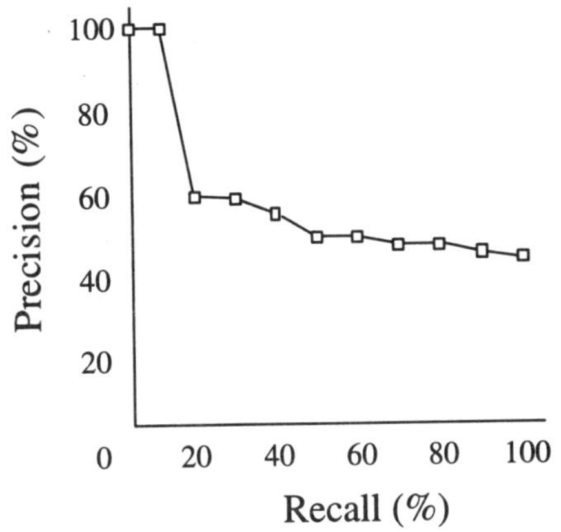

Figure 6.5: A recall-precision curve example for document retrieval.

biometric search engine to return *all* the enrolled biometric identifiers $B_{i1}, B_{i2}, \ldots$ in database M that are associated with an input query biometric B'_ℓ (as we would with documents). The requirement is usually only that *at least one* correct identifier is returned. We now look at some performance measures more specifically associated with either threshold-based or rank-based identification systems.

6.3 Threshold-based identification

A threshold-based identification system compares the query biometric B' to each of the identities in the database, as though that identity had been claimed by the subject in a verification system. In essence, the system is built around an internal black box that directly responds to the question: "Is this subject individual ID_i?" From a implementation viewpoint, systems can often parallelize this operation efficiently with distributed databases (as in [62]) so that it is not implemented as m separate, serial operations. But the effect will be the same: m "Yes/No" decisions.

Thus we have an internal biometric matcher that is optimized around some operating point t_o, with FAR and FRR estimated on biometric databases as described in the previous chapter (though t_o, FAR and FRR would probably be different to those chosen for a verification system). Then for each enrolled identity ID_i, we carry out a two-way hypothesis test. An ideal system will return m "No" (H_a) answers when the test subject is not enrolled, or a single "Yes" (H_o) answer for the correct enrollee when the test subject is in the database. However, in practice, a number of other situations will arise, giving us a variety of error conditions:

1. More than one candidate may exceed the threshold, giving an ambiguous candidate list and no definitive identification.

2. A single False Match may exceed the threshold, giving a misidentification.
3. No candidate exceeds the threshold even though the test subject is enrolled, giving a False Reject.

We will derive the False Accept Rate, $\mathrm{FAR}(m)$, and False Reject Rate, $\mathrm{FRR}(m)$, for an identification application with m enrollees in database $\mathbf{M}$. In particular, it is interesting to see how these are related to the 1:1 FAR and FRR (or equivalently, $\mathrm{FAR}(1)$ and $\mathrm{FRR}(1)$), the corresponding error rates of the underlying verification matcher.

6.3.1 Simple $\mathrm{FAR}(m)$ and $\mathrm{FRR}(m)$

Previous researchers [47,75,156] have simplified the situation by ignoring the case where multiple (correct or incorrect) candidates are matched. They declare an imposter to be falsely accepted if one or more scores for incorrect candidates exceeds the threshold. Under this assumption the chance of *correctly* rejecting an imposter is

$$Prob(\text{correct reject}) = \prod_{i=1}^{m} (1 - \mathrm{FAR}_i). \tag{6.12}$$

Here the FAR_i are the separately measurable False Accept Rates for each identity ID_i in database $\mathbf{M}$. This equation just computes the probability that the system will *not* falsely accept the imposter as any of the m identities in the database. Although the FAR_i are non-identically but independently distributed random variables, we substitute the expectation of $\mathrm{FAR}_i = \mathrm{FAR}$ (the overall system performance parameter) to obtain

$$Prob(\text{correct reject}) = (1 - \mathrm{FAR})^m. \tag{6.13}$$

This implies that the probability of a *false* accept is just the complement:

$$\mathrm{FAR}(m) = 1 - Prob(\text{correct reject}) = 1 - (1 - \mathrm{FAR})^m \tag{6.14}$$

For FAR small, $(1 - \mathrm{FAR})^m \approx 1 - m \times \mathrm{FAR}$. Thus we obtain the major result that $\mathrm{FAR}(m)$ is approximately linear in *m:*

$$\mathrm{FAR}(m) \approx m \times \mathrm{FAR} \tag{6.15}$$

A correct identification is considered to occur when the proper candidate score is matched (e.g., its score exceeds the threshold), regardless of what happens with the other candidates. Thus,

$$Prob(\text{correct identification}) = 1 - \mathrm{FRR} \tag{6.16}$$

To find the probability of *failed* identification, we take the complement and get

$$\begin{aligned} \mathrm{FRR}(m) &= 1 - Prob(\text{correct identification}) \\ &= 1 - (1 - \mathrm{FRR}) \\ &= \mathrm{FRR}. \end{aligned} \tag{6.17}$$

Note that $\mathrm{FRR}(m)$ is thus independent of m; it is just the same as the FRR of the underlying matcher.

6.3.2 Unambiguous answers

These are useful measures in a supervised application, where the candidate list is presented to some expert for review, whether the expert is a human or a subsequent biometric identification system, since $\text{FAR}(m)$ is the chance that an imposter generates one or more spurious candidates in the list and $\text{FRR}(m)$ is the chance that the correct enrollee is missed off the candidate list. *Reliability* and *Selectivity* (Section 6.2.1) are performance measures specifically designed for this situation. However, for an autonomous identification system that must determine a person's identity and act upon it, we can refine $\text{FAR}(m)$ and $\text{FRR}(m)$, recognizing that an ambiguous answer is a failure of the system, since it can take no definite action. An "Acceptance" will thus only occur when exactly one candidate scores above the threshold, and is either Correct or False depending on whether this candidate is the correct answer or some other enrollee.

First consider the case where the test subject is an imposter (not in $\mathbf{M}$). A False Accept occurs when exactly one database entry is falsely matched while all the others are not matched:

$$\begin{aligned}\text{FAR}(m) &= \binom{m}{1}\ \text{FAR} \times (1-\text{FAR})^{m-1} \\ &= m \times \text{FAR} \times (1-\text{FAR})^{m-1}. \end{aligned} \tag{6.18}$$

If $\text{FAR} \times m \ll 1$, $(1-\text{FAR})^{m-1} \approx 1$ so this reduces to

$$\text{FAR}(m) \approx m \times \text{FAR}. \tag{6.19}$$

Next, the chance of clearly rejecting the imposter is, as before, the probability of correctly rejecting all the entries in the database $\mathbf{M}$:

$$Prob(\text{correct reject}) = (1-\text{FAR})^m. \tag{6.20}$$

The only other alternative is to return an ambiguous answer, whose likelihood can be found as the remaining probability:

$$\begin{aligned} Prob(\text{ambiguous imposter}) &= 1 - Prob(\text{correct reject}) - \text{FAR}(m) \\ &= 1 - (1-\text{FAR})^m - m \times \text{FAR} \times (1-\text{FAR})^{m-1} \\ &= 1 - [(1-\text{FAR}) - m \times \text{FAR}]\,(1-\text{FAR})^{m-1} \\ &= 1 - [1 - (m+1)\,\text{FAR}]\,(1-\text{FAR})^{m-1}. \end{aligned} \tag{6.21}$$

We can perform a parallel analysis of cases for the situation where a person who really *is* in the database is being identified. The chance of being correctly and uniquely identified is the probability of matching the correct record, but none of the $m-1$ others:

$$Prob(\text{correct identification}) = (1-\text{FRR}) \times (1-\text{FAR})^{m-1}. \tag{6.22}$$

Since this is the only case that counts as a non-ambiguous identification, we can define the False Reject Rate of the system as

$$\begin{aligned}\text{FRR}(m) &= 1 - Prob(\text{correct identification}) \\ &= 1 - (1-\text{FRR}) \times (1-\text{FAR})^{m-1}. \end{aligned} \tag{6.23}$$

Notice that this is higher than the $\text{FRR}(m)$ derived in the previous section, and again, when $\text{FAR} \times m \ll 1$, this reduces to

$$\text{FRR}(m) \approx \text{FRR}. \tag{6.24}$$

A misidentification occurs when a single answer is returned, but it is the wrong answer. For this to happen, the correct record must be falsely rejected, while exactly one of the $m - 1$ other records is falsely accepted (the rest being correctly rejected):

$$\begin{aligned} Prob(\text{misidentification}) &= \text{FRR} \times \binom{m-1}{1} \text{FAR} \times (1 - \text{FAR})^{m-2} \\ &= (m-1) \times \text{FRR} \times \text{FAR} \times (1 - \text{FAR})^{m-2}. \end{aligned} \tag{6.25}$$

The remaining alternative is to return an ambiguous answer with more than one identity in the list (which might or might not include the true identity). The probability of this can be found as the remainder after subtracting from 1 the probability of the other two cases:

$$\begin{aligned} & Prob(\text{ambiguous genuine}) \\ & = 1 - Prob(\text{correct identification}) - Prob(\text{misidentification}) \\ & = 1 - [(1 - \text{FRR})\,(1 - \text{FAR})^{m-1}] - [\text{FRR} \times \text{FAR} \times (m-1)\,(1 - \text{FAR})^{m-2}] \\ & = 1 - [1 - \text{FRR} - \text{FAR} + m \times \text{FRR} \times \text{FAR}]\,(1 - \text{FAR})^{m-2}. \end{aligned} \tag{6.26}$$

For an identification system to work autonomously, $\text{FAR}(m)$ and $\text{FRR}(m)$ must be low, but also the probabilities of ambiguous answers must be low. Ambiguous answers can be dealt with by:

1. Running an exception procedure (calling in a human supervisor).

2. Considering them rejects—increasing $\text{FRR}(m)$ above its value in (6.24) of FRR.

3. Passing them on to some other biometric identification system, using the same or another biometric.

4. Accepting the highest-scoring candidate as the correct answer. This amounts to the "hybrid" system we described earlier (and requires access to sorted scores, not just the "Yes/No" decisions of the matcher).

To summarize, for this type of identification system to work autonomously, not only must the FAR and FRR of the underlying matcher be low, the probabilities of ambiguous answers must also be kept under control. Just as with the ROC for verification systems, there is a trade-off between the probabilities of error conditions, i.e., $\text{FAR}(m)$ and FRR, which corresponds to a trade-off of convenience against security of the search engine. To characterize fully the error conditions of an identification system, the properties of the exception handling mechanism must also be specified, e.g., the operating threshold of a secondary matcher, t_s.

6.4 Rank-based identification

A rank-based identification system returns the top K most similar enrolled candidates to the presented biometric B'_ℓ in some order. Some other secondary decision-making process (secondary matcher), for example, a human operator decides on the actual strength of the similarities. Such a system generally operates as a system that uses filtering and/or binning (Section 6.1.1). Only under very restricted conditions can such a rank-based identification system truly do identification (determine the unique identity of each query biometric). These conditions are—

1. The output candidate vector or list must be of length one ($K = 1$); otherwise ambiguous answers will always be returned.

2. The closed-world assumption must hold true; i.e., the application scenario must be constrained in such a way that only enrolled users present their biometrics to the system. If the closed world assumption is relaxed and the query biometric B'_ℓ is not in $\mathbf{M}$, ideally, the candidate list should be empty. In practice, by design, the rank engine is still required to return a list of length K.

Under these conditions, a rank-based system can *only* make one kind of error—a mis–identification—when the true identity is ranked lower than one or more of the other enrolled candidates. In practice since it is not really possible to *guarantee* a closed world, performing identification by always taking the highest-ranked candidate is not ideal for applications where security is very important.

6.4.1 Rank-order statistics

There are biometric "evaluation scenarios" where rank-order statistics are used to evaluate biometric match engines. In particular, they have been used in a number of evaluations of face recognition systems (FERET and FRVT [15,164–166]; see Section 7.7). Here, the "biometric scenario" (sometimes called the "watch-list scenario") is to pop up a small selection of $K > 1$ mug-shots showing possible matching biometrics in biometric database $\mathbf{M}$ of templates $\mathbf{C}_K = (B_{(1)}, B_{(2)}, ..., B_{(K)})^T$ to query biometric B'_ℓ. This watch list scenario is really a negative identification application; and following our convention, the identification database should be denoted as $\mathbf{N}$. However, denoting the database as $\mathbf{M}$ of m (say) most-wanted terrorists is equivalent; the important thing to note here is that m could be relatively small, (m could even be 1). In general, the output list of K items is smaller than m, however. The question then becomes how is $\mathbf{C}_K$, the "optimal" output candidate vector of "optimal" length K determined, and how does one measure performance when vectors $\mathbf{C}_K$ instead of unambiguous answers are returned?

Earlier in Section 6.1.4 we defined the rank $r(B'_\ell, B_i)$ of a particular enrollee i when the database is *sorted* by similarity to a query biometric B'_ℓ. When testing against a known biometric database (and enforcing the closed-world assumption), we know the true answer and can define $r(B'_\ell)$ as the rank of the correct answer for this query:

$$r(B'_\ell) = r(B'_\ell, B_i) = i \text{ when } \mathcal{B}'_\ell \equiv \mathcal{B}_i. \tag{6.27}$$

An effective system should have r always low, and ideally $r = 1$ (this ideal behavior is impossible, of course). This rank-based output could be extended to open-world scenarios by defining

$$r(B'_\ell) = 0 \text{ when } \mathcal{B}'_\ell \notin \mathbf{M}; \tag{6.28}$$

a rank-based system could just output rank $k = 0$.

6.4.2 Rank Probability Mass (RPM) function

The ranking behavior of the identification system can be characterized by the probability distribution of r, which we term the *Rank Probability Mass* function $P(r)$:

$$P(r) = Prob(r), \quad r = 1, ..., m. \tag{6.29}$$

That is, for any input query, B'_i corresponding to an enrolled identities, the probability that the correct identity is ranked in position r is $P(r)$. Of course, we have $\sum_{r=1}^{m} P(r) = 1$. In the extension to the open-world case, the probabilities $P(1)$ ... $P(m)$ are scaled down by the prior probability of a genuine user. $P(0)$ is then the probability of correctly detecting an imposter.

The RPM can be estimated empirically as $\hat{P}(k)$, using a database $\mathbf{L} = \{B_\ell, \ell = 1, ..., L\}$ of test biometrics and counting the test cases $\mathbf{L}_k$ for which the rank of the correct answer is k. If we define

$$\mathbf{L} = \{B_1, B_2, \ldots, B_L\}, \text{ and} \tag{6.30}$$

$$\mathbf{L}_k = \{B'_\ell;\ r(B'_\ell, B_i) = k, \text{ i.e., when } \mathcal{B}'_\ell \equiv \mathcal{B}_k\}, \tag{6.31}$$

then

$$\hat{P}(k) = \frac{|\mathbf{L}_k|}{|\mathbf{L}|}, \quad k = 1, ..., m. \tag{6.32}$$

If we have sufficient samples of a particular biometric $\mathcal{B}'_\ell \in \mathbf{M}$, sampled over a long time interval and representative set of circumstances, then we could even estimate the RPM $\hat{P}_i(k)$ for a particular person ID_i, as shown in Figure 6.6. This gives us a way of finding individuals who are hard to recognize because they lead to many high-scoring False Matches—the "lambs" defined in Chapter 9). We continue this discussion in Section 16.5, where we examine what can be estimated about a search engine, given a database of biometric samples.

Given the estimates $\hat{P}_i(k),\ k = 1, ..., m$ and if each $\hat{P}_i$ is estimated with the same number L of rank scores, the average RPM can be calculated as the average over the enrolled individuals (which might be weighted if we have a prior distribution over the users of the system):

$$\hat{P}(k) = \frac{1}{m} \sum_{i=1}^{m} \hat{P}_i(k), \quad k = 1, ..., m. \tag{6.33}$$

If we can construct a generic intruder model, then we could also estimate an RPM for this model and assess the model's performance on imposters and enrolled candidates.

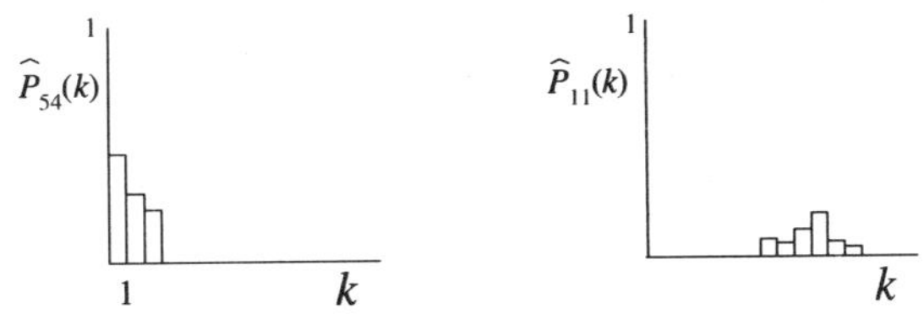

Figure 6.6: Estimated Rank Probability Mass function for biometric template B_{54} is "good" the one for biometric template B_{11} is much worse.

6.4.3 Choosing the list length

The main question to be determined for a rank-based identification system is, What is the optimal candidate list or vector $\mathbf{C}_K$? There are several possible ways of answering this. In many cases an *a priori* fixed K is chosen based on application logistics. For instance, there might only be space on the display screen to show 8 mug-shots with good resolution, or there might only be enough human labor available to examine two alternate fingerprints for every passenger screened. In a hybrid system K might be changed dynamically for each query, which we analyze with Reliability-Selectivity curves.

The situation of most interest here is when a fixed K is chosen according to performance requirements for the system—K is chosen so that the probability that the correct answer is contained in the shortlist is greater than some specified performance goal. Note that setting $K = 1$ will not be a good choice if there is much probability mass in the RPM away from $k = 1$. The maximum likelihood rank may not even be one if the matcher has a large database, and if there are enrolled biometrics which easily match many others (the "wolves" of Chapter 9).

Given (6.29), we are now in the position to introduce the Cumulative Match Curve (CMC), which is sometimes used as a performance statistic of biometric search engines. The CMC is the cumulative sum of the RPM:

$$\mathrm{CMC}(k) \;=\; \sum_{i=1}^{k} \hat{P}(i), \;\; k = 1, ..., m. \tag{6.34}$$

By examining the RPM, or by calculating the CMC, i.e., estimating the expected rank,

$$E(r) \;=\; \hat{r} \;=\; \sum_{k=1}^{m} k\, \hat{P}(k), \tag{6.35}$$

we can evaluate the performance of a rank-based identification system. Certainly we would expect $K > E(r)$ to give a good chance of finding the correct answer in the candidate list, given the expectation that correct identities have lows ranks (high scores).

The CMC shows (as in Figure 6.7), for any given list size k, the probability of a test subject's correct identity showing up on the shortlist (if the test subject has been enrolled). With the CMC, we can choose K so that this probability meets a performance goal, or

Figure 6.7: A Cumulative Match Curve of a biometric identification system that returns a ranked list.

determine this probability for a K determined by other factors. The CMC increases with k, and when $k = m$ (the whole database) this probability is one. We can see that the CMC is very similar to the Reliability-Selectivity curve—with almost identical axes—and it can be used in the same way. On the y-axes, the Reliability and CMC(k) are both the chance of the correct answer (the subject is enrolled only once) being on the candidate list. These are both measures of the size of the candidate list. In the CMC the x-axis is the chosen parameter K, whereas in Reliability-Selectivity curve for the same rank-based system it is the expected number of False Accepts = $(K - Rel)$. For a given underlying scoring function operating as a rank-based or a threshold-based identification system, we would expect the corresponding CMC and Reliability-Selectivity curve to be very similar.

7

Performance Testing

Biometrics is an emerging science and technology with fierce competition and manufacturers continually refining the technology and claiming high accuracy. "High accuracy" may be claimed in many different ways: simply as a system that works, or as a system that makes no or very few errors, or a system that is "100 percent" accurate. Obviously, such loose definitions of accuracy are undesirable and there is a need for precise measurements and well-defined biometric system testing procedures.

Two biometric capabilities (*matching* and *ranking*) and biometric system errors rates have been introduced and defined in Chapters 5 and 6. The former chapter is concerned with 1:1 *biometric matching*, the latter is concerned with 1: m *biometric searching*. In this chapter we continue the split. Sections 7.1–7.6 we attempt to relate the "error quotes" that can be found in the biometric literature to the error definitions of Chapter 5, the so-called trade-off ROC errors for 1:1 biometric matching. In Section 7.7, we discuss some of the practices in testing 1: m search systems and quote some of the error estimates. As noted in Chapter 6, rank-order statistics, like the CMC, are a good tool to evaluate search engine accuracy.

All this is not an easy task and therefore we proceed with caution and look at the different types of testing scenarios that are used in the biometrics research community in Section 7.1 first. We then, in Sections 7.3–7.5 look at accuracy numbers for our six most common biometrics as best as we can, and reconstruct and interpret them from the biometric literature (this is summarized in Section 7.6).

7.1 Measuring performance

There are three prominent testing methodologies for biometric matcher evaluation, *technology evaluations*, *scenario evaluations*, and *operational evaluations* [132]. Evaluations are often organized in the form of a competition between vendors and developers of biometric matchers that compute the similarity $s(B', B)$ between samples or templates B and B'. Because biometrics is such a relatively young research area, the contestants in these evaluations are often both commercial vendors and academic institutions. Furthermore, biometric system testing is focused on the 1:1 verification systems. The fundamental error rates FAR (FMR) and FRR (FNMR) are derived from an ROC using correctly labeled databases of biometric samples. Such tests are also called *benchmarks* in that various biometric matchers can be compared on "*frozen*" biometric databases.

7.1.1 Technology evaluations

In this type of evaluation the contestants have a certain period of time to train their biometric match algorithms on some "training data" $\mathbf{A}$. This database is often made up of several separately collected databases $\mathbf{A} = (\mathbf{A}_1 \cup \mathbf{A}_2 \ldots)$ which are made available by the contest organizers.

In the next step of the evaluation, the matching algorithms are tested on newly made available, labeled sequestered "test" data $\mathbf{Q}$. We call this $\mathbf{Q}$ because this data is meant to mimic in the best possible way of "opening up" a biometric match engine, essentially developed using closed-world data $\mathbf{A}$.

Hence, a technology evaluation of biometric matchers consists of at least two phases:

1. *Training phase:* This phase is depicted in Figure 7.1. Database $\mathbf{A}$ with correctly labeled biometric samples is made available at the start of Phase 1. To achieve optimal performance, the system would be trained on the database. To maximize the benefit of training, the training database $\mathbf{A}$ is split into two parts: the actual data used to train $\mathbf{C} \cup \mathbf{A}$ and a validation set $\mathbf{A} \backslash \mathbf{C}$ as shown in Figure 7.1.

   However, how to train their matchers is up to the contestants and unless prohibited by competition rules, each contestant will supplement the training database with private databases also.

   Each individual database $(\mathbf{A}_1, \mathbf{A}_2, \ldots)$ may have been collected to be representative of a different user population, and with a corresponding test set $(\mathbf{D}_1, \mathbf{D}_2, \ldots)$, in which case a separate training phase may be carried out with each database separately, in preparation for testing with the corresponding test set.

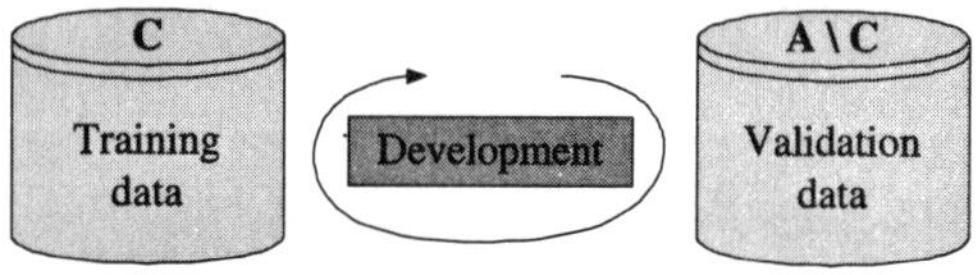

Figure 7.1: Phase 1 of a technology evaluation: Data is made available and the contestants train the algorithms.

The biometric sample databases $\mathbf{A}_1, \mathbf{A}_2, \ldots$ are collected off-line and the samples need to be properly labeled with unique subject (or real-world biometric) identifiers (e.g., numbers); this is called *ground truth*. In general, the samples are obtained from volunteer test subjects and great care should be taken to label the data correctly. In general, it is a good practice to record as much information as possible about the samples. For instance, if multiple biometrics (say, several fingers) are sampled per subject, this information should be retained because (as we will see in Chapters 15 and 16) such samples are not independent and influence the accuracy of the error estimates.

Good biometric data is important, both for system testing and system training. Errors in labeling result in data that are not "clean"; this can have rather dramatic effects on the estimated error rates when such databases are used for testing. Similarly, if poorly labeled databases are used to train the matcher this will result in a faulty match engine. Moreover, all databases evolve over time as individuals are added into and deleted from the database. It is important to be careful about the integrity of the database over its entire lifetime; otherwise a good database may go bad.

2. *Competitive testing phase:* For this phase the performance statistics that are to be measured are defined and it is determined how these parameters will be estimated from test data. The sequestered data **D** is made available and the competing biometric matchers are tested on this data as shown in Figure 7.2.

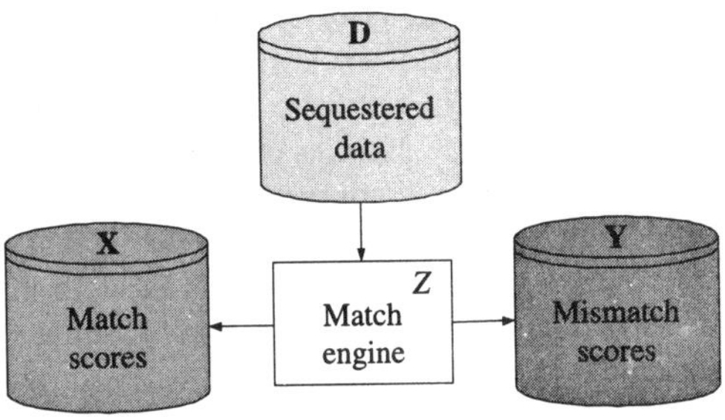

Figure 7.2: Phase 2 of a technology evaluation: The matcher Z is tested on newly made available data **D**.

Using database **D** for each matcher Z, a set of match (genuine) scores $\mathbf{X} = \{X_1, X_2, ..., X_M\}$ and a set of non-match (imposter) scores $\mathbf{Y} = \{Y_1, Y_2, ..., Y_N\}$ is generated. The set **X** is obtained by matching mated samples; the set **Y** by matching samples from different biometrics. These scores can then be used to estimate the FMR and FNMR (FAR and FRR for positive authentication) as a function of decision threshold T. This is equivalent to estimating the match score distribution $F(x) = \text{FRR}(x)$ and non-match score distribution $G(y) = (1 - \text{FAR}(y))$ for matcher Z. The probability distributions F and G specify matcher behavior at all operating points and functionally describe the error rate.

The data **D** are sometimes sequestered and the algorithms are tested after submission of the competing biometric matchers to some independent testing agency. The face biometric has a long history of testing face-matching algorithms in this way [15,164,165]; such testing has been started more recently in the fingerprint area [13,130]. Influenced by the rigorous testing procedures of speech recognition algorithm, the area of speaker recognition also has a good history of competitive algorithm testing (see [33,174]).

In technology evaluations, the match capabilities of core 1:1 matchers are estimated and compared and not the FAR/FRR or FPR/FNR of some biometric application. Strictly speaking, these technology evaluations estimate the FMR of (5.5) and the FNMR of (5.6) as a function of T, the decision threshold.

Two common and easy ways to make mistakes in biometric system design are—

1. *Testing on the training set*—the test score databases **X** of and **Y** are somehow used to train a biometric match engine.
2. *Overtraining*—The training databases **X** and **Y** are used too extensively to optimize the associated FAR/FRR ROC curve. A sign that a system may be overtrained is exceptionally good performance on just *one* particular data set.

Classical books like [56] warn of these pitfalls.

All these databases are from samples obtained from a finite number of volunteer subjects. As long as all this data is considered to be acquired from the user population of the biometric application, this data is a *closed* database of samples. When these databases are also used to generate False Match error conditions, the associated error estimates like the FAR and FNR are so-called "zero-effort" error estimates, as introduced by the signature research community (see Section 7.5). It is unclear, in general, how these zero-effort estimates differ from the true (False Accept and False Negative) error rates. Since biometric data is often not available from the true enemies of an application, it is hard to generate true False Accepts and False Negatives. Estimates of these immeasurable error rates, of course, are impossible to compute without somehow modeling this population. The error conditions that are easy to detect are a False Reject and a False Positive, and therefore the FRR and FPR estimates are more meaningful estimates of the true False Reject Rate and true False Positive Rate.

When examining ROC trade-off error curves, one should always be suspicious about how the False Accepts and False Negatives are determined (from the available samples or otherwise).

7.1.2 Scenario evaluations

This type of evaluation takes place in a testing facility, e.g., as shown in Figure 7.3. The biometric authentication devices (1:1 matchers) are installed in an "office environment" as in the figure (loosely the test facility layout used in [131]) and the biometric devices are tested.

A group of volunteer subjects is drafted, and this group uses the system periodically over some duration (ideally months or even years to capture the temporal variability of the biometrics) while statistics are collected. Many such evaluations will try to compare different biometrics or different vendors' technologies at the same time, so the test facility will contain all the systems being tested, and each volunteer will generally use all of them in sequence. This of course creates a database of biometric samples that can subsequently be used for technology evaluations. Any estimate of a FAR should be scrutinized carefully because the False Accept error condition may be somehow artificially inferred.

7.1.3 Comparison of the methods

Biometric test scenarios are described in the excellent report "Best Practices in Testing and Reporting Performance of Biometric Devices," by Mansfield and Wayman [132]. Quoting

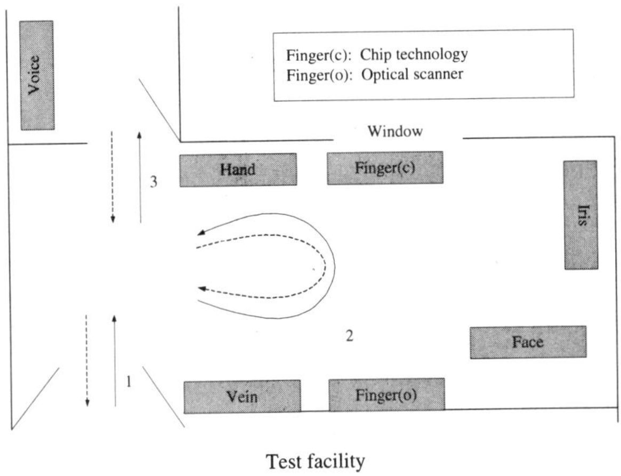

Figure 7.3: Scenario evaluation where volunteers are drafted to use several biometric systems on a regular basis. Adapted from [131].

from the report—

> Our aims are:
>
> 1. To provide a framework for developing and fully describing test protocols.
> 2. To help avoiding systematic bias due to incorrect data collection or analytic procedures in evaluations.
> 3. To help testers achieve the best possible estimate of field performance while expanding the minimum of effort in conducting their evaluation.
> 4. To improve understanding of the limits of applicability of test results and test methods.

The report covers both user tests (scenario evaluations) and testing on databases (technology evaluations). A host of recommendations is given in this report, ranging from how best to select a crew of volunteers to recommendations on what fingerprint sensors to use (sensors that comply with certain standards and technical specifications as described in [42]).

For volunteer selection, which is perhaps the most challenging aspect of biometric testing, the report notes that volunteer selection may bias the tests:

> Individuals with objections against the technology and individuals with physical challenges may be less likely to participate in the testing. It then may be necessary to unevenly select subjects from the volunteers to make the crew as representative as possible and to make sure that problem cases are not underrepresented.

Unfortunately, the authors do not give many practical recommendations on this issue. They note:

> Our understanding of the demographic factors affecting biometric system performance is so poor, that target population approximation will always be a major problem limiting the predictive value of our tests.

Both evaluation methods are used by the biometric industry, but it seems that academia tends to use (often public) databases, i.e., technology evaluations. Of course, during scenario evaluation, databases of biometric samples are processed (and may be collected) and the estimation of error rates becomes a matter of computing estimates based on sample data.

A scenario evaluation using "end-to-end" biometric authentication systems is very different from a technology evaluation. In technology evaluations, biometric matchers are tested on databases with no control of the acquisition procedure (possibly collected from volunteers); in scenario evaluations of the entire end-to-end biometric systems, including acquisition, are tested on volunteers. It is unclear how the test results obtained with such disparate testing procedures are to be compared.

The first problem with most scenario evaluations is that the user population is a closed set and during the test the biometric devices are not really attacked in the true sense of the word. As stated previously, one should always be suspicious about how the False Accepts and False Negatives are determined (from the available samples or otherwise). Estimates of these immeasurable error rates are, of course, impossible to compute without somehow modeling the enemy population. The error conditions that are easy to detect are a False Reject and a False Positive, and therefore the FRR and FPR estimates are more meaningful estimates of the true False Rejects and positives.

Another significant problem with both technology, but more so with scenario evaluations, is that such tests are not "*double blind*" tests as used in drug discovery [41]. It would, first of all, be necessary that *both* the subject *and* the testers do not get to know the matcher decision, because this will influence the results of the scenario evaluation. For instance, if the system returns a wrong result, the tester may be tempted to have the subject submit his biometric a second time.

7.1.4 Limits to evaluation

Some basic requirements for good biometric device testing are—

1. The biometric authentication application should be mandatory to the whole user population or rules for the selection of a subset of the volunteers need to be developed, as hinted at in [132].

2. A related requirement is that the user population should be fairly represented in the test.

3. The subjects themselves should probably be unaware of the matching decision.

Clearly, to measure accuracy fairly and in an unbiased fashion one may find other requirements, but it seems that, because of the fact that people are involved, biometric matcher evaluation is a wide open area of research. However, at a minimum, the evaluation should be performed under the most realistic circumstances and maybe the only place to perform biometric testing is at the actual installation; this type of testing is called an *operational evaluation.*

In the meantime, when reporting realistic error rates for the various biometrics, we have to deal with what we have. The biometrics face, finger, and voice are ahead in terms of technology evaluations. Public databases are collected and maintained by universities (e.g., the University of Bologna) or by government agencies such as NIST and DoD, and comparative evaluation reports are available as already mentioned. On the other hand, for estimating error rates of commercial and academic authentication systems that use iris or hand, no public databases are available and no open competitions are being held. This could be attributed to the lack of competing systems using these biometrics. Therefore, for these biometrics, we have to resort to public reports on biometric technology commissioned by government agencies or other institutions that are mainly scenario evaluations. The signature biometric is a class in itself because much of the signature verification testing is done with technology evaluations but not in a very systematic way.

Incidentally, while an operational evaluation of a biometric system is clearly desired, one cannot measure the *true* False Accept Rate for positive authentication systems and the *true* False Negative Rate for negative or screening biometric authentication systems (watch lists, see Section 6.2.1). Ironically, an operational evaluation cannot measure how well the biometric system can detect the very event that a biometric system is trying to prevent. The error condition could be granting unauthorized access or granting access to undesirable subjects. If it happens, nobody except the actual subject will ever know. Statistics that an operational evaluation can measure are the False Reject Rate as function of a threshold for positive authentication, the False Positive Rate as function of a threshold for negative identification systems, or rank-order statistics that are related to the expected length of candidate lists.

Any attempt at measuring these "hidden" system parameters (i.e., true False Accept Rate for positive authentication) will be by actively trying to defeat a biometric system. These are forgeries, impersonations, disguises, etc., and it is not clear what error rates would actually be measured. Are we measuring the skills of the intruders, or the system error rates? The use of Rank Probability Mass (RPM) statistics to evaluate this type of *operational data* is further explored in Section 7.7.

7.2 Implications of error rates

Before we look at accuracy of the various biometrics, let us look at the implications of various estimates of these 1:1 error rates. O'Gorman [156] gives two examples below that point out problems with the state of the art.

Biometric Authentication—"Why does it reject me?"

Verification protocol: Frequent flyer smartcard with biometric:

1. Assume a system where each person is verified with a fingerprint template on a smartcard or stored in a central database.

2. Each passenger uses a unique frequent flyer number and a fingerprint sample as credentials.

3. Use a typical False Reject Rate (FRR) for finger of 0.03 (= 3 percent).

4. If 5,000 people per hour are requesting access (Newark airport hourly volume) in a 14 hour day,
   roughly *2,100 people will fail to be verified (FR)*
   (3 percent of 5,000 $\times 14 = .03 \times$ 70,000=2,100).
   Of course one could pick a lower FR operating point, but of course, this would adversely affect FA.

5. These people have to be verified through some exception handling procedure (which is part of the protocol). Note that this might be as simple as a retry in some cases.

O'Gorman [156] then points out that throughput (the number of 1:1 biometric matches that can be performed by a biometric matcher per unit of time,—as defined in Section 10.3). Here is the problem: somewhere this verification exception handling has to be done by someone. But the throughput problem is compounded if some form of biometric screening, say, with face images, is performed. O'Gorman notes,

> Even if the probability of a False Positive is set really low, "to inconvenience as few as possible," [at the expense of high probability of False Negatives] there may still be many False Positives.

Biometric Screening—"Why does it point to me?"

Screening protocol: Match passenger face images with government face image database:

1. Assume a system that checks each person's face against a negative database **N** of $n = 25$ alleged terrorists.

2. Use a best-case False Positive Rate (FPR) for face of: 0.001 (= 0.1 percent).

3. If 300 people are requesting jumbo jet access for a flight:
   7 of those will likely match suspected terrorists
   ($25 \times 300 = 7{,}500$ matches are performed,
   which gives $0.001\times$ 7,500 $\approx$ 7 False Positives).
   That is, for each terrorist added to the database, on average 0.3 people per flight will be falsely flagged.

4. Again, the False Positives have to be screened through some exception handling procedure (which is part of the screening protocol).

Note here that face screening is done at a False Negative Rate (FNR) rate of 10 to 20 percent on the ROC. This means that a person in negative database **N** (the most wanted) has an 80 percent to 90 percent chance of being identified; of course, this also means that a person in database **N** (the most wanted) has a 10 percent to 20 percent chance of *not* being identified. However, it is here that the deterrence factor comes in. Publication in the popular press of these type of error numbers may be hard to explain, since just like running a verification system at low FAR (high security), one would desire to run a screening system at low FNR (high security).

O'Gorman then continues with the question, How does a 1 in 1,000 rate result in 7 in 300 False Positives? The short reason is given in Point 3 above. The number of False Positives can be approximated as (see Section 6.3.1, expression (6.15)):

$$\text{FPR}(n) \approx n \times \text{FPR}(1).$$

Thus, matching a positive data set **M** of m subjects, the population of a jumbo jet $m = 300$ now requires m matches against a database **N** of n terrorists. Each passenger has to be matched against each (n) terrorist, where the probability of a False Positive is FPR(n). The number of False Positives is then given by

$$\#\text{ False Positives for plane} = m \times \text{FPR}(n) = m \times n \times \text{FPR}(1). \tag{7.1}$$

7.3 Face, finger, and voice

Let us start with some illuminating numbers on error rates of biometric authentication systems. The numbers in Table 7.1 are directly taken from [13,150,164], the most recent competitive and comparative testing of face, finger, and voice engines. All these competitions are technology evaluations. In this table, finger, face, and voice are all operating at around

Authentication	**False Reject**	False Accept
Screening	False Negative	**False Positive**
Finger	3 to 7 in 1,000 (0.3–0.7%)	1 to 10 in 100,000 (0.001–0.01%)
Face	10 to 20 in 100 (10–20%)	100 to 1,000 in 100,000 (0.1–1%)
Voice	10 to 20 in 100 (10–20%)	2,000 to 5,000 in 100,000 (2–5%)

Table 7.1: Best- and worst-case error rates for face, voice and finger.

10 percent False Reject; best-case False Accept Rates are then 1 in 100,000 ($10^{-5} = 0.001$ percent) for finger, 100 in 100,000 ($10^{-3} = 0.1$ percent) for face and 2,000 in 100,000 (1 percent) for voice, respectively. An approach to infrared face recognition using infrared imaging technology is described in [204] as reporting over 90 percent recognition rates with 100+ individuals. This means that the internal 1:1 matcher runs at FAR, say, roughly 10^{-3} and at FRR around 10^{-1}, 10 percent, which means infrared face recognition may be a close competitor to visible-light face recognition.

7.4 Iris and hand

There are standard biometric evaluations, both technology and scenario evaluations procedures, proposed by Mansfield and Wayman [132] ("Best Practices in Testing and Reporting Performance of Biometric Devices"), maintained by National Physics Laboratory in the UK. The tests we found on iris and hand were done according to these recommendations.

Iris

The test scenario in [131] is verification within "a normal office environment" with cooperative, habituated users. The tests were conducted with 200 volunteers, over a three-month period with 1–2 months between enrollment and verification. The subjects were selected by invitation and, according to the study, few dropped out during enrollment. A test facility was used with the biometric authentication systems positioned roughly as in Figure 7.3. (Other biometrics in addition to iris and hand were tested in [131].)

	False Reject	False Accept	FTE	FTA
Iris	0.0%	2.0%	0.5%	0.0%
Explanation	Two different iris images are falsely matched	Two images of the same iris fail to match	Iris image cannot be acquired for enrollment	Iris image cannot be acquired for verification

Table 7.2: The iris verification error rates from [131].

The iris authentication error rates found in [131] are summarized in Table 7.2 for a factory selected operating point. The FTE (Failure to Enroll), FTA (Failure to Acquire), FAR, and FRR are as defined as in Chapter 5. Also introduced in that section is the FTU (Failure to Use) rate, the probability that a person will not use a voluntary biometric authentication system and will stick to the legacy system. (Compare this to the old lady who insists on dealing with human tellers at the bank.) The volunteers were selected by invitation and most of them accepted the invitation. It is unclear from [131] what the FTU is.

A test of iris identification that dates back to 1996 is the one from Bouchier et al. [24]. In this test the iris authentication system was running not in *verification* mode but in *identification* mode. That is, the authentication protocol is just a single biometric with no claimed identity. Subjects just present the iris and the system makes the correct or incorrect decision, where an incorrect decision (error condition) can only be a misidentification. For such identification systems, the FAR is a function of m, the number of subjects in database M. Again, following (6.15) we have

$$\text{FAR}(m) \approx m \times \text{FAR}(1).$$

The FAR is affected by m; the FRR, however, is not.

In a first test by Bouchier et al. [24], 895 iris identification transactions were recorded with 106 actual rejections, i.e., a FRR of 11.8 percent. The authors observe that many people have problems enrolling with *one* of their eyes and can only enroll with their "dominant eye." These people have less chance of correct identification opposed to people enrolled with two eyes, who have two chances of correct identification. If users that were able to enroll with two eyes were allowed to do so, this FRR therefore drops to 10 percent. These figures are contested by the iris identification system manufacturer (see attachment to [24]); nevertheless, since so few independently estimated error rates for iris can be found, we adopt the 10 percent FRR as worst case, though it is rather anecdotal. Neither the tests by Mansfield et al. nor the tests by Bouchier et al. show any False Accepts.

Wildes in [232] reports on an iris system developed at the Sarnoff Corporation that is not used in a commercial authentication system. The iris database was collected using 60 irises from 40 people, including identical twins and a range of iris colors, blue, hazel, green, and brown. No False Accepts and no False Rejects were reported. However, as the author notes himself, the test is of limited statistical significance because the data set is too small (Table 7.3). Recent testing performed by King [112] for DoD a test with 258 participants and 186,918 identification attempts, reporting 6 percent FRR and 2 potential False Accepts (FAR $\approx$ 0.001 percent). Because the data were not collected during testing, the two False Accepts cannot be reproduced. Sixty people felt that the identification took too long, which is probably due to the repeated image sequence identification attempts. Using the 30-rule [171], to observe a 10^{-5} FAR one needs to produce 30 errors. Hence, $30 \times 10^5 = 3$M attempts are really needed, which implies that the 10^{-5} FAR is therefore very anecdotal. It means, however, that the FAR for iris is not zero.

	False Reject	False Accept
Iris	2 to 10 in 100 (2–10%)	$\geq 10^{-5}$ (0.001%)

Table 7.3: Best- and worst-case error rates for iris from reports [24,112,131].

Why the error rates for iris are as reported is unclear. When running at such an excellent False Accept Rate (FAR $= \epsilon$), running at an equally excellent False Accept Rate (FAR $= 2\epsilon$) should result in lower False Reject Rates (than FRR $=$ 2 to 10 percent). This should be possible to do, no matter how the distance between two biometric samples is defined. That is, the operating point should really be shifted for this application.

Boucher et al. [24] gave a number of reasons for the false iris rejects (which may or may not apply to current systems):

1. User or environmental error: Presenting the wrong eye, not keeping the eye open, reflections from external light sources that obscure the iris.

2. Reflection from glasses: Glasses produce glare in the image. This could be avoided by users, except when the glasses are very thick.

3. User difficulty: Hair obscuring the iris image, difficulty focusing, problems with identifying based on non-dominant eye.

Obviously, contact lenses in the subject's eyes may also negatively affect iris identification.

The error numbers (FAR and FRR) for iris are controversial. The Failure to Enroll (FTE) rate is perhaps the most controversial, there is anecdotal evidence that a segment of the population just cannot enroll with the iris biometric. Some systems require the subject to move around until a series of lights appear to be aligned in order to get a good closeup of their iris. Perhaps some of the FTE is due to this poor user interface. In general, obtaining a high-resolution image of an iris within difficult ambient lighting conditions and with unpredictable specular reflection conditions is a nontrivial task. Computer vision experts are of the opinion that if the developers have solved this task, they have accomplished quite a feat. This is especially true if the imaging system works when the subject wears small, hard contact lenses that may even obscure the iris.

Hand geometry

An experimental hand geometry authentication system is described in [102] and [241], the only two hand verification system published papers in the open literature. Here the hand geometry authentication system was trained and tested using a group of 50 users, resulting in 500 images; 140 images were discarded due to incorrect placement of the hand by the user. The authors note that this will create Failure to Enroll (FTE) problems and False Rejects because of non-habituated users and conclude that user adaptation to this biometric is necessary.

The system is tested by enrolling each person with two hand images and matching mated pairs to estimate the FRR and non-mated pairs to estimate the FAR. This is a technology evaluation of Figures 7.1 and 7.2 with results as in Table 7.4. Note that four different operating points are reported in this table. Acceptable FRR of 3 percent may be achieved but this is at a FAR of 15 percent, which is not a reassuring number from a security point of view.

Hand geometry	
False Reject Rate	False Accept Rate
1 in 33 (3%)	1 in 7 (15%)
1 in 20 (5%)	1 in 10 (10%)
1 in 10 (10%)	1 in 20 (5%)
1 in 3 (30%)	– (0%)

Table 7.4: False Reject Rate, False Accept Rate pairs for the system in [102].

It should be mentioned that these are the numbers achieved by an experimental university prototype, and do not reflect the accuracy of commercially available systems. However, the error rates reported in [102] are the only estimates obtained by a technology evaluation. Other error rate estimates for hand geometry authentication systems that are found in the literature are based on scenario evaluation (user tests). We look at two studies, a 1991 Sandia study [86] and the above mentioned Mansfield report [131].

The 1991 Sandia report "A Performance Evaluation of Biometric Identification Devices" by Holmes et al. [86] describes an "office environment" as in Figure 7.3: "The verifier tests at Sandia were conducted in an office-like environment; volunteers were Sandia employees and contractors. A single laboratory room contained all of the verifiers [including the hand geometry system]. Each volunteer user was enrolled and trained on all verifiers." In another part of the report we find that "nearly 100 volunteers attempted many verifications on each machine."

However, in short, what the study finds is that the Equal Error Rate, where the FAR equals the FRR, for the tested hand geometry verification system is approximately 0.2 percent.

Hand geometry is also tested using scenario evaluation in [131] (as mentioned above, the tests were conducted with 200 volunteers, over a three-month period with 1–2 months between enrollment and verification). This report finds a FTE = FTA = 0.0 percent; it further seems to estimate the hand geometry EER at roughly 1 percent. Putting these numbers together, we get Table 7.5.

	False Reject	False Accept
Hand geometry	2 to 10 in 1,000 (0.2–1.0%)	2 to 10 in 1,000 (0.2–1.0%)

Table 7.5: Estimated error rates for hand geometry from reports [86,131].

Sandia [86] also reports on a user survey regarding the verification systems that were tested. Questions like *Which machine do you feel ..., ... rejects you the most often?* and *... is most friendly/fun?* were asked. The answers for hand verification were remarkably favorable. The manufacturer subsequently used the number of positive responses and the number of negative responses to define an *acceptance ratio* for each biometric device. Hand geometry, based on that measure, is *more than 16 times* more accepted than the second runner-up, a retina verification system; a number which is of course quoted by the manufacturer [241].

7.5 Signature

There are also static off-line and dynamic on-line technologies, but the results are scattered throughout the literature. Static approaches are published in *document analysis* while dynamic approaches are published in *signal processing* or time signal analysis literature.

Signature is commonly called a biometric but it misses one of the necessary conditions as prescribed in [40]; signature does not have the characteristic of *permanence* in that a person can change his or her signature pretty much at will. One could, if one wished, classify signature as a knowledge authentication method but we will continue to refer to signature as a biometric.

Authentication decisions in general in terms of accept and reject are called *genuine* and *forgery*, respectively (see Figure 7.4). The signature verification area has over the years developed an interesting way of classifying the quality of a forgery as

1. *Zero-effort forgery:* A zero-effort forgery is not really a forgery but a random scribble or signature of another individual.

   The FAR and FRR of a verification algorithm can be measured by just taking other non-matching signatures as forgeries (as is usually done for FAR estimation).

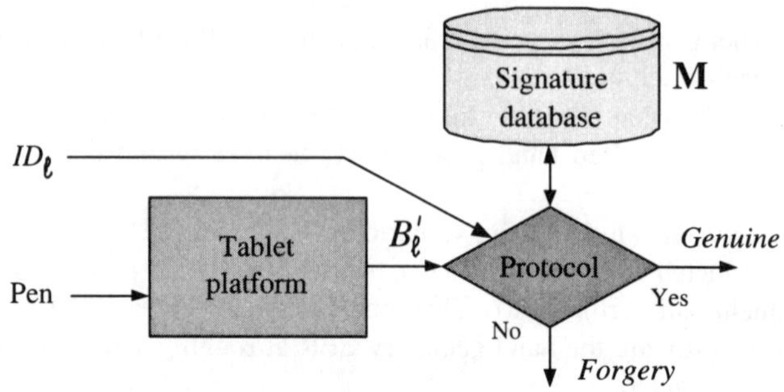

Figure 7.4: Signature verification and decision terminology.

A special case of zero-effort forgery is the writing the correct name in the wrong handwriting.

2. *Home-improved forgery:* These are forgeries which are made when the forger has a paper copy of the signature in possession and has ample time at home to practice and copy the signature at home. The imitation is based on just a static image of the signature.

3. *Over-the-shoulder forgery:* Here the forger is present while the genuine signature is written. The forger does not just learn the spatial image of the signature but also the dynamic properties by observing the signing process, which could be a curve in a five-dimensional space (see Section 3.6).

   Over-the-shoulder and home-improved are not distinguished when testing signature verification matchers. These are called amateur, semi-skilled, or even skilled forgeries [168].

4. *Professional forgery:* These are forgeries produced by individuals that are skilled in the art of handwriting analysis. These people do not necessarily have special skills in forging, but they do know how to circumvent obvious problems and they can use their knowledge to produce high-quality forgeries.

This terminology and methodology probably traces back to the long history of handwriting authentication in the judicial system. The motive also has been the strong desire to detect forgeries right from the start of research in automated signature verification, simply because there have been many instances of the need for signature forgery detection and writer authentication over the ages.

In that sense, the signature verification literature makes an attempt to define levels of severity of an attack on the verification system, i.e., the impersonation attacks described elsewhere in this book. That is, a distinction is made between *passive False Accepts* and *active False Accepts* (discussed in Chapter 9, as attacking the "lambs"). So from the beginning the area defined different sets of error numbers associated with a signature verifier, loosely defined by the type of attack: attack by random signatures, attacks by amateur forgers, and attacks by professional forgers. For none of the other biometrics has much effort been devoted to the skill level of the impersonator. In the voice verification area Doddington et al. [51] have made an initial attempt at qualifying the skill level of the attacker by introducing a category of speakers called *wolves,* those speakers that are particularly good at imitating other voices (Chapter 9).

The signature research area selected more complicated ways of testing a verification system. Moreover, signature as a biometric has made rapid advances over the years, mainly perhaps because of rapid advances in the input devices [57]. These rapid developments probably impeded the development of a standardized testing and evaluation culture in this technology area. Quoting Dolfing [52]: "In general, the comparison of signature verification approaches is difficult due to the different equipment, algorithms, models and databases. There is no standardized benchmark or database. Additionally, the type and quality of forgeries used to estimate the verification accuracy is different for each approach. This leads to False Acceptance, False Rejection and Equal Error Rates which make no sense without information about other system parameters."

Automated signature verification has a long history, starting with approaches to static signature verification as described as early as 1977 [141]. An early survey paper from 1988 by Plamondon and Lorette [168] covers both static and dynamic approaches to signature verification. They note that signature outperforms initials and handwritten passwords. They further note dynamic signature is superior and that (in 1988) static writer verification (from a full page of text) performs about as well as static signature verification. Judging from [25], where it is concluded that signature and writing offer complementary (rather than correlated) information, the issue of whether writer verification is superior to static signature verification is still unresolved.

Unfortunately, the study [131] does not include a user study (scenario evaluation) of dynamic signature; and we have to go back to the 1991 Sandia report [86] to find some numbers from a user study of a commercial dynamic signature verification system. Error numbers for several verification protocols (one-try, two-try, and three-try) are given, which shows a large improvement from the one-try to the two-try numbers (Table 7.6). This may indicate that a subject is much better able to put the signature so that it can be matched in the second try, perhaps because habituation of the biometric on that particular device is poor.

Dynamic signature		
	False Reject	False Accept
Three-try	2.06%	0.70%
Two-try	2.10%	0.58%
One-try	9.10%	0.43%

Table 7.6: The signature error numbers from [86] allowing one or more tries.

Dolfing [52] does give comparative error estimates of a number of dynamic signature verification systems and we include in this list also the system described in [121]. The IBM system described by Worthington et al. [238] does best with FRR = 1.77 percent and

FAR = 0.28 percent based on the first signature. Given that the proprietary database of signatures that is used to test the system of [238] probably consists of habituated signatures, it may be best to compare the two-try protocol of the system of Table 7.6 to the IBM system; the system [121] has error rates less than FRR = 0.2 percent and FAR = 0.2 percent. This means that we find some consistency between signature accuracy numbers obtained by a technology evaluation and scenario evaluations. We roughly combine the one-try numbers of both tests and arrive at Table 7.7.

	False Reject	False Accept
Signature	2 to 10 in 100	2 to 5 in 1,000

Table 7.7: Best- and worst-case error rates for *dynamic signature* [52,86].

The FAR for static signature seem to be about a factor 10 less, when the FRR is in the range 2–5 percent.

Interestingly enough, the matching technology that is used in [238], referred to as "regional correlation," is more akin to approaches found in fingerprint and face representation technologies.

Signature, as a man-made biometric, is doing surprisingly well in comparison to the other biometrics. The reported error rates that we find are consistently in the range of EER = 0.2–0.5 percent; however, the signature verification has different evaluation criteria in that error rates for different types of forgeries, like *zero-effort* forgery, *home-improved* forgery, and *professional forgery* are measured separately, and it is even more difficult to relate the different studies.

7.6 Summary of verification accuracies

Table 7.8 summarizes *the best error rates* that we found in the literature and as they are described in Sections 7.3, 7.4, and 7.5. The last column gives the type of testing, done for the particular biometric; T is technology and S is scenario evaluation. It cannot be emphasized enough that face, finger, and speech are tested with technology evaluations; iris and hand are mainly tested with scenario evaluations; and signature is tested with both technology and scenario evaluations (Figures 7.1–7.3).

The problem in general with any biometric authentication system, again, is one of *volume* (see Section 10), especially if using one-attempt authentication protocols. The more matching attempts for verification results the fewer False Rejects. But, the higher the number n of terrorists in the most wanted list (the negative database N), the more False Positives. This carries through to (positive) identification of m enrolled identities in positive database M. In an identification application a False Positive is like a False Accept for security purposes. Hence, when security is involved, the immediate conclusion is that it is *never* a good idea to use "pure identification" because both empirical and theoretical error rates of biometric identification (1: m matching) are high, in which case the exception handling needs much attention [156].

	False Reject / (FN)	False Accept / (FP)	Evaluation method
Fingerprint	3 to 7 in 100 (3–7%)	1 to 10 in 100,000 (0.001–0.01%)	T
Face	10 to 20 in 100 (10–20%)	100 to 1,000 in 100,000 (0.1–1%)	T (S)
Voice	10 to 20 in 100 (10–20%)	2,000 to 5,000 in 100,000 (2–5%)	T
Iris	2 to 10 in 100 (2–10%)	$\geq 10^{-5}$ ($\geq$ 0.001%)	S
Hand	1 to 2 in 100 (1–2%)	10 to 20 in 1,000 (1–2%)	S (T)
Signature	10 to 20 in 100 (10–20%)	2 to 5 in 100 (2–5%)	T & S

Table 7.8: Roughly the error rates that can be found in the literature, based on scenario (S) and technology (T) evaluations.

However, when using any of the biometrics in Table 7.8 for screening (or negative identification), the only acceptable FPR (if any of the FP rates are acceptable) is obtained by running the authentication systems at (roughly) the FNR in the table. This means that fingerprint will then run at 3–7 percent and face and voice run at 10–20 percent and a person in the screening database **N** has anywhere from a 3 percent to 20 percent chance of *not getting caught.*

Finally note that the tabulated values were measured at operating points that might not be suitable for a particular application (decreased FAR can be traded for increased FRR to some extent). As a rule of thumb, an authentication system would in practice be run with a FRR around 5 percent (which merely inconveniences the user), while a screening system would typically require a FRR of 0.1 percent (for subjects who require exception handling).

7.7 Identification system testing

Biometric 1:1 matchers are tested using (sequestered) databases of biometric test samples **D** as in Figure 7.2. Biometric 1:m search engines are also tested using databases **D** of biometric test samples. Unfortunately, very few biometrics have been used in 1:m identification systems and, of these, even fewer have been tested competitively using rank-order statistics as the CMC. One primary exception is face recognition.

7.7.1 Biometric data and ROC, CMC

Biometric capabilities such as ranking and matching need to be developed by modeling biometric data and training using biometric data. We have seen specific uses of data already, for example:

1. In Section 7.1 we already saw that many data sets are used for different biometric functions and tests.

2. We saw in Chapters 5 and 6 that there are two different biometric statistics, the ROC and the CMC —score- and rank-based statistics, respectively.

The latter can be summarized as follows:

1. *ROC:* This statistic measures the capabilities of a match engine $s(B', B)$ with some fixed t_o or as a function of some operating threshold T.

2. *CMC:* This statistic measures the capabilities of a rank engine $R((B_1, B_2), B'_\ell)$ with ordered entries $(B_1, B_2) \in \mathbf{M}$ and some unknown sample B'_ℓ.

   However, the desired capabilities of a rank engine are not well understood.

The way the CMC is estimated today assumes that rank $R(B_1, B_2)$ is determined by scores $s(B'_\ell, B_1)$ and $s(B'_\ell, B_2)$. Hence, it is assumed that better ranking is achieved using a better biometric matcher. The modular design of a system as in Figure 7.5 allows for better use of data in training/testing of the biometric subsystems.

The statistics ROC and CMC are intricately related but distinct statistics. The ROC is good for scores, continuous random variables; the CMC is good for ranks, discrete random variables.

7.7.2 Biometric search engines

In this section, we explore a hybrid approach (see Section 6.1.2) to the $1\!:m$ problem: *ranking* followed by *scoring*. We assume here that ranking is an important biometric capability, which is *different* from biometric matching. One might also build a hybrid identification engine by performing matching first and ranking next, or both simultaneously using special mechanism like indexing. There surely is no widespread consensus about this assumption in the biometrics research community.

The input to a $1\!:m$ search engine is a biometric sample B'_ℓ, the output is a (sorted list) vector $\mathbf{C}_K(B'_\ell) = (\text{ID}_{(1)}, ..., \text{ID}_{(K)})^T$. This reduction process from m possible identities in $\mathbf{M}$ to the sorted shorter vector $\mathbf{C}_K$ of K identities can be done in 2 steps:

1. Depending on input query B'_ℓ a (more-or-less) complete reordering of a portion of vector $\mathbf{M}$ to vector $\mathbf{C}_{m'}$ is determined; i.e., $\mathbf{C}_{m'}$ is just a permutation of a portion of $\mathbf{M}$. This is achieved by a *biometric rank engine*. It *does not imply* that all m entries need to be ranked!

2. A shorter candidate list $\mathbf{C}_K$ is derived from $\mathbf{C}_{m'}$ by using a *biometric match engine* to make threshold-based match decisions (some threshold t_o).

Basically, this is somewhat of a speculation, because existing $1\!:m$ systems are highly proprietary.

Let us accept two biometric capabilities as *basic* biometric modules:

1. *Biometric ranking:* Determine if the order of two biometric templates (B_1, B_2) with respect to a third biometric template B'_ℓ is correct or not.

2. *Biometric matching:* Determine, using a scoring function $s(B, B')$ and decision threshold t_o, if two biometric samples B and B' match or not.

Let us also accept that a search engine in its simplest form is implemented as the cascade of a rank engine, followed by a match engine as shown in Figure 7.5.

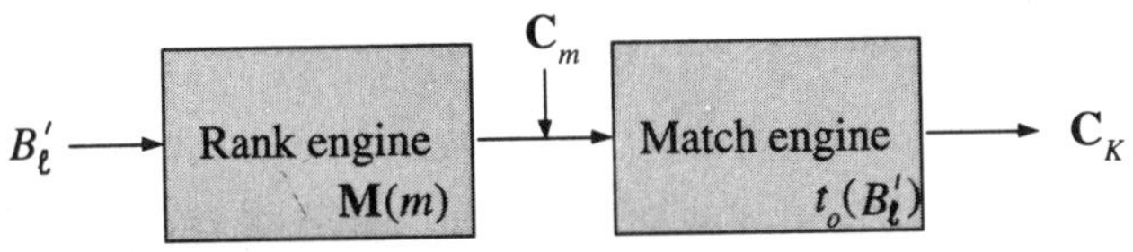

Figure 7.5: A (1: m search engine) is a rank engine followed by a match engine; manual labor may be needed at all stages of the search.

The 1: m search engine with an enrollment database $\mathbf{M}$ is defined as a process of reducing vector $\mathbf{M}$ to a vector $\mathbf{C}_K$, given an input query B'_ℓ:

$$\mathbf{C}_K = (B_{(1)}, B_{(2)}, ..., B_{(K)})^T = (\mathrm{ID}_{(1)}, \mathrm{ID}_{(2)}, ..., \mathrm{ID}_{(K)})^T.$$

If $B'_\ell \notin \mathbf{M}$, C_k should be $\emptyset$. A possible architecture to achieve this is proposed in Figure 7.5:

1. *A biometric rank engine:* Determine some reordering $\mathbf{C}_m$ of vector $\mathbf{M}$ by repeatedly applying *biometric ranking*, followed by—
2. *A biometric match engine:* Determine, using a scoring function $s(B'_\ell, B_{(k)})$ and decision threshold $t_o(B'_\ell)$, a short candidate vector $\mathbf{C}_K$ of the K top candidates in order of decreasing likelihood.

7.7.3 1: m search engine testing

A big distinction of a 1: m search engine, compared to a 1:1 matcher is the prerequisite of an enrollment database $\mathbf{M}$. The available test samples are therefore used differently, since an identification system by necessity implies an enrollment database $\mathbf{M}$ of m entries:

$$\mathbf{M} = (B_1, B_2, ..., B_m)^T. \tag{7.2}$$

This is the enrollment database and we cannot stress enough that the quality of the enrollment database $\mathbf{M}$ in real-world systems needs to be carefully controlled and monitored over the lifespan of a system. For 1: m search engine testing, this should be kept in mind, but in (7.2) we just select the first m samples in $\mathbf{D}$ as database samples.

Then for all other samples, denoted as $\{B'_\ell,\ \ell = m+1, ...\} = \mathbf{D}\backslash\mathbf{M}$, a rank $\hat{r}(B'_\ell)$ is estimated as follows:

1. Compute the sets of scores

$$\mathbf{X}_\ell = \{s(B'_\ell, B_i);\ i = 1, ..., m\} \text{ for } \ell = m+1, ...$$

2. Sort these scores:

$$\tilde{\mathbf{X}}_\ell = (s(B'_\ell, B_{(1)}), s(B'_\ell, B_{(2)}), ..., s(B'_\ell, B_{(m)}))^T, \tag{7.3}$$
$$\text{such that } s(B'_\ell, B_{(k)}) > s(B'_\ell, B_{(k+1)}),\ 1 \le k < m. \tag{7.4}$$

Hence, the set $\mathbf{X}_\ell$ is rewritten as a vector $\tilde{\mathbf{X}}_\ell$ of scores, from high to low, of B'_ℓ matched with $B_i \in \mathbf{M}$.

3. If $(B'_\ell, B_{(k)})$ is the mated pair, i.e., if $B_i = B_{(k)}$ matches B'_ℓ, $\hat{r}(B'_\ell) = k$.

The result of such a test is just a set $\mathbf{K}$ of estimated ranks. This $\mathbf{K}$ could be rewritten as a set of sets $\mathbf{K} = \{\mathcal{K}_i,\ i = 1, ..., m\}$. A set of ranks $\mathcal{K}_i$ can be obtained by selecting those B'_ℓ that are impressions in $\mathbf{D}\backslash\mathbf{M}$ mated with database biometric $B_i \in \mathbf{M}$.

Note, however, that (7.3) assumes that sorting the scores is *the way* to find the most similar $B_i \in \mathbf{M}$ to input query B'_ℓ. The location k of mated pair $(B'_\ell, B_{(k)})$ in the sorted list is then the output rank estimate $\hat{r}(B'_\ell)$, as shown in Figure 7.6.

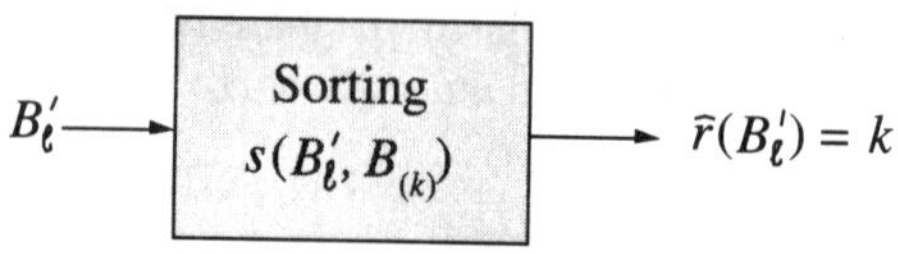

Figure 7.6: An identification system could be designed by sorting the scores $s(B'_\ell, B_{(k)})$ and defining rank $\hat{r}(B'_\ell) = k$ so that $(B'_\ell, B_{(k)})$ is a mated pair.

In that sense this is a closed-world test in that only samples from enrolled user population $\mathbf{M}$ are used. The biometric search problem is (currently) best modeled by a finite discrete probabilistic ranking model that determines the sorted score (rank) $\hat{r}(B'_\ell)$ of an input query biometric B'_ℓ with respect to $B_i, i = 1, ..., m \in \mathbf{M}$. The preferred performance statistic for comparing the different matchers is the Cumulative Match Curve (CMC) as defined in Section 6.4.2, with an estimate:

$$\text{CMC}(k) = \frac{1}{|\{B'_\ell\}|} \sum_{i=1}^{k} (\# \hat{r}(B'_\ell) = i) = \frac{1}{|\{B'_\ell\}|} (\# \hat{r}(B'_\ell) \le k),\ \ k = 1, ..., m. \tag{7.5}$$

That is, CMC(k) is the fraction of the B'_ℓ that have rank (or are sorted) $\hat{r} \le k$. The CMC corresponds to a Rank Probability Mass (RPM) function

$$\hat{P}(k) = \frac{1}{|\{B'_\ell\}|} (\# \hat{r}(B'_\ell) = k),\ \ k = 1, ..., m, \tag{7.6}$$

which is an estimate of (6.29). The estimate given in (6.30) is another way to estimate the Rank Probability Mass function of the 1: m search engine with enrollment database $\mathbf{M}$.

The desired output of a 1: m search engine is that the output rank of B_ℓ is 1 if (B'_ℓ, B_i), $B_i \in \mathbf{M}$ is a mated pair. This, of course, only happens when $B_i = B_{(1)}$ after sorting the scores $s(B'_\ell, B_i)$. In fact, if B'_ℓ is associated with an identity ID_i or sample B_i in **M**, rank $\hat{r}(B'_\ell)$ should be 1. Equivalently, identity ID_i should be associated with rank 1, i.e., $\text{ID}_{(1)} = \text{ID}_i$, the very first one of the sorted identities. This is not achievable in practice, and it is desired that rank $\hat{r}(B'_\ell)$ be as low as possible. Given a $1\!:\!m$ search engine, we have—

1. An enrollment database $\mathbf{M} = (B_1, ..., B_m)^T$,
2. The desired rank $\hat{r}(B'_\ell) = 1$, if B'_ℓ is a sample of $\text{ID}_i \in \mathbf{M}$,
3. The $1\!:\!m$ search process produces low ranks for all samples of $\text{ID}_i, i = 1, ..., m \in \mathbf{M}$.

The Rank Probability Mass estimate $\hat{P}(k)$ of (7.6) of a $1\!:\!m$ search engine should have high probability values $\hat{P}(k)$ for low values of k.

7.7.4 Face Recognition and Verification Test 2000

FRVT 2000 (Face Recognition and Verification Test) [15] was the first attempt to characterize performance measures for assessing commercially available face identification systems. The five participating vendors had to compute an all-against-all match of a database of 13,872 face images with varying parameters of compression, image distance, and facial expression. For each vender, similarity scores were thus obtained between all pairs of images in the database, yielding more than 192 million scores per identification system. As outlined in Section 7.7.3, many rank estimates $\hat{r}$ are computed for many B'_ℓ and the $\text{CMC}(k)$ is estimated as in (7.5).

FRVT reports performance in terms of identification rate for most of the experiments. Based on the CMC, the test report characterizes each of the five algorithms by the sorted scores returned by the algorithms compared to the known ground truth. CMCs for several different test parameter configurations are included. The impact of factors such as compression, distance, resolution, expression, and media on the system performance were also carefully studied. As a bonus, the report provides performance measures of the simpler verification task in terms of ROC curves.

Some of the results of the report are—

- Compression of the face image does not impact performance adversely.
- Pose changes up to 25 degrees can be handled by the algorithms, but beyond 40 degrees the performance degrades sharply.
- Recognizing face images taken 12 or more months apart is difficult.
- Performance of the system decreased when the distance between the camera and the person increased.
- Identification is more sensitive to expression changes than verification is.
- Matching indoor images with outdoor images will require further research to achieve adequate performance.

It should be noted, however, that FRVT in reality tests the sorting capabilities of the five particular participating 1:1 face-matching algorithms. The implicit assumption here is that sorting match scores is the best approach to $1\colon m$ search.

7.7.5 FRVT 2002

FRVT 2002 extended the FRVT 2000 test program with a much broader test suite. The database size increased from 1,196 in FRVT 2000, to 37,347 in FRVT 2002. For the first time, the standard error ellipses were introduced in reporting the errors. The errors as a function of gallery size m and sorted list size K were comprehensively covered in FRVT 2002. In terms of the accuracy numbers, the general conclusion of the FRVT 2002 report was that the identification errors have been reduced significantly. The difference in results for the plain *verification* task are summarized in Table 7.9. Note that different databases were used in these experiments, so care should be taken when directly comparing these numbers, but they give a rough indication of the improvement in face recognition.

	FRVT 2000	FRVT 2002
FAR	FRR	FRR
10^{-4}	–	27%
10^{-3}	74%	17%
10^{-2}	58%	10%
10^{-1}	32%	4%

Table 7.9: Verification error rate (FRR) as a function of FAR in both FRVT 2000 (M34) and FRVT 2002 (HCINT).

FRVT 2002 [164] provides performance measures for assessing the capability of commercially available face recognition systems. Ten participating vendor software algorithms were tested extensively in three main regimes. The high computational intensity (HCINT) test consisted of matching 121,589 images of 37,437 persons. Computation of the all-against-all scores requires matching of more than 15 billion pairs. The best error rates for the *verification* part of the HCINT test are summarized in Table 7.9. The medium computational intensity (MCINT) test consisted of still image and video face recognition. In the still images category, the query set consisted of 7,722 face images that required the computation of 56 million scores. Finally, the video face recognition test consisted of 1,008 facial video sequences as well as 882 still facial images. A special evaluation of 3D morphable models was also made in this category. This test showed that a 3D morphable model used to reconstruct a frontal view, as a preprocessing stage, independent of the recognition algorithm, significantly improved the recognition rate of most of the vendors' systems on oblique views of faces.

The upshot of the FRVT 2002 testing is that, under controlled lighting, face recognition performance can be very good. The best *identification error* performance is summarized in Table 7.10a, as the percentage of trials in which the correct identity appears in the list of the top K candidates. Note that this identification test was a closed-world test; hence, we only have the identification rate as a function of the short-list size K, and no intruders are

K	Identification error
1	35%
10	18%
100	10%

(a)

m	Identification error
25	7%
100	8%
1,000	17%
10,000	23%

(b)

Table 7.10: FRVT 2002 Identification performance.

expected as input queries. Table 7.10b summarizes the error rates for the single top-ranked candidate $K = 1$ as a function of the database size m.

The sorting capabilities of the face-matching systems were also tested in an "open" mode as watch list systems. Here, the $K = 1$ best performance ranged from 53 to 73 percent with the FAR ranging from 10^{-3} to 10^{-1} using a watch list database of $m = 3{,}000$ persons. More details of these test results can be obtained from [164].

FRVT 2002 was a significantly more elaborate program than FRVT 2000. FRVT 2000 consisted of face images taken indoors on the same day, indoors on different days, and images with expression and illumination changes. FRVT 2002 expanded upon this set by adding images taken outdoors on the same day, outdoors on different days, and also examined pose (position and orientation) variations. However, a few analyses have been dropped from the FRVT 2000, assessing the impact of compression, image resolution, and distance from the camera.

Some of the other interesting results provided in the report include—

- Identification performance as a function of the database size m and short-list size K.
- Estimation of variance of performance across different groups of people (by sex and age groups).
- Performance characterization as a function of elapsed time difference between enrollment and verification.

The report also documents the performance of face recognition from videos, and the impact of 3D morphable face models (for video) and score normalization in terms of accuracy.

FRVT 2002 also compares the performance of face recognition systems to that of a specific fingerprint matcher. However, the comparison is questionable since it is unclear how well the fingerprint system used performed relative to available commercial systems. Also the population and acquisition conditions used for the fingerprint database may have been substantially different from those used for the face database.

7.8 Caveats

In this section we have reported discrete FAR and FRR values for verification and identification. Generally for verification systems the FRR rate is set to the maximum allowable (typically around 10 percent) and then the systems are tuned to give the best corresponding FAR rate. For identification systems, the short-list size K is constrained to be only a single candidate, or some other fixed small number of candidates and the best identification rate is reported for these scenarios. In many ways it would have been more useful to show representative ROC and CMC curves, respectively, to illustrate better the range of operating points available. However, systems using different biometrics (e.g., fingerprint versus hand) are often tuned for, and evaluated over, different ranges of operating points and error rates.

Also, while we have reported on the efficacy of face recognition as an identification method, we have not mentioned identification by fingerprint. There are a huge number of AFIS systems in operation worldwide. These systems are apparently providing satisfactory identification functionality but their true operating accuracies are seldom reported (possibly for proprietary or security reasons).

Finally, lots of sample data exist for the various biometric identifiers (finger, face, etc.). However, if not used carefully, the existence of all this (legacy) biometric sample data may well be more of a liability than an asset. Legacy and acquired biometric data is used for training, enrollment, matcher design, rank-engine optimization, testing, match- and search-engine comparisons, etc. The biometric sample data needs to be well annotated and documented because of the many ways this data is being used.

8

Selecting a Biometric

The choice of biometric is *not* just based on error rates. Many factors are involved in selecting a biometric, given some application. The cost, or added cost, and security of the installation, no doubt, depend on the choice of biometric; therefore, selecting the appropriate biometric for an application is of prime concern. Accuracy is an important factor in selecting a particular biometric, but by no means does it have to be the most important factor (see Figure 8.1).

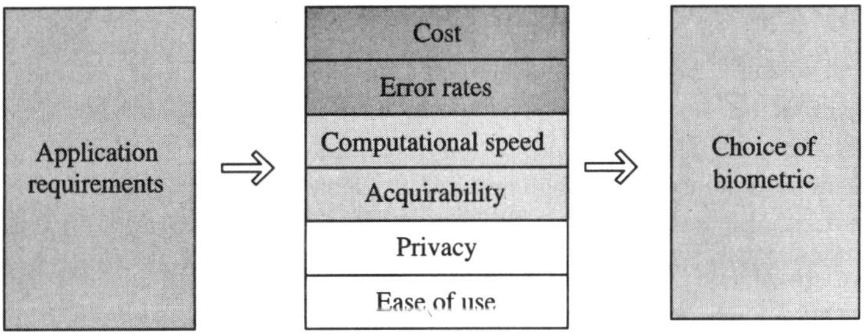

Figure 8.1: Selecting the "right" biometric is a complicated problem that involves more factors than just accuracy.

In the first part of this chapter we take a comparative look at the six most common biometric identifiers discussed in Chapter 3, but now we compare the biometrics in terms of additional important properties and attributes other than just the error rates. In the second portion of the chapter we examine the requirements of particular application scenarios and how these impact the choice of biometric. Finally, we discuss how to go about selecting a biometric for a particular application.

Here we are concerned with the more-or-less objective evaluation of biometric systems. However, this is not necessarily what drives sales. From a manufacturer's point of view, aesthetics can be particularly important [65]:

	Finger	Face	Voice	Iris	Hand	Signature
Maturity	very high	medium	medium	medium	high	medium
Sensor type	contact	non obtrusive	non obtrusive	non obtrusive	contact	contact
Sensor size	small	small	very small	medium	large	medium
Sensor cost	< \$200	< \$50	< \$5	< \$300	< \$500	< \$300
Template size	< 500	< 1,000	2,000	256	< 100	200
Scalability	high +	medium	low	very high	low	high –

Table 8.1: Comparison of the attributes of the six popular biometrics.

"Good looking devices outsell ugly ones regardless of reliability," Mr. Argenbright said. "It kills me."

8.1 Biometric attributes

Several intrinsic and extrinsic attributes of the biometrics and the sensors that are needed to acquire the biometric signal can be used to compare the six common biometrics, as shown in Table 8.1.

One of the more important, though not directly quantifiable, properties is—

- *Maturity:* This property deals with the stage of development of the technology of acquiring and recognizing a biometric, as well as the understanding of the processes that affect the variations in the biometric samples. A well-studied science behind a biometric that explains the uniqueness of the biometric, makes the biometric more mature than a less well-understood biometric.

  Another, related aspect is the use of the biometric in human expert biometric systems, where the question as how to handle the "abnormal" cases has been studied over large samples of diverse collections.

While the processing times for performing biometric matches is certainly a significant factor, in this book we do not consider them as a biometric property, per se. These processing times are much harder to pin down because they depend on how efficiently the code was written and what sort of processor it runs on. Moreover, computing time is becoming less of an issue as processors get ever faster. What we do discuss are processing times as a function of m, the database size, as in Chapter 10.

8.1.1 Sensor properties

As listed in Table 8.1 there are several sensor-related issues to consider when selecting a biometric system.

- *Sensor type:* A contact or non-contact type of sensor can be used to acquire the biometric signal. Non-contact sensors can be useful in covert, surveillance applications, whereas contact sensors require the cooperation and knowledge of the subject to provide the biometric. In general, the use of non-contact sensors is much less obtrusive than contact sensors.

- *Sensor size:* Some biometric sensors like microphones or hidden cameras can be very small. For some, a complex optical setup is needed to acquire the signal reliably (e.g, for iris), possibly resulting in a larger sensor. Sometimes the physical security of the sensing device is important in unmanned installations involving biometrics. This demands that the sensors be physically strong to be able to handle physical attacks. Such extra requirements can make the sensor large (and costly). Of the sensors used for the six biometrics, hand geometry sensors by nature are largest, as they have to acquire the image of the full hand length, width, and height.

- *Sensor cost:* An issue related to the sensor size is the sensor cost. Even though the costs of biometric sensors decrease every year, the relative cost of one sensor compared to another remains more or less the same. Note that the cost is the retail cost of the basic sensor and does not include the cost of other associated materials such as cables, kiosks, user interfaces, and maintenance needed to run an installation smoothly.

8.1.2 Template sizes

A template is a machine representation of a biometric sample. Hence a template in some way describes the acquired biometric sample so that automated matching of biometrics can be done as precisely as possible. The size of this representation, in terms of bytes or kilobytes (KB), can be an important factor that decides the type of biometric that can be used in an application. Smaller templates enable the use of smaller storage devices like magnetic stripe cards and the implementation of truly distributed biometric databases.

In addition to the template storage, often other associated data such as ID number needed by the matching algorithm may be required. Sometimes compression techniques can be used to make the stored templates compact, or encryption and digital signatures might make them larger. To address this, we define an *on-line* storage (template) as the run-time storage requirement of a biometric representation and *off-line* storage requirement of the compressed biometric records in the biometric database. The approximate template size numbers quoted in the last row of Table 8.1, and throughout this book, are on-line storage sizes. While these are issues to be aware of, we only report template sizes (on-line storage) for the biometrics of interest in this book.

Template sizes for the various biometrics vary a lot, from very small for iris to larger for voice; there is also variation in the reports on template sizes. For example, Ruggles in a study [193] prepared for the California State Legislature reports (in bytes) iris = 256, finger

= 512–1K, hand geometry 9 bytes. Using Ruggles, and roughly following the advances in the state of the art, we summarize storage requirements as the last row of Table 8.1.

Template sizes are a very poor indication of individuality, bit strength, or False Accept Rates of a biometric. In general we would expect a small template to have less discriminant information than a larger one, but if the representation is inefficient, a large template may contain less information than a smaller one, and may have worse False Accept Rates.

As noted before, for privacy or security reasons, it may be desirable to store the authentication credentials, including the biometric sample or machine representation, on a distributed database, for example, held on smartcards in the possession of the enrolled population. This way, the biometric sample never leaves its owner. For central databases, there has to be some kind of trust established between the users of an application and the database owner and operators—the trust that the biometric sample are held in confidence.

When looking at biometrics from a storage point of view, a first issue in biometric authentication systems is *verification* versus *identification*. Identification involves 1:m matching of biometric representations while authentication only involves 1:1 matching. This means that for identification, large centralized databases $\mathbf{M}$ are needed and that in general an order m times more computer operations are required per match.

There exist fingerprint matching algorithms that find matches databases with large m, using transform clustering [209]. Extensions of these approaches can be implemented using geometric hashing schemes [237] that deal with increasing m by using more storage for a larger machine representation of the entire database. These are approaches that basically use space-time trade-offs. Sometimes the storage requirements depend almost exponentially on m (for example, [75]).

Storage may also become an issue when the algorithms use the notion of cohorts (see Section 9.5.4) because then if the individual templates are stored in a distributed database of biometric templates, for each individual biometric representation somehow the cohort or the world (if this is the cohort model) needs to be included in the template.

8.1.3 Scalability

The issue *scalability* in the fifth row of Table 8.1 is a little different. The scalability of a biometric is tied to the intrinsic individuality property of the biometric and is very much related to the error rates as discussed in Chapter 7.

A highly scalable biometric can be deployed to identify people in a large population without incurring unacceptable error rates or throughput times. A poorly scalable biometric can only handle small databases well, and would have larger errors while dealing with a large database. In fact, large database sizes cannot and should not be handled with the weak biometrics as enrollment and authentication processes are likely to become clogged with the frequent need for manual intervention and exception handling as we will see in Chapter 10.

Face and fingerprint, and indeed the other biometrics, may be much less accurate than is commonly believed. Still the accuracy numbers of the table indicate that there is a group of strong biometrics (iris and finger) and a set of weak biometrics (hand, face, voice). Signature is a wild card here, in that its scalability is hard to judge because of the lack of large scale tests.

It is important to understand that breakthrough technology would be needed to elevate the weak biometrics into the class of strong biometrics; i.e., simple tuning and incremental improvements are not sufficient. The current rate of improvement is too slow to expect adequate accuracy and successful deployment in the high-security protocols in the near future, and accelerated research programs in these "acceptable" biometrics is warranted.

It should be noted that, in some cases, scalability is actually an undesirable characteristic, since it can add to a lack of acceptability because of privacy issues. Using a strong biometric for a verification system such as dormitory meal plan, or employee time and attendance requires the presentation of a biometric that could be used to identify the user in some other database (such as searching a police records), raising users concerns about privacy.

8.2 Application properties

It is difficult to answer what the "best" biometric is without thinking about what type of an application the biometric is going to be used in. Here we will briefly examine the categories of biometric applications typically encountered.

There are plenty of places where biometric authentication can bring additional security. There are access control points of various security levels, and there are authorizations of all sorts of transactions. What is important in a biometric authentication application varies among the applications. Salient questions about applications are—

1. what is the protected asset;
2. who are the authorized users and operators;
3. who are the adversaries of the application;
4. what can be the costs of security violations;
5. what is the actual cost of the security; and
6. what is the added "cost" of inconvenience?

We will look at various applications in relation to these types of questions.

8.2.1 Wayman's application taxonomy

Wayman in [229] introduces a useful taxonomy of biometric applications and it is best to simply quote Wayman verbatim:

> Each technology has strengths and (sometimes fatal) weaknesses depending upon the application in which it is used. Although each use of biometrics is clearly different, some striking similarities emerge when considering applications as a whole. All applications can be partitioned according to at least seven categories.

Cooperative versus non-cooperative: This refers to the behavior of the "wolf,"[1] (bad guy or deceptive user). In applications verifying the positive claim of identity, such as access control, the deceptive user is cooperating with the system in the attempt to be recognized as someone s/he is not. This we call a "cooperative" application. In applications verifying a negative claim to identity, the bad guy is attempting to deceptively not cooperate with the system in an attempt not to be identified. This we call a "non-cooperative" application. Users in cooperative applications may be asked to identify themselves in some way, perhaps with a card or a PIN, thereby limiting the database search of stored templates to that of a single claimed identity. Users in non-cooperative applications cannot be relied on to identify themselves correctly, thereby requiring the search of a large portion of the database. Cooperative, but so-called "PIN-less," verification applications also require search of the entire database.

Overt versus covert: The second partition is "overt/covert." If the user is aware that a biometric identifier is being measured, the use is overt. If unaware, the use is covert. Almost all conceivable access control and non-forensic applications are overt. Forensic applications can be covert. We could argue that this second partition dominates the first in that a wolf cannot cooperate or non-cooperate unless the application is overt.

Habituated versus non-habituated: The third partition, "habituated versus non-habituated," applies to the intended users of the application. Users presenting a biometric trait on a daily basis can be considered habituated after short period of time. Users who have not presented the trait recently can be considered "non-habituated." A more precise definition will be possible after we have better information relating system performance to frequency of use for a wide population over a wide field of devices. If all the intended users are "habituated," the application is considered a "habituated" application. If all the intended users are "non-habituated," the application is considered "non-habituated." In general, all applications will be "non-habituated" during the first week of operation, and can have a mixture of habituated and non-habituated users at any time thereafter. Access control to a secure work area is generally "habituated." Access control to a sporting event is generally "non-habituated."

Attended versus non-attended: A fourth partition is "attended/unattended," and refers to whether the use of the biometric device during operation will be observed and guided by system management. Non-cooperative applications will generally require supervised operation, while cooperative operation may or may not. Nearly all systems supervise the enrollment process, although some do not [231].

[1] Here Wayman refers to users who by intention behave like a "Wolf" from the Doddington zoo of Chapter 9 [51].

Standard environment: A fifth partition is “standard/non-standard operating environment.” If the application will take place indoors at standard temperature (20° C), pressure (1 atm.), and other environmental conditions, particularly where lighting conditions can be controlled, it is considered a “standard environment” application. Outdoor systems, and perhaps some unusual indoor systems, are considered “non-standard environment” applications.

Public versus private: A sixth partition is “public/private.” Will the users of the system be customers of the system management (public) or employees (private)? Clearly attitudes toward usage of the devices, which will directly effect performance, vary depending upon the relationship between the end-users and system management.

Open versus closed: A seventh partition is “open/closed.” Will the system be required, now or in the future, to exchange data with other biometric systems run by other management? For instance, some state social service agencies want to be able to exchange biometric information with other states. If a system is to be open, data collection, compression, and format standards are required.

Wayman continues, saying the “list is open, meaning that additional partitions might also be appropriate. We could also argue that not all possible partition permutations are equally likely or even permissible.”

Note that *open/closed* relates to standards, and not to the user set as such terminology sometimes implies. The partition: open set versus closed set systems does not really deserve full standing in Wayman’s taxonomy, because a closed set biometric authentication system is an unrealistic scenario.

8.2.2 Weighting the factors

Table 8.2 looks at various applications and weights the importance of the various aspects of the different biometrics (as discussed in Section 8.1). Attributes like *Failure to Enroll* and *Template size* are listed in the first column and the importance of these attributes to the application are weighted. Drawbacks of biometrics and biometric attributes are listed in the left column of the table, while the different applications are listed along the top row:

Weighing	Application A	Application B
Drawback	*low*	*medium*

The value *low–high* determines the weight (importance) of a drawback in light of the particular applications, A and B. Here the drawback is of *low* importance to application A while of *medium* importance to application B.

Accuracy as a biometric attribute is handled by using two different error specifications. That is, the table evaluates the performance of a system at two different operating points instead of a single point, like the Equal Error Rate. This gives a rough idea of the trade-off between convenience and security that can be expected for each method. Be advised that the ratings in this table are purely subjective and are for example purposes only.

1. **#FA per 10,000 (@FRR = 10%):** The number of False Accepts per 10,000, when the False Reject Rate is a standard 10^{-1}.

   A value of 100 would mean accepting 100 out of every 10,000 intruders or, equivalently, that 1 percent of attempted break-ins would succeed. This might be an operating point of "medium" interest for building access control systems which typically undergo few challenges, but it is of "low" interest to applications that require significant security.

2. **#FA per 10,000 (@FRR = 1%):** The number of expected False Acceptances for every 10,000 attacks when the system FRR rate is set to a very forgiving 1 percent of all trials.

Typically the numbers in this row are of "high" interest to the high-volume applications with large user groups. These credit card and ATM-type applications need good security but still a reasonable amount of user-friendliness. Such a low False Reject Rate is of "low" interest to other applications where the users will tolerate much more of a burden.

Importance weighting	**Physical access**	**Stock trading**	**Airport access**	**Credit card**	**ATM use**
Intrinsic properties					
Required cooperation	low	low	high	low	medium
Social stigma	medium	low	medium	high	high
Intrusiveness	medium	low	medium	high	medium
Population missing	low	low	medium	medium	high
Sampling properties					
Inconvenience	medium	low	medium	high	high
Required proximity	low	low	high	medium	medium
Acquisition time	high	medium	high	medium	medium
Failure to Enroll	medium	low	medium	high	high
Failure to Acquire	medium	medium	high	high	high
1:1 Matching properties					
#FA per 10K (@FRR=10%)	medium	high	high	high	medium
#FA per 10K (@FRR=1%)	medium	low	high	medium	high
Template size (bytes)	low	low	medium	high	high
Technology properties					
Installation cost	medium	low	medium	high	high
Continual run cost	medium	low	medium	high	high
Cost per match	low	low	medium	medium	medium

Table 8.2: Importance weighings for various applications.

To put the weighting of the factors and attributes of Table 8.3 in a little more perspective let us describe our applications further. The particular applications we look at roughly divide into three groups with different requirements:

1. Physical access control and authorization systems for small user groups;

2. physical access control for a large user group; and

3. transaction authorization for a large user group.

Physical access and stock trading

These are the applications with typically a small number of enrolled users. The monetary consequences of authentication failures in terms of damages and losses can easily be on the order of one million dollars. The user of the application has every reason to cooperate with the application because this type of authentication is a daily job requirement. Hence, this authentication application is not voluntary and there exists a relatively static access control list. These are typically the applications where the negative convenience aspects of any of the biometrics are relatively unimportant.

Airport screening

This application is in a class of its own because the consequences of authentication errors can lead to direct loss of life. This is also an application for which the user community is composed of all sorts of constituents. This in itself makes the design of airport biometric authentication difficult and costly.

However, here the higher costs of installing and running the installation is of medium importance, as opposed to a more specialized application such as stock trading where the costs are of lower importance.

Credit card and ATM transactions

This application scenario is largely the opposite of physical control and stock trading. Credit card transactions and ATM transactions are applications with typically a larger number of users enrolled in the application and the monetary consequences of authentication failures in terms of damages and losses are only on the order of \$500 or so; i.e., there is an order of 10^{-4} lower monetary risk compared to the stock trading application.

Public acceptance, possibly not much enhanced by press coverage, of the biometric in the application is key here. The user of the application does not have many reasons to cooperate with the application if an alternative system offered by a competitor (perhaps without a biometric altogether) is simpler to use. Moreover, because of lack of liability of the user for ATM fraud and credit card fraud (currently—this may change), there is little incentive to use and cooperate with a difficult-to-use biometric system. These are typically applications where all the negative convenience aspects of any of the biometrics are weighed heavily because the requirements are high. Moreover, the cost of the installation has to be low because the applications may potentially be low-margin businesses. Therefore, the higher cost of biometric authentication for these applications is an important negative decision factor.

8.3 Evaluating options

Now that we have talked about the relative properties of various biometrics and the requirements imposed by several classes of applications, we return to the *core question* of How does one choose a biometric? Of course the answer is not simple; there are a large number of factors to be considered, as already pointed out. In this section we examine a simple evaluation metric for matching biometrics to applications. This is not meant to be a rigorous analytical tool, merely a starting point for focussing in on what are the most important details of particular applications.

8.3.1 Mismatch calculation

Let us consider a simple, yet somewhat naive, method for choosing the "best" biometric for an application. The basic idea is to create a table similar to Table 8.2 but listing the different biometrics in the top row:

Value	Biometric X	Biometric Y
Drawback	*low*	*medium*

The entries with values *low-medium-high* weight the importance of the particular drawback or negative attribute in light of the particular biometric, X or Y. The drawback in this example is of *low* importance to biometric X while of *medium* importance to biometric Y.

Table 8.3 shows an example of such a companion table with the various biometrics in the top row. The part associated with accuracy is numeric because we have some idea of the ROCs of the various biometrics.

For example, let us look at the top two numbers in the **Finger** column:

1. When the False Reject Rate (FRR) $= 10^{-1}$, the FAR $= 10^{-5}$, so the number of False Accepts in 10,000 trials is $10^{-5} \times 10^4 = \mathbf{0.1}$.

2. But when we set FRR $= 10^{-2}$ for added user convenience, FAR rises to 10^{-3} and the number of False Accepts becomes $10^{-3} \times 10^4 = \mathbf{10}$.

And for the corresponding top two numbers of the **Face** column—

1. When we set FRR $= 10^{-1}$, we have $FRR = 10^{-3}$ and consequently **10** False Accepts in 10 thousand trials (since $10^{-3} \times 10^4 = 10$).

2. At FRR $= 10^{-2}$, the FAR $= 10^{-1}$, yielding $10^{-1} \times 10^4 = \mathbf{1{,}000}$ False Accepts.

The numbers here are just "ballpark" figures to show the use of such a table—they are **not** the result of any sort of rigorous testing regime, nor are they related to any particular algorithm or technology.

To actually use this as a matrix in a calculation, the textual entries would have to be converted to numerical "penalties" or costs. For instance, assume "high" carries a penalty of 10, "medium" a penalty of 3, and "low" a penalty of 1.

Although there are already some true numerical values in the table, we also need to convert these to penalty values such that they have magnitudes comparable with those of the textual attributes. One such penalty (cost) function might be

$$C = v/100$$

Biometric drawbacks	**Finger**	**Face**	**Voice**	**Iris**	**Hand**	**Signa-ture**
Intrinsic properties						
Cooperation required	high	low	low	medium	high	high
Social stigma	high	low	low	medium	medium	low
Intrusiveness	medium	low	low	medium	medium	low
Population missing	low	low	medium	low	medium	medium
Imaging properties						
Inconvenience	low	low	low	medium	medium	medium
Proximity required	high	low	low	medium	high	high
Acquisition time	low	low	medium	medium	medium	medium
Failure to Enroll	medium	low	medium	high	low	low
Failure to image	medium	medium	medium	medium	low	low
1:1 Matching properties						
#FA per 10K (@FRR=10%)	0.1	10	300	0.001	10	300
#FA per 10K (@FRR=1%)	10	1,000	1,000	0.1	100	1,000
Template size (bytes)	500	1,000	2,000	250	100	200
Technology properties						
Installation cost	low	low	low	medium	medium	medium
Continual run cost	low	low	low	medium	low	low
Cost per match	medium	low	low	low	medium	low

Table 8.3: Drawbacks of various biometric technologies. These are only approximate values; the error rates are "ballpark" figures and do not relate to any particular algorithm or system.

for the template sizes,

$$C = 10 \times log_{10}v$$

for expected breaches at 10 percent FRR, and another function,

$$C = max(0, 10 \times log_{10}v + 10),$$

for violations at 1 percent False Reject.

One could then line up a column of the weights (Table 8.2) with a column of the drawbacks (Table 8.3) and compute the mathematical dot product to yield a mismatch score. Below we do this for the **Physical access** and **Finger** columns.

It can be seen that the mismatch of this application and biometric is 135. For comparison, performing the same calculation with the **Face** column yields a mismatch score of 201. Since lower scores correspond to a better match between the application and the biometric, fingerprints would be preferred over face recognition for this particular physical access application.

Note, however, that there are many ways to adjust the system to give totally different results. One can change the relative weighting of various attributes for a particular application (change Table 8.2). One could assign different levels of demerit to various attributes of

Mismatch score calculation	Physical access		W	×	P		Finger
Intrinsic properties							
Cooperation required	low	→	1	**10**	10	←	high
Social stigma	medium	→	3	**30**	10	←	high
Intrusiveness	medium	→	3	**9**	3	←	medium
Population missing	low	→	1	**1**	1	←	low
Imaging properties							
Inconvenience	medium	→	3	**3**	1	←	low
Proximity required	low	→	1	**10**	10	←	high
Acquisition time	high	→	10	**10**	1	←	low
Failure to Enroll	medium	→	3	**9**	3	←	medium
Failure to image	medium	→	3	**9**	3	←	medium
1:1 matching properties							
#FR per 10K (@FRR=10%)	medium	→	3	**0**	0	←	0.1
#FR per 10K (@FRR=1%)	medium	→	3	**30**	10	←	10
Template size (bytes)	low	→	1	**5**	5	←	500
Technology properties							
Installation cost	medium	→	3	**3**	1	←	low
Continual run cost	medium	→	3	**3**	1	←	low
Cost per match	low	→	1	**3**	3	←	medium
SUM				**135**			

Table 8.4: Computing a mismatch score by assigning numeric values and summing: *W* denotes weight, *P* denotes penalty and '×' denotes product.

the biometric (change Table 8.3). The interpretation of symbolic weights as numerical factors could be adjusted, for instance, by using $(10, 5, 2)$ instead of $(10, 3, 1)$. Or one could use one such mapping for the weights table and a different mapping for the drawbacks table. Finally, the conversion of probabilities and template sizes to penalty values is largely arbitrary. With proper adjustment of even some of these factors, a vendor could likely show that his particular biometric was, in fact, the best for every single application! **Beware.**

Perhaps a better use for this sort of calculation is to point out system sensitivities. This can be done by looking at the individual numbers in the "product" column. High numbers show up for those attributes of the biometric that do not fit the application well, and essentially tell what the weak points of the system might be. In the case of Table 8.4 the value 30 is associated with *Social Stigma* and *#FA per 10K (@FRR = 1%)*. This suggests that one problem when using fingerprints for physical access is that people might feel that they are being treated as criminals. A second problem would be that, when run at a very low FR rate, the overall FA rate is not as good as one might like. So one might consider running at a higher FR rate such as 10 percent (thus inconveniencing the users more) in order to get an acceptable level of security (good FA rate). Similarly, when performing the same calculation for **Face** one finds a 60 in the product column next to the entry *#FA per 10K (@FRR = 10%)* and 90 next to *#FA per 10K (@FRR = 1%)*. This implies that the biggest liability for face recognition in this application is its low security. To mitigate this, one might try to carefully limit the size of the user population through other means.

8.3.2 Zephyr^{TM} charts

One way of visualizing a comparative analysis of different biometrics is with Zephyr charts. These charts use four factors to compare the different biometrics. Although originally very general criteria like effort, intrusiveness, accuracy, and cost were proposed, one can in principle select very specific sets of criteria and assign values for each of the biometrics [143]. Deciding on an overall criteria for selection, one can eliminate the biometrics that do not qualify, based on thresholds on the area A (see Figure 8.2) of the Zephyr graph.

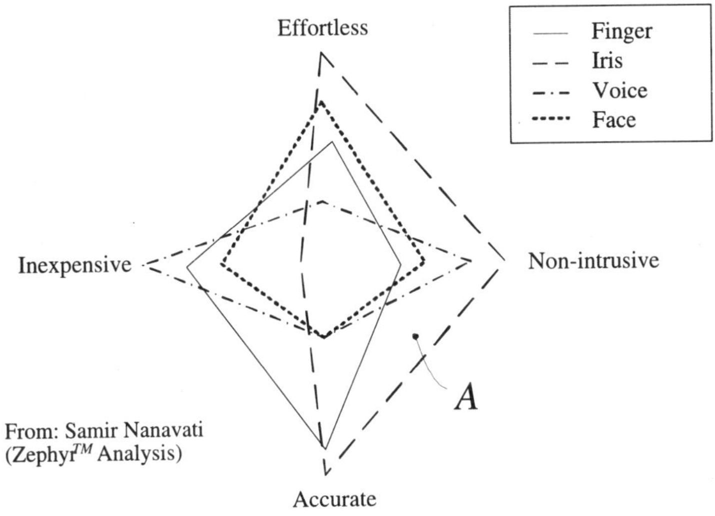

Figure 8.2: An example of a Zephyr chart (courtesy Samir Nanavati).

Figure 8.2 shows in a Zephyr graph how the biometrics finger, iris, voice, and face compare—the larger the area A, the "better" the biometric. This Zephyr analysis is about four variables, positive attributes of the biometric, that somehow express the "value" of a biometric:

1. *Effortless:* A rating of how easy is it to offer a sample of the biometric.

2. *Non-intrusive:* A measure that expresses how little intrusion or inconvenience the biometric causes.

3. *Accurate:* The accuracy that can be achieved by the biometric.

4. *Inexpensive:* A measure of how cheap the hardware associated with the biometric is.

A Zephyr analysis implies weighting these four variables in light of a certain application. It can be seen from Figure 8.2 that for the above four measures the biometrics are ordered as follows:

Effortless	Iris	>	Face	>	Finger	>	Voice
Non-intrusive	Iris	>	Voice	>	Face	>	Finger
Accurate	Iris	>	Finger	>	Face	$\approx$	Voice
Inexpensive	Voice	>	Finger	>	Face	>	Iris

Note that this is an importance weighting of the four characteristics of the four biometrics in light of some application and a user population.

It is sometimes argued that the area within a curve is a measure of a biometric's suitability, but the following counterexample shows that this is too simplistic. For a moment, consider the Zephyr graphs of Figure 8.3. Four attributes N, E, S, and W of a biometric in a given application are defined. We see from this figure that the biometric on the left (configuration $(2, 10, 10, 2)$) is better than the one in the middle, which in turn is better than the one on the right. However, now consider swapping the meanings of the N and the E axes. In that case, the biometric on the right is the best one. Hence, *just swapping two axes in the Zephyr graph changes the ordering of the biometrics,* and evaluating and eliminating biometrics based on these type of graphs may not be that meaningful. The area interpretation of these figures implies that some special benefit is obtained by having high values on two adjacent axes.

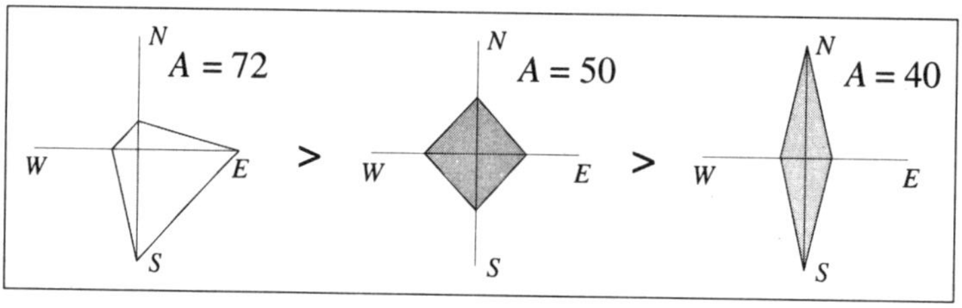

Figure 8.3: The middle labeling is $(5, 5, 5, 5)$. In one labeling of the axes (left), the configuration is better than $(5, 5, 5, 5)$; in another labeling (right) the same configuration is less than $(5, 5, 5, 5)$.

8.3.3 Open competitions for comparing biometrics

Often, external test agencies or competition coordinators specify methods of their evaluation. We describe two such examples.

Voice

The NIST speaker recognition contest SPEAKER99 [33,174] uses the following cost to declare a winner based on a single number:

$$Cost = \sum_{error\_cond} Cost_{error\_cond} \times P_{error\_cond} \times P_{cond}. \tag{8.1}$$

Here $Cost_{error\_cond}$ is the cost associated with an error condition, P_{error_cond} is the probability of this error condition, and P_{cond} is the prior probability of condition *"cond."* The P_{cond} attempts to specify the likelihood of the errors. For example, the probability of an intruder's trying to break into a system in a crime-prone area (e.g., a city) will be higher compared to a less crime-prone area (e.g., rural town or village). For example, in the case of speaker verification task, NIST identified the following values:

C_{miss}	C_{false_alarm}	P_{cond}
10	1	0.01

The cost of a *miss* (False Reject) is 10 times the cost of a False Alarm.

Even though (8.1) is used to evaluate speaker recognition algorithms, one can in principle use the same cost function in comparative studies of other biometric matchers.

Finger

The commendable recently published contests in fingerprint verification FVC 2002 & 2000 [130] has identified several error parameters and declares a winner for each. The organizers adopt the False Match Rate (FMR) and False Non-Match Rate (FNMR) terminology and the error parameters that are evaluated are—

1. The Equal Error Rate (EER), the operating point or threshold t_e where FMR = FNMR.
2. The *zero False Match Rate,* the lowest FNMR at which no False Matches occur.
3. The FNMR for which FMR $\leq 1\%$.
4. The FNMR for which FMR $\leq 0.1\%$.
5. Average time to enroll a fingerprint.
6. Average time to match two fingerprints.

Not all of these matcher evaluation parameters are equally important, but in order to be able to compare these numbers correctly, valid and appropriate statistical tests need to be developed, which is the subject of Chapters 15 and 16. We need, at a minimum, a database to test each biometric matcher using the performance criteria.

8.4 Affordability and cost

Aside from the properties listed in Table 8.2, there are a number of meta issues, mostly related to cost, to consider. We give some examples but do not claim to be exhaustive here.

While discussing the cost of a biometric sensor, we mentioned that the basic cost of the solution is quite different from the total cost of ownership (TCO). The total cost of ownership includes cost of maintenance of the sensor, cost of running the facility, and other related costs. Often this will be much higher than the basic cost of the equipment, which is only a portion of the upfront cost of installation.

A very first cost issue in biometric authentication systems relates to *verification* versus *identification*. Identification involves $1{:}m$ matching of biometric representations while verification involves 1:1 matching. This means that identification requires large centralized databases; a secure communication infrastructure; and, roughly speaking, m times more computational power than is involved for verification solutions. Another important cost factor is the choice between single or multiple biometric authentication protocols.

There are many costs related to adding biometrics to the authentication protocol. A large portion of the upfront investment for biometric authentication is system training. Enrollment of the population can be included in system training cost or in run-time support cost. In reality, as we discuss in Chapter 9, much of the training (fine tuning) takes place during operation and (especially for large-scale systems) represents a major portion of lifecycle support cost. So we define total training as follows:

- *System training:* The improving, refining, and tuning of the biometric matcher while the system is installed. This training task almost always has to be performed during operation of the authentication system at the site of operation and therefore could be expensive, but this training might be needed to reach the required False accept and False Reject rates, or to improve system accuracy during operation.

  Reaching these specifications may further involve quality control of biometric acquisition, which could be application dependent. This may affect the convenience of the overall solution and negatively affect the throughput, thereby increasing costs, especially enrollment costs.

- *Enrollment:* System training is related to enrollment. Enrolling a subject is the process of training a system with a biometric representation of the subject. As noted, this is an often underemphasized task. When security is the prime concern of an application, the enrollment should be at least as secure as the authentication. Usually, the enrollment has to be performed in a more secure and trusted environment than the authentication itself. As will be seen in Chapter 9, enrollment is a process that has to be administered very carefully and therefore can be a significant cost factor.

Besides this, there are a number of critical continuing cost factors:

- *Continued enrollment:* When the enrolled population of a biometric application reaches a steady state, the enrollment costs may level off (most of enrollment occurs immediately after commissioning a system). However, as discussed in Chapter 9, maintaining the enrollment database's integrity is a difficult, ongoing task because of all the problems of an enrolled population coming from all categories of Doddington's zoo [51], as well as deliberate imposters and clerical errors. It is essential that the acquisition of the biometric samples, especially during enrollment, be done with high quality standards, especially if m becomes large.

- *Failure to Enroll (FTE):* The FTE rate is something that is hard to estimate prior to system development and installation. FTE requires exception handling and can be a significant extra hidden cost factor.

  When a biometric authentication system operates at high levels of FTE rates, conversion of the system from voluntary to mandatory may be very costly.

- *User education:* A certain level of understanding of a biometric application is required from the user. Misunderstandings in the use of the application can have disastrous acceptance or throughput consequences. Marketing cost of the biometric installation might be significant portion of the investment.

- *Supervisory labor:* Depending on the various job levels among the inspectors and operators of a biometric authentication system, certain skill levels in biometric applications are required.

  Many biometric applications are supervised by an inspector. These inspectors typically control the quality of the presentation (acquisition) of the biometric and perform exception handling in the case of, say, *Failure to Acquire* (FTA, see Chapter 5). But in addition, very importantly, these operators may need to perform *visual matches* of the biometrics. (See Chapter 2 for a discussion of the contributions of operator errors to the overall error rates.)

- *Maintenance labor:* Human labor is needed to maintain the application. This labor ranges from highly skilled computer operators to less skilled personnel to perform important tasks as cleaning the biometric scanning devices.

Cost is always a concern, and it is not only the initial cost of the sensor or the match engine that is involved. The life-cycle support cost of providing system administration services and enrollment/authentication operator can quickly overtake the initial cost of the hardware.

In summary, the above-mentioned cost factors are influenced by the particular biometric in a different way for each biometric. Further, the total cost is affected much by the properties of the individual biometric as discussed in Section 8.5, i.e., properties like maturity, scalability, sensor cost/size, and template sizes of the biometrics. This makes the complete profit and loss statements for biometric systems extremely hard and beyond the scope of this book.

8.5 Positives and negatives of the biometrics

For a given application, there may be several biometrics that could work well in terms of objective and subjective parameters. Each biometric has certain advantages and disadvantages that should be kept in mind to pick the best biometric. A technology and application exploration of the area of biometrics can be found in [17]. This report looks at today's existing installations and application aspects of each biometric. A number of positives and negatives for each of the six most popular biometrics are enumerated.

8.5.1 Fingerprint

pros:

- There is a long tradition of the use of fingerprint as immutable identification in law enforcement. However, there are some recent challenges to the premise of uniqueness. See Chapter 14, which tells the judicial story behind fingerprints.
- There exist large legacy databases of fingerprints, albeit largely of criminals. However, among others, the California, Colorado, Florida, and Texas Departments of Motor Vehicles are working to establish a fingerprint biometric system for driver's licenses and record data.
- Fingerprint lends itself well to forensic investigation, i.e., the study of latent fingerprints [66]. Criminals often leave a trail of fingerprints (e.g., car, door knob, glass, weapon) that allows for reconstruction of facts after the events take place.
- A fingerprint is easily sampled using low-tech means, and the size and price of fingerprint readers are still declining. The conversion of fingerprints into digital images is getting easier, better, and cheaper. There are low-cost fingerprint scanners available (under $100) that are already in widespread use in many access control applications.

cons:

- Fingerprints suffer from poor public reputation in some countries because of the very strong relationship between fingerprint and criminal history. This relation has its advantages but does not help in the acceptance of fingerprint as a biometric.
- In some cultures there is an additional stigma because fingerprints are associated with illiterate people, who use them in place of signatures.
- There is a large variation of the quality of the fingerprint over the population. The appearance of a person's print depends on age, grease, and cut or worn fingers, i.e., on occupation and lifestyle in general.
- Sampling an image of a fingerprint is a matter of pressing the finger against the platen of a fingerprint reader. This creates technical problems as discussed in Chapter 3 and potential problems with cleanliness of the sensor and public hygiene.
- In some very rare cases there are people without fingers, or without a full set of fingers. Obviously these individuals cannot be fingerprinted.

8.5.2 Face

pros:

- Photos of faces are widely used in passports and driver's licenses where the possession authentication protocol is augmented with a photo for manual inspection purposes; therefore there is wide public acceptance for this biometric identifier.
- Face recognition systems are the least intrusive from a biometric sampling point of view, requiring no contact, nor even the awareness of the subject.

- The biometric works, or at least works in theory, with legacy photograph databases, videotape, or other image sources.
- Face recognition can, at least in theory, be used for screening of unwanted individuals in a crowd, in real time.
- It is a fairly good biometric identifier for small-scale verification applications.

cons:

- A face needs to be well lighted by controlled light sources in automated face authentication systems. This is only a first challenge in a long list of technical challenges that are associated with robust face authentication.
- Face currently is a poor biometric for use in a pure identification protocol (see Chapter 7). It performs better in verification, but not at the accuracy rates that are sometimes claimed.
- An obvious circumvention method is disguise, which will easily cause False Negatives in screening applications; i.e., the undesirable but disguised person is not identified.
- There is some criminal association with face identifiers since this biometric has long been used by law enforcement agencies ("mug-shots"). However, face is widely used and accepted in more mundane scenarios such as passports, season tickets, and the like.

8.5.3 Voice

pros:

- Like face, voice is a "natural biometric" (one that people use instinctively to identify each other) under certain circumstances (phone), and machine decisions can be verified by relatively unskilled human operators.
- Like face sampling, voice sampling can be accomplished quite unobtrusively.
- There exists great public acceptance of this biometric partly because of its naturalness and partly because there is little association of voice with identifying criminals.
- The voice biometric requires only inexpensive hardware and is easily deployable over existing, ubiquitous communications infrastructure (the telephone system). Voice is therefore very suitable for pervasive security management.
- Voice allows incremental authentication protocols. For example, the protocol prescribes waiting for more voice data when a higher degree of recognition confidence is needed.
- The voice biometric may achieve high accuracy and flexibility when combined with knowledge verification, in the authentication protocol called *conversational biometrics* (see Section 11.2.3).
- The voice biometric allows for checking identity continuously (Section 2.5.4); i.e., in a passive way the voice can be authenticated throughout a conversation.

cons:

- Imitation by skilled impersonators may be possible. This potential susceptibility has not been studied in any detail yet; unlike with signature, there is not much of a discipline for voice testing using real forgeries.

- With the text-to-speech technology improving, e.g., [206], it becomes possible to create nonexistent identities with machine voices (when enrollment and authentication are remote), and trainable speech synthesis may make it possible to create an automatic system that can imitate a given person saying anything.
- Voice recognition is dependent on the capture of a high-quality audio signal. Speaker identification systems remain susceptible to background noise, channel noise (from phone lines, wireless transmission, or severe compression) and unknown channel or microphone characteristics.
- Some people cannot speak at all due to injury, physical or mental abnormalities, or deafness. It is also possible to lose your voice temporarily.

8.5.4 Iris

pros:

- Iris is currently claimed and perhaps widely believed to be the most accurate biometric, especially when it comes to FA rates. Iris has very few False Accepts (the important security aspect). Therefore iris might be a good biometric for pure identification applications.
- Given that the iris sample acquisition process is solved using unobtrusive and distant cameras, the sensing of the biometric is without physical contact and without too much inconvenience.
- Iris has received little negative press and may therefore be more readily accepted. The fact that there is no criminal association helps.
- The dominant commercial vendor claims that iris does not involve high training costs. As discussed above, although, enrollment is the major portion of the system training and this may not be as much of an advantage as it is purported to be.

cons:

- There are no or few legacy iris databases. There consequently does not exist much legacy infrastructure, which will make the worldwide upfront investment too high initiate an iris ID system. Though iris may be a good biometric for identification, large-scale deployment is impeded by lack of installed base.
- Since the iris is small, sampling the iris pattern requires much user cooperation or complex, expensive input devices.
- The performance of iris authentication may be impaired by glasses, sunglasses, and contact lenses; subjects may have to remove them.
- The iris biometric, in general, is not left as evidence on the scene of the crime; so it is not useful for forensic applications.
- Some people are missing one or both eyes, while others do not have the motor control necessary to reliably enroll in such a system.

8.5.5 Hand

pros:

- There seems to exist a certain amount of public acceptance for the hand biometric because it is already used at Disney World, INSPASS (The United States

Immigration and Naturalization Service Passenger Accelerated Service System [90]), and at various universities for verification in, e.g., the University of Georgia.

- It is claimed that hand geometry measuring is an easy "do-it-yourself" operation.
- There exists *at least one* scenario evaluation of hand geometry as a biometric [131], which shows that hand is a good biometric for verification. But little is known about error rates for hand geometry recognition.
- Hand geometry, being a relatively weak biometric, can be used for verification in circumstances where a stronger biometric might raise public fears about misuse of the data for identification.

cons:

- As with fingerprint, hand geometry is measured when a subject presses the biometric against a platen (although there are ways around this). Such contact may be cause for some public hygiene concerns.
- Hand geometry is not very distinctive from one person to the next. Therefore, as discussed in Chapter 7, this biometric is a poor one to select for pure identification.
- *Moreover,* there exists *only one* scenario evaluation of hand geometry as a biometric. There is debate as to whether hand geometry is truly a biometric *or not.*
- Again, some people do not have hands or measurable fingers for various reasons.

8.5.6 Signature

pros:

- Signature is a man-made biometric where forgery has been studied extensively, therefore forgery is detected even when the forger has managed to get a copy of the authentic signature or even more characteristics of a signature.
- At enrollment stage there is already some possibility of detecting a forged signature. This can be used to detect the impersonation (forgery) of existing identities (signatures) at enrollment time.
- Training is intuitive and fast, and people understand (from speech recognition technology) that the system needs to be trained. They intuitively understand what it means to enroll in a way that will not cause False Accepts on signature.
- Signature verification in general has a fast response and low storage requirements.
- A signature verification is independent of the native language of the user.
- Signature is inherently a combination of information and biometric, the information component (what is written, and how it is written) can be chosen, and indeed changed, by the user.
- Very high compression rates do not affect shape of the signature (100–150 bytes).

cons:

- There is much precedence for using signature to authenticate documents. This may give rise to the perception that signature is not secure enough for protecting airports, etc.
- The study of the individuality of signature is a little different in that it is rich in ideas and has a much better understanding of personalization (forgery) efforts. Unfortunately, the community is not very disciplined in their system testing and it is hard to get a feeling for scaling properties and conditions for the signature identifier.
- A five-dimensional pen (see Chapter 3) may be needed to arrive at the desired accuracy. The five-dimensional pen as input device records pressure and angles too. This makes the hardware costly. The effectiveness of signature for access control using state-of-the-art input devices is unknown.
- Some people have palsies, while others do not have enough fine motor coordination to write consistently.

8.6 Biometric myths and misrepresentations

We offer a list, in no particular order, of some of the most prevalent misinformation and misunderstandings related to biometrics, some of which is possibly deliberately promulgated by proponents and opponents of biometrics.

Biometric X is the "best" for all applications

There is no single "best" biometric. Each application and scenario will call for a particular combination of many factors, including price, accuracy, usability, and user acceptance. In this chapter we looked at these factors with respect to applications and biometrics. Chapters 3 and 4 describe different biometrics and indicate the strengths and weaknesses of each.

Biometric X is unique for each individual

The appearance of the traits of living organisms that we measure as biometrics is the result of genetic predisposition modulated by stochastic morphological developmental processes. The amount of variation due to each of genetics and development varies from biometric to biometric, but the effect of the random developmental process ensures that each person's trait is unique *if analyzed with sufficient detail.*

However, in this book we are dealing with practical identification technology, with limits on the resolution and modalities of the sensing, limitations on data storage, and practical limitations on the ability to compare the sensed data, as well as inherent intra-personal variations over time. While it is next to impossible for any two people to have identical biometric representations in any reasonable system, it is also extremely unlikely that two measurements of the same person would give identical representations. The necessary "tolerances" introduced to overcome this problem, which allow matching of biometrics despite measurement noise and temporal variation, necessarily result in significant probability that one person's data will be found to match with another person's representation, thus banishing the notion of uniqueness *for all biometrics*.

Chapter 14 further discusses the claim that some biometric X may truly be unique, by defining the "individuality" of a biometric. Individuality has to do with how different one machine representation can be from another, and how many different representations there are.

A single number quantifies system accuracy

If this were true, it would make comparing matchers extremely easy. However, in Chapter 5, we show that the design of a biometric system is very much a trade-off between two different types of errors that can be made. Consequently, there are operating areas for different types of matchers. We show that one matcher can be better than another matcher in one operating area, while the reverse is true in another operating area.

This single number could be, for instance, the Equal Error Rate (EER), the operating point where FAR = FRR = EER.

Our system is "plug and play"

A biometric authentication system is designed using databases of biometric samples and trained on databases of biometric samples. Such a system may function at a certain accuracy when installed. Almost always, though, tuning a system to the sensing devices and sample acquisition circumstances and training the system on samples of the actual user population will increase a system's accuracy. Think of, for instance, a speech recognition application that is trained for a specific speaker population and application.

Real accuracy performance can be predicted

As noted above, a system needs to be trained on biometric training data. It is always best to train the system on biometric data that is obtained from the true user population, and in the circumstances in which the biometric system will truly be used.

Accurate performance prediction would mean that one is able to accurately model the user population and the collection environment *a priori*. As seen in Chapter 9, such models are not easily computed from available databases, and even pilot studies cannot hope to model all the factors and variations present in a large-scale deployment.

The vendor reporting best FAR and FRR has the "most accurate system"

Vendors and technology enthusiasts too often report system accuracy results which may be misleading.

First, the data used for evaluating the accuracy performance may be collected under unrealistic and controlled conditions. Seeking evaluation results on standard public domain databases (evaluations as described in Chapter 7) may be a quick way to establish a realistic measure of performance. Second, when the results are reported on standard data sets, they are often not reported for the entire set but on a subset of the data, or without following the standard test protocol. Ascertaining that this subset is not selected based on unreasonable criteria will ensure that the results reported on the subset are applicable to the entire data set.

This topic is addressed in Chapter 5. It is good practice to give confidence intervals around accuracy estimates (see Chapter 15) to give an idea of how meaningful the figures are.

Multiple biometrics outperform single biometrics

While biometric person identification may potentially benefit from multiple biometrics and multiple samples per identification, it does not necessarily imply that doing so will always result in better speed/cost/accuracy performance (Chapter 11).

Our biometric system does not use a decision threshold

Any biometric needs to define a criterion for determining when two biometric samples are from the same subject. This needs to be somehow defined by using testing and training sets (Chapter 5). Training of a system, in itself, *is* really establishing an optimal decision threshold.

Our feature extractor can be used with any match engine

This question arises particularly for fingerprint recognition systems, many of which use similar representations of minutiae for identification. Despite this apparent commonality, feature extraction algorithms will generate slightly different features. Chapter 9 describes probabilistic enrollment, which involves modeling the data in terms of features and defining match probabilities in terms of templates constructed from the features. Consequently, feature extraction algorithms and match engines are best designed in tandem, with each adapted to the biases of the other.

Large templates mean better accuracy

The biometric identifier representation size rarely constitutes a good metric for accuracy performance; the single most critical determinant of the accuracy of the system is the extent to which the variations of biometric identifier from different identities overlap and the extent to which in the individual identifiers remain invariant. (For example, recording a person's height any more precisely than to the nearest centimeter will not help accuracy in identification.)

Chapters 3 and 7, respectively, discuss the six most commonly used biometrics and their published accuracies.

Face recognition prevents terrorism

The current state of face recognition is not advanced enough. Face scanning of passengers against "most wanted" lists simply will result in too many False Positives as discussed in Chapter 7. In fact, there have been several face recognition tests at US airports and public places [3,196], which show that the face biometric needs much further study.

Biometrics means 100 percent security

Typical biometric matchers are imperfect and this statistically guarantees that with a sufficiently large number of intruder input presentations, the system *will* let in some of the intruders (see Chapter 5). No system is 100 percent foolproof, especially if you take into account attacks directed at the system from professional hackers using diverse techniques such as social engineering, collusion, and coercion.

Chapter 12 lists possible attack points of biometrics systems and offers solutions to many of the problems.

Biometric systems are no threat to privacy

Privacy can be bolstered by technology but not guaranteed by it. Privacy is a policy matter. This is not to say that protection of privacy rights cannot be facilitated by biometrics. In fact, a well-designed biometric system operating under well-thought-out procedures for accountability truly protects the rights of otherwise vulnerable individuals.

Biometric systems invade our privacy

Many people are afraid that biometrics, with its promise of perfect identification, inherently means a "Big-Brotherish" complete erosion of privacy. While biometric systems could be exploited by totalitarian societies, it is the implementation of biometrics systems and policies and legal framework that will determine their effect on our privacy. Biometrics technology can in fact be used to protect our privacy in many ways—for instance, by giving improved security and accounting trail on medical records, or by providing authentication without the revelation of personal information—compared with the current widespread use of social security number, mother's maiden name, or driver's license (which additionally reveals address and date of birth) for identification.

Biometric sensors are unhygienic or otherwise harmful

Many biometric sensors require human contact for their operation but interacting with the sensors is no more a threat to public hygiene than are doorknobs, keyboards, money, and so on. Retina and iris scanners do not shine lasers into the eye, and in those systems that do provide illumination there is no indication that this is in any way harmful.

Part III

System Issues

9

Creating and Maintaining Databases

Biometric enrollment is a serious biometric research topic because it asks an individual to give out even more private information (e.g., fingerprint) about his or herself. However, enrollment is just a process directed by some *enrollment policy.* This policy needs to be acceptable to the public, since (almost by definition) policies are public documents and at least must be understood by the public. Obviously, part of the enrollment process should be a clear statement of how, where, and when this private information will be used.

Technically, enrollment is the process of selecting (trusted) individuals through some enrollment policy E_M and storing machine representations of these m enrolled members in a verification database $\mathbf{M}$. Trusted individuals are authenticated from seed documents, such as birth certificates, proofs of citizenship, and passports; we call this "positive enrollment." For criminal identification systems, on the other hand, enrollment is the process of determining the undesirable individuals through some enrollment policy E_N and storing (much more elaborate) machine representations of the n selected individuals in the screening database $\mathbf{N}$. This is achieved through the use of existing criminal database systems somehow and we call this "negative enrollment" (inspired by [231]).

In the following, the target population of an application is taken to be the world population $\mathbf{W}$ because *any* subject d_k in $\mathbf{W}$ could attack an authentication system. During enrollment a "*ground truth*" is available in the form of legacy databases, whether civilian or criminal. Because of errors and fraud, these legacy databases are of course not necessarily an accurate and precise reflection of the ground truth; there are *fake* and *duplicate* identities. Analysis of the situations involving fraudulent enrollment using stolen biometric identifiers is beyond scope of this book and will not be considered.

A *fake identity* can be one of two cases, created and stolen identities:

1. A *created identity:* Some subject d enrolls in $\mathbf{M}$ as d'_k using documents for a non-existent identity, either fake documents or documents legitimately issued by an authority in which the fake ID had been enrolled earlier. Hence then subject d'_k leads an existence as if he or she had a valid birth certificate, passport, and so on.

   When biometric samples are securely associated with these seed documents, bio-

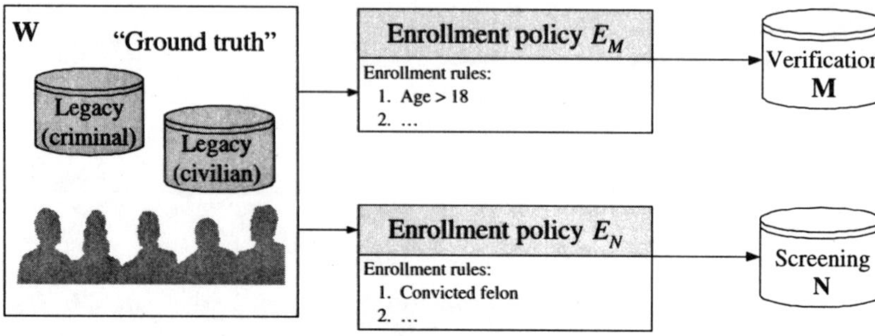

Figure 9.1: Enrollment consists of one or two processes, or both: Building a member database M and building screening databases N of undesirable people, like "most wanted" lists.

metrics can make fraud more difficult here. Biometric screening against criminal databases N may help in detecting fakes since subject d may be enrolled in these databases.

2. A *stolen identity:* A fake identity can also be a falsely enrolled subject d'_k as subject d_k, the stolen identity. It is a machine identifier ID_k in which the biometric sample *does not* correspond to the true existing subject d_k as recorded in trusted civil databases (the true world model) but rather corresponds to some other subject d'_k. The biometric samples of subject d'_k have been enrolled as if they were the samples of d_k.

A *duplicate identity* is a pair of identities or identifiers ID_1 and ID_2, with $ID_1 \neq ID_2$, in database M that are associated with *one* particular subject d with biometric $\mathcal{B}$. Note that the duplicate identity implies a fake identity (but the reverse is not necessary). Say, for example, that subject A who is not yet enrolled under his true identity I_A manages to enroll as an another existent identity I_B, who does not happen to be enrolled in the database yet. This is a case of false identity. Now if A tries to enroll again as either I_A or any other identity, we also have a duplicate identity.

A biometric authentication system is only as secure as its enrollment subsystem. In this chapter we describe and probabilistically model the enrollment process.

9.1 Enrollment policies

The enrollment processes are shown in Figure 9.1. Enrollment policies are in a way very similar to authentication protocols. The difference is that an enrollment policy makes use of manual or machine authentication protocols, and *not* vice versa. An authentication protocol does not execute enrollment policies and in terms of workflow is a faster process. We define the enrollment policies E_M and E_N as follows:

E_M : Positive membership enrollment, this is a process of the registration of M trusted subjects $d_m, m = 1, ..., M$ in some database **M**. The enrollment could, for instance, be based on some already enrolled population **W** of an application such as credit card holders. Then of course the integrity of the database **M** is no better than the original database **W**.

When for an application, database **M** is created anew, positive enrollment policies should ideally include biometric authentication. In any case, a policy for enrolling subjects into **M** needs to be developed and defined and a subject needs to be authenticated by one or more authentication protocols. These policies are more likely to involve visually supervising enrollment personnel; in such situations, enrollment needs to take place at physical enrollment stations.

A membership enrollment policy may include matching trusted subjects d_m against screening databases **N**. In any case, the positive enrollment policies E_M should be made public so that subjects d are aware of such negative identification practices during positive enrollment. Often membership in these positive databases is voluntary.

E_N : Negative enrollment is a process of registration of N questionable subjects $d_n, n = 1, ..., N$ by storing machine descriptions of these subjects in database **N** (here **N** could be a collection of databases). Such databases will contain much more specific and detailed descriptions of subjects d_n. The policies here are about when a subject d in a legacy database is included in the screening database **N** and what information about a subject d is to be stored along.

The policies for negative enrollment are ideally completely public in that a subject d is enrolled in **N** if and only if he or she breaks the law, which is a public document. In that sense, negative enrollment policies specify those behaviors that *should not* be exhibited in order to avoid enrollment. It can be easily seen that enrollment is not voluntary for these databases.

The databases, **N**, collected using enrollment policy E_N, have a dual use:

1. They are used to screen subjects d during verification.
2. The databases are used in an enrollment policy E_M; i.e., in order to be enrolled each trusted subject needs to be matched against databases **N**.

The process of enrollment assumes that there exist universal databases such as passport and birth certificate databases.

9.1.1 Enrollment for positive authentication

Careful enrollment policies and quality control of the enrollment are central to the ultimate success of biometric authentication. Much care is needed during enrollment as explained in the rest of this chapter. This is needed because of technical reasons and, perhaps more importantly, it is needed for sociological reasons to ensure no issues arise like the creation of "second-class" citizens.

Regarding enrollment process and policies E_M, we expand on things we already said in the introduction. Enrollment policies that determine who are the eligible subjects d for database **M** are to be defined for each application. Issues with enrollment policies E_M include—

- The ultimately "true" identity of a subject d has to be (manually) verified or established somehow. What proofs of identity are accepted at enrollment time?

- A definition of membership requirements for the prospective members $d_m, i = 1, ..., M$ of **M** needs to be defined. That is, it needs to be defined who the customer is. Or, more precisely, it needs to be defined who the "trusted identities" among the customers are.

  Unfortunately, this is sometimes called the *target population,* a poor choice of words because this population *is also* (part of) the target population of an attacker.

- Conversely, it needs to be exactly specified and agreed upon what type of information about a subject d documented in legacy databases, such as FBI, INS, DMV, and credit reports excludes subject d from database **M** and thereby from the access control application.

- What credentials are issued and what credentials are shared with the application (biometrics, personal knowledge) for subsequent authentication? Choices made here will limit the ultimate flexibility of the authentication protocols as we saw in Chapter 2. Retroactive collection of additional information or biometric samples from the population **M** is an expensive exercise.

- Personnel has to be trained to perform the enrollment and initial authentication of the identity of the subjects d.

- What definitely should not be forgotten is the training of the user population where an understanding of the technologies and policies is crucial.

- Physical enrollment stations need to be designed and locations for these stations need to be determined.

- It has to be ensured that through the enrollment system no security holes are created.

Again, during enrollment, other databases may be available, indicated by the databases *legacy civilian* and *legacy criminal* in Figure 9.1. Then, during enrollment a subject can be screened against the databases to determine whether a person has some undesirable credentials: these "credentials" can be civilian, i.e., credit report records or driver's license records; or these credentials can be criminal, like previous arrests.

9.1.2 Enrollment for screening

Enrollment policies that determine who are the undesirable subjects $d_n, n = 1, ..., N$ to be registered in screening database **N** need to be defined for each application. These policies include—

- It needs to be exactly specified what type of information about a subject d makes this subject eligible for inclusion in $\mathbf{N}$. Conversely, it needs to be explicitly stated what criteria are used for removing an identity from the database (either because they have undergone correction process or because the insertion was accidental). It needs to be exactly specified and agreed upon what the type of information about a subject d documented in legacy databases, such as FBI, INS, DMV, and credit agencies, is sufficient for the inclusion in $\mathbf{N}$.

  The details of the policies for running such enquiries of legacy databases may or may not be publicly made available. Nevertheless, the enrollment policy somehow has to define this.

- It needs to be determined who has access to the database $\mathbf{N}$, i.e., which parties are allowed to query database $\mathbf{N}$ and to which fields of records, the enrolled subjects, these parties may query. One also needs to address issues such as how the parties are authenticated/authorized before they can access the database and how the integrity of the transaction logs and database is maintained.

Enrollment policies can only be developed through an open national and international debate. Consequently, this debate will probably be chaotic, like everything else in the biometrics area. The resulting enrollment policies will be also, like so many other things in biometrics, a complicated matter of compromises.

Unenrollment

Unenrolling an individual d from database M is governed by the enrollment policy E_M, too, of course. Public opinion has identified this as a concern about biometrics. However, the biometric research community, surprisingly, has given little attention to unenrolling individuals from biometric database M. Valid technical questions to ask here are—

- Is it really just a matter of erasing some database entry in M?
- How does this continuous enrollment and unenrollment effect the performance of a biometric system?
- Can this "obsolete" biometric data be used for other purposes, e.g., monitoring the performance of the biometric system over time?

9.1.3 Social issues

In general, many unanswered questions about *how* to make biometric authentication work without creating additional security loopholes, and without infringing on civil liberties, need to be answered. For national authentication applications, like travel, difficult questions will have to be answered about who will be eligible and therefore who will be enrolled in automated authentication systems. Other questions that need to be answered are, Who will administer and maintain databases of authorized subjects and How will the data integrity of these databases be protected. Perhaps the largest issue with data integrity is keeping the databases "clean" with strict criteria for enrollment and strict criteria for continued enrollment of subjects.

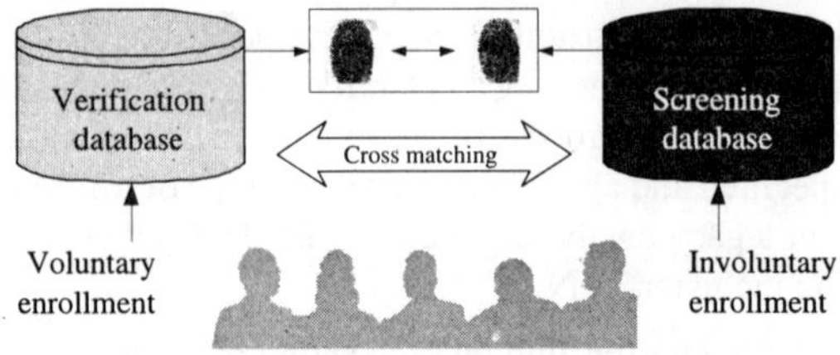

Figure 9.2: Cross-matching of large databases is a privacy concern.

A new government bureaucracy may be created to maintain the enrollment database. Cross-matching as in Figure 9.2 is then a real possibility, especially for keeping the authentication database current. This matching of voluntary databases with databases collected through involuntary means, such as criminal databases, may be very controversial. Therefore, another possibility is that one or more private entities maintain one or more of these databases and keep them current, very much like today's system of credit rating. The question then is how these databases are kept current, which can only be achieved by somehow monitoring the behavior of the enrolled subjects.

When it comes to security, especially physical security, most everyone seems to be willing to give up some privacy. Privacy concerns are mainly with *information security*, i.e., data confidentiality and integrity and have to do with trust in the application, the technology, and the authentication system operator. Unauthorized access to biometric authentication databases may lead to "privacy attacks." The use of authentication methods will lead to impersonation attacks (Chapter 12) that can be viewed as the ultimate invasion of privacy. The stealing of identities is a more personal violation when biometric samples are also stolen.

9.2 The zoo

Assume for a moment that the database does not contain duplicate or fake identities. Because of the fuzzy or probabilistic matching of biometric templates (e.g., non-zero error rates), an enrolled database $\mathbf{M}$ of biometric identifiers brings with it the notion of subjects with strong or weak biometric identifiers. That is, for the same biometric, one subject may be easy to authenticate reliably because he/she has a very distinctive pattern, while another subject may be hard to authenticate because his/her biometric pattern is not very distinctive. Early on, the speech recognition community realized that there are subjects d whose speech is easy to recognize and there are subjects that are hard to recognize; these subjects are called *sheep* and *goats*, respectively.

Doddington et al. [51] applied this notion of animals to the area of speaker recognition and added some animals. The *sheep* and *goat* are the traditional categories:

Sheep: This is the group of subjects that dominate the population, and authentication systems perform reasonably well for them simply because their real-world biometric is very distinctive and stable.

Goats: The group of subjects that are particularly difficult to authenticate because of a poor real world biometric that is not distinctive, perhaps due to physical damage to body parts involved in a biometric identifier or due to large spurious variability in the biometric measurements over time.

This is the portion of the population that generates the majority of False Rejects.

Classifying the enrolled subject in terms of animals is a good metaphor for understanding the enrolled population of a particular installation with database **M**. This classification was originally intended for voice recognition and for speaker identification, but is applicable to other biometrics as well. It also stresses the importance of careful enrollment, because if enrollment is neglected it can really become "a zoo out there," to quote Doddington et al., who added two more subject categories:

Lambs: These are the *enrolled* subjects who are easy to imitate. A randomly selected speaker from the general population is highly likely to be erroneously authenticated as an enrolled lamb. Equivalently, in the signature area, a *zero-effort forgery* (in essence a scribble) is highly likely to be erroneously authenticated as an enrolled lamb, simply because lambs have easy-to-forge signatures.

Lambs are the subjects who, once enrolled in a database, are the cause of most False Accepts because they are imitated by wolves.

Wolves: These are subjects that are particularly good at imitating, impersonating, or forging a particular biometric. For example, in speaker identification their speech is likely to be accepted as another enrolled speaker's speech. Consequently wolves make successful intruders and cause False Accepts. Note the distinction in the underlying mechanisms generating the False Accepts due to goats and due to wolves. Equivalently, in signature a professional forger is a wolf and if the forger has knowledge about signatures of enrolled lambs, the easiest thing to do for the intruder is simply to forge the signature of a known lamb.

Chameleons: These are the subjects who are both easy to imitate and are good at imitating others. They are a source of passive False Accepts when enrolled and of active False Accepts when being authenticated.

The classes are summarized in Table 9.1.

A possible way to distinguish these two types of False Accepts is to think of lambs as causing "passive" False Accepts and wolves as causing "active" False Accepts. This notion of the two types of False Accepts is introduced in [19]. The passive False Accept Rate is measured by randomly matching random biometric samples to enrolled and other samples. However, each enrolled subject $d_m, m = 1, ..., M$ has its own False Accept and False Reject rate and hence attacking an installation by impersonating the right lamb will result in very high error rates, but this requires prior knowledge of the database **M**. Actively attacking a system with unenrolled wolves will still result in a significantly higher (active) False Accept rate than attacking the system with random samples. Note that the distinction between lambs and wolves exists because (apparently) the match measure between templates

Doddington's zoo [51]	
Sheep	Well-behaved subjects in the population
Goats	The subjects that cause the False Rejects
Wolves	The subjects that attack other subjects in the population, causing active FA
Lambs	The subjects that are attacked by other subjects, causing passive FA
Chameleons	The subjects that both attack and are being attacked

Table 9.1: Subject categories and their positive and negative impacts on authentication system performance.

is not symmetric. What happens if the measure *is* symmetric? In such cases, *chameleons* were introduced in [19] because a matcher can be designed to be symmetric, i.e., distance template T_a to template T_b equals the distance from template T_b to template T_a. We arrive at the strange situation that wolves $\equiv$ lambs. This "crossing" between wolves and lambs, we call chameleons.

The classification of individuals in terms of animals depends on the way the match score probability is defined [19] and is still poorly understood. Such a classification of fingerprints in terms of their matching behavior is attempted in [21].

Hence, every biometric implementation has its own zoo $\mathbf{M}$, which is composed of population groups in terms of the above animals. The exact composition of the target population in terms of thcsc animals greatly influences the accuracy of a particular application. Conversely, strict enrollment policies can result in the disqualification of the undesirable animals (resulting in a high Failure to Enroll rate) and the enrolled population will consist, as desired, mostly of sheep. Biometric sample quality improvement by acquisition quality control, is a first place to improve performance of a matcher.

9.3 Biometric sample quality control

Many *random* False Rejects/Accepts occur because of adverse signal acquisition situations. Poor input control perhaps constitutes the single most important reason for high False Reject/Accept Rates. Apart from designing better user interfaces, there are two solutions to this. One can either probabilistically model and weigh all the adverse situations into the feature extraction/matching system or one can try to dynamically and interactively obtain a desirable input sample. The latter is possible in interactive overt authentication protocols involving cooperative users and overt screening protocols involving uncooperative users (see Figure 9.3).

Automatic implementation of either strategy needs an algorithmic characterization of what a *desirable* pattern sample is, some quality measure. The term *quality* is then somehow related to "processability" of the sample. A system faces difficulty in analyzing poor-quality samples and performs well when presented with good-quality samples. It is important to quantify the concept of quality so that consistent actions can be taken for different

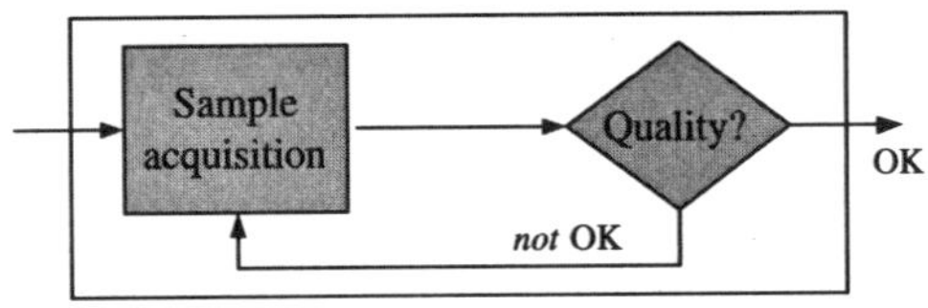

Figure 9.3: Dynamically improving the biometric samples by selecting high quality samples for acquisition.

types of undesirability, so that an appropriate corrective action can be taken by the system (e.g., apply image enhancement) or suggested for the subject (e.g., present the biometric in a different, "better" way). Finally, it is desired that assessment of quality be fast and not involve actually matching the given sample either with its mates or with its non-mates. Often quality may be used to pick one single best example for enrollment out of a set of biometric samples.

Conveniently handling biometric samples of diverse quality is important to any practical biometric authentication system. However, in theory, for almost all applications it is possible to compromise some convenience (ease-of-use or low FRR) by accepting only a certain quality of input. Not surprisingly, almost all operational biometric systems have an implicit or explicit strategy for handling samples of poor quality. Some of the simplest measures for quality control include the provision for presenting multiple samples to the system. In other schemes, the system provides a user with live visual display of the biometric that has been sampled (which is of course not practical for all biometrics).

This input quality control will result in higher Failure to Enroll (FTE) rates; accepting low-quality biometric samples lowers FTE. Quality control of the input samples (especially during enrollment) is important to maintain a certain level of quality; i.e., with higher FTE, only good biometric samples will join the enrolled population. This is important because the number of users that, in one way or another, are subject to some exception handling (c.g., a human gatc agcnt) can bc an addcd cost in a particular biometrics installation. Unfortunately, these subjects are hard to recognize, thereby increasing the FR; but perhaps more importantly, these subjects will also increase the FA rate. Hence, the FTE rate is an extremely important variable of an application because if it is too high, an application simply becomes too costly. Consequently, the optimal FTE for a biometric application is a trade-off between upfront investment and the continuing operating cost of a biometric installation, with the FTE of an installation greatly affecting the operating cost.

In Chapter 5 we discussed the ROC and its use as a tool for comparing biometrics, but we see here that the enrollment policy affects the FAR and the FRR in a trade-off with the FTE rate. Before we assumed that the *true ROC curve* of a matcher was known. To estimate the ROC curve of an application, a carefully designed experiment needs to be performed. However, the test practices in the biometric community are only in the beginnings of standardization and the ROC curve is to a certain extent under control of the system designer. For example, by excluding some of the most difficult data (e.g., the "goats," the hard to match subjects) an ROC can be made to look better. In general, the overall ROC estimate can be improved as shown in Figure 9.4 by eliminating poor data.

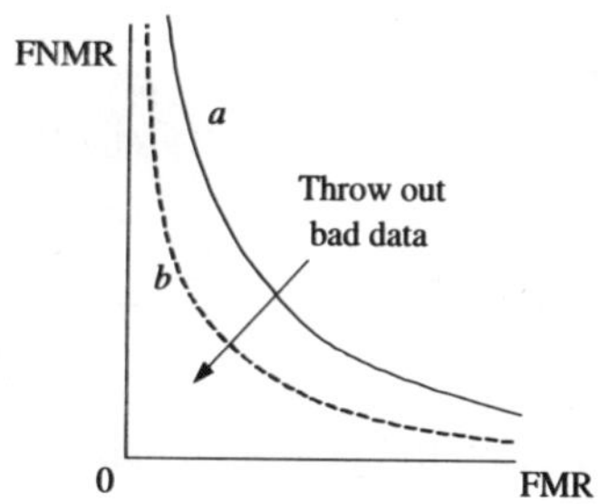

Figure 9.4: An ROC curve estimate can be made to look arbitrarily good.

Enrollment of a subject in a negative authentication AFIS application with database N, is often done in an involuntary fashion, e.g., when the subject is under arrest, and enrollment can be done with great care. On the other hand, subjects might also be enrolled with pieces of covertly obtained (biometric/demographic) information, in which case high quality of enrollment is very hard to achieve. Negative authentication databases are rarely built from scratch, generally being mixtures of existing legacy databases. Depending upon the integrity of these individual legacy databases and the amount of care taken to build them, significant variations in the accuracies can be expected. Collating poor-quality legacy databases may often introduce significant authentication costs due to an excessive amount of manual intervention needed for exception processing.

9.4 Training

Training of a biometric authentication system is conceptually similar to "machine learning." It is not the training of system operators or system users, though that type of training, of course, also greatly improves system performance.

Why does a biometric system need to be trained? There are many things peculiar to biometric authentication systems. One thing is that a biometric match engine does not produce 1 and 0 decisions per se; but rather the match engine computes similarity scores $s(B', B)$ between biometric samples B and B'. This score is a random variable, which is in general high if B and B' are from the same biometric and is in general low if the samples are from different biometrics. The goal of system training is to make the average difference between these match scores and mismatch scores as high as possible. This is a difficult problem because the biometric identifiers of the users are unknown *a priori*, unlike, for example, identifiers like account numbers.

Instead of a score s, a match engine might alternatively compute a distance measure $Dist(B', B)$ that indicates how "close" sample B' is to sample B. Distance is a dissimilarity measure, so when the measures are properly scaled and normalized $s(B', B) = 1 - Dist(B', B)$ and there is a one-to-one relation between score and distance. Therefore, we only concentrate on scores in this chapter. The score $s(B', B)$ is related to the probability that the samples are from the same biometric. When the score is high, the probability that the samples B and B' are from the same real world biometric $\mathcal{B}$ is likely to be high too.

For many match engines the score is not a probability. Even if the score is computed in a probabilistic fashion, assumptions and approximations are made and the score will not be the probability that one really would like to compute. Assume we have samples B' and B from real world biometrics $\mathcal{B}'$ and $\mathcal{B}$, respectively. Then, given these samples, one would really like to compute the probabilities

$$\mathit{Prob}(\mathcal{B}' = \mathcal{B} \,|\, B',\ B) + \mathit{Prob}(\mathcal{B}' \neq \mathcal{B} \,|\, B',\ B) = 1. \tag{9.1}$$

If $s(B', B) \equiv \mathit{Prob}(\mathcal{B}' = \mathcal{B} \,|\, B',\ B)$, we would be done and there would be no reason to delve into much of the material presented in this chapter. Unfortunately, $s(B', B)$ is *not* the probability that B and B' are samples from the same biometric. To score $s(B', B)$ needs to be normalized somehow to satisfy the equality in (9.1). To do this one would need to compute

$$\mathit{Prob}(\mathcal{B}' \neq \mathcal{B} \,|\, B',\ B) \quad \text{or} \quad \mathit{Prob}(\mathcal{B}'' = \mathcal{B} \,|\, B',\ B), \tag{9.2}$$

with $\mathcal{B}'' \neq \mathcal{B}$. If $\mathcal{B}' \neq \mathcal{B}$, there must exists some biometric $\mathcal{B}''$ that is equal to $\mathcal{B}$. So we need to compute not only how well $\mathcal{B}'$ fits the data, but also we need to compute if some $\mathcal{B}''$ fits the data better. This is the dilemma: we need to compute the probability that "something else" possibly fits the data better.

9.4.1 Training and modeling

The probabilities of (9.1) and (9.2) might be computed by meticulous modeling of $\mathcal{B}$ and the sampling process $B = S(\mathcal{B})$. This is an impossible task in general; and, therefore, one has to resort to training the biometric authentication system, while only partially modeling $B = S(\mathcal{B})$. Using properly labeled sample data, the algorithms for automated biometric matching are trained to handle efficiently and effectively the customer population of an application while protecting the application against potential intruders. Hence, an authentication system is trained as to what the user population looks like but, ideally, should also be trained as to what the nonuser (unwanted) population looks like.

As a consequence, the particular error statistics, as discussed in Chapters 5, 15, and 16, depend on the user population and on the population of potential abusers. These parameters cannot be computed and can only be statistically estimated. Testing of a biometric authentication system is done through technology, scenario, and operational evaluations as discussed in Chapter 7 (more about this later).

9.4.2 There are two aspects to training

First we look at two aspects of biometric system training in general and of the training of biometric authentication applications in particular. Here system training refers to the training of a particular biometric application and not so much to the training of a core biometric matcher. Let us first look at the application of biometric verification and identification, i.e., positive authentication. The training of the application involves issues like the *enrollment policies* and the *authentication protocols*, that is—

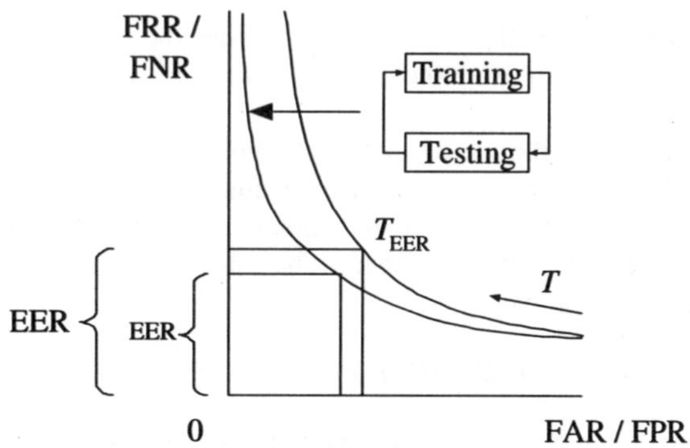

Figure 9.5: Improving the ROC curve by training a biometric authentication system.

1. *Enrollment of subjects:* A first training issue is the enrollment of a subject where a system is trained to authenticate a person by verifying a match with their biometric $\mathcal{B}$, or rather a sample B of that biometric instance. During enrollment one or more samples B of a subject's biometric $\mathcal{B}$ are acquired and biometric samples or templates derived from the samples B are stored in some member database $\mathbf{M}$.

   The care with which the process of enrollment is administered greatly influences the overall accuracy of the biometric authentication system both for positive and negative authentication.

2. *Tuning to a specific user population:* In addition to the enrollment subsystem, a biometric authentication system itself needs to be trained. This is achieved through refining and enhancing the signal or image to match the user population characteristics and incrementally improving the match engine. Here the cycle of training using training data followed by testing using separate test data is a standard procedure for optimizing pattern recognition systems. The objective, as shown in Figure 9.5, is to iteratively improve the accuracy, i.e., minimize *either* the FAR and FRR or minimizing *both* the FAR and FRR by attempting to move the ROC curve as close as possible to the coordinate axes. (In the case of a negative authentication system, the error rates to be minimized are the FPR and FNR, of course.)

In particular, in Section 5.2 we discuss what is involved when a biometric application is trained using the ROC curve as in Figure 9.5. First, though, we look at the enrollment process, which is also system training. After all, when a subject d is enrolled, the subject's biometric $\mathcal{B}$ is "shown" to the system and a machine representation B (often called a template) is constructed so that $\mathcal{B}$ can be recognized at later times.

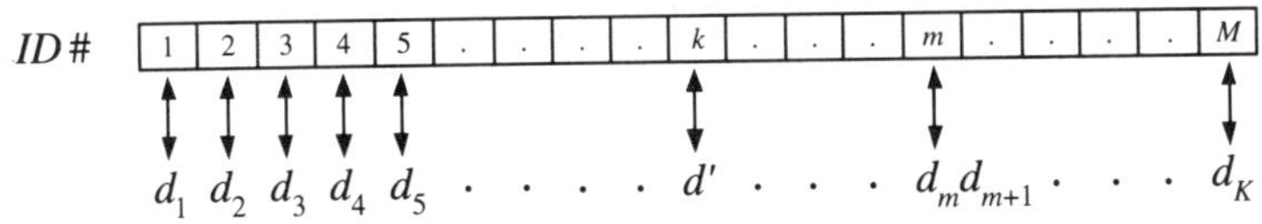

Figure 9.6: Sequentially enrolling subjects by assigning unique ID numbers.

9.5 Enrollment is system training

Enrollment of a biometric authentication system with database $\mathbf{M}$ is achieved by selecting subjects d from the world population $\mathbf{W}$ and assigning an identifier *ID* to each subject. In a sense, enrollment is a process of labeling the population. This process is shown in Figure 9.6: sequentially m identifiers ID_i, $i = 1, ..., m$ are assigned to the sequence of subjects d_k, $k = 1, ..., K$.

When a certain subject d_k enrolls with ID number $ID = k$, there are really three possibilities:

1. Most commonly, subject d_k is somehow correctly "linked" to *truth data* and legitimately assigned $ID = k$.

2. Subject d_k is in reality a subject d_j, with $j < k$, i.e., d_k is "duplicate" in some sense of subject d_j. The result of enrollment is that ID numbers ID_j and ID_k are *duplicates*, representing the same individual.

3. Subject d_k is in reality a subject d_j, with $j > k$. That is, subject d_k is *faking* (impersonating) unenrolled subject d_j, a legitimate subject; alternatively, subject d_k is *creating* some d_j as a seemingly legitimate subject. Either way, the result of this enrollment is that ID number ID_k corresponds to a "fake" identity.

Any subject d in $\mathbf{M}$ may be any of the three above and we have non-zero probabilities

$$P_D = Prob\,(d \text{ is a duplicate}) \ \text{ and } \ P_F = Prob\,(d \text{ is a fake}).$$

These are the probabilities that some subject $d \in \mathbf{M}$ is also enrolled under a different ID number and the probability that subject $d \in \mathbf{M}$ is a fake identity, respectively.

The enrolled population $\mathbf{M}$ is of course a function of time. Given an application, at any point in time t, let again $\mathbf{M} \subset \mathbf{W}$ denote the enrolled population. The enrollment status at time t is then the development of an ideally one-to-one mapping between the set $\mathbf{M} \subset \mathbf{W}$ and a set $\mathbf{I}$ of identities, or less ambiguously "identifiers," $\{ID_i,\ i = 1, ..., \#IDs = m\}$. Hence, enrollment can be modeled as a process that randomly selects a sequence of subjects $(d_1, d_2, ..., d_m)$ where there is a finite probability P_D that some d_i is a duplicate and there is a finite probability P_F that this d_i is a fake identity, a created or stolen identity.

A result of the enrollment process is shown in Figure 9.7: when forming set $\mathbf{M}$ by drawing samples from set $\mathbf{W}$ there is a finite chance of a repeat and there is a finite chance of a fake identity.

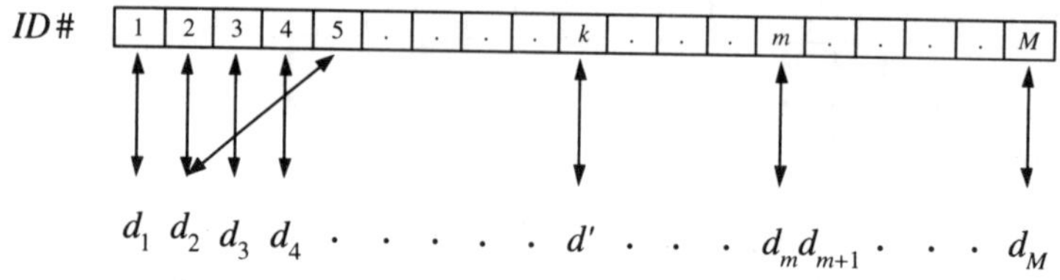

Figure 9.7: The result of enrolling subjects by assigning unique ID numbers: subject d_2 is enrolled two times with $ID = 2$ and $ID = 5$; subject d' is enrolled as a fake with $ID = k$.

Incidentally, the capability of detecting a duplicate identity depends on the False Negative Rate, the probability that $\mathcal{B}(d_k) \equiv \mathcal{B}(d_j)$ for some j is *not* detected. This is because duplicates are usually discovered by screening the biometric of subject d_k against the existing database $\mathbf{M}$.

With the set $\mathbf{M}$, $\{d_i,\ i = 1, ..., m\}$ the registered subjects of the population $\mathbf{W}$ and the set $\mathbf{I}$, $\{ID_i,\ i = 1, ..., m\}$ the set of unique, logical identifiers, we have the associations

$$\begin{array}{rcll} d_i & \Leftrightarrow & ID_i & \text{for} \quad i = 1, ..., m \\ \mathcal{B}(d_i) & \Leftrightarrow & ID_i & \text{for} \quad i = 1, ..., m. \end{array}$$

Equivalently, we have the pairs $\{(\mathcal{B}_i, ID_i) = (\mathcal{B}(d_i), ID_i),\ i = 1, ..., m\}$ where $\mathcal{B}_i = \mathcal{B}(d_i)$ is the true, real world biometric of subject d_i. However, there are the following instances:

- $(\mathcal{B}(d), ID_i)$, where d is the same as some other subject d_j, hence a duplicate entry.
- $(\mathcal{B}(d), ID_i)$, where d is a stolen (from some unenrolled subject d_j) or created identity.

9.5.1 Database integrity

We introduce a number of issues associated with the integrity and quality of the enrollment biometric database $\mathbf{M}$. By "integrity" of $\mathbf{M}$ we mean how well the database reflects the truth data of the seed documents used for enrollment; by "quality" of $\mathbf{M}$ we mean the quality of the biometric sampling.

The overall process used for checking the database integrity and purging the detected duplicates is called *database consolidation*. Due to non-zero error rates of matching, this process is not perfect; consequently duplicates may exist in the database after consolidation. The database integrity when it comes to duplicates is determined by P_D, the probability of duplicates

$$P_D = \text{FNMR}_E \times P_{DEA}. \tag{9.3}$$

Here P_{DEA} (Double Enroll Attack) refers to the probability that an already enrolled subject d_j wishes to re-enroll in the database as a different identity d_k. Expression (9.3) assumes that the database is consolidated by an all-to-all match of the enrolled subjects d_i in $\mathbf{M}$. The FNMR_E is the probability that a match between two samples of the same biometric is not detected, i.e., is missed. Note that P_D may not be constant over a period of time as more and more identities get enrolled; the number of duplicates in $\mathbf{M}$ is $P_D \times m$, with m the number of entities in $\mathbf{M}$.

The enrollment integrity is further determined by P_F, the probability of a fake enroll as d_k (created using stolen biometric identifiers of some d_j or created using fake biometric identifiers for a fake d_j). Let FMR_E be the probability that a match between two different biometric samples is falsely declared during enrollment, then

$$P_F = \text{FMR}_E \times P_{IA}, \tag{9.4}$$

where P_{IA} is the probability of impersonation attack. The number of fake identities in $\mathbf{M}$ equals $P_F \times m$. The concepts of P_{DEA} and P_{IA}, granted, are nebulous but as we will see, give valuable insights for analyzing enrollment.

Expressions (9.3) and (9.4) contain the FNMR and the FMR of the enrollment process, FNMR_E and FMR_E. These terms denote effective error rates of individual comparisons, had the negative identification in the system been implemented using an exhaustive comparison of all identities enrolled in the system. In other words, FNMR of enrollment is the probability with which a pair of matching subjects (d_k, d_j) is not detected by all-against-all matching, which is the FNMR of match engine used to detect duplicates (should be low to keep the number of False Positives low, since these need manual inspection). The FMR of enrollment, on the other hand, is the probability with which a pair of mismatching subjects (d_k, d_j) is not detected by the (human) enrollment process. These are the error rates of the enrollment matcher, and so, not unexpectedly, there is a trade-off between the error rates P_D and P_F. The optimal operating point might be determined by minimizing a weighted sum of the FNMR and FMR, operating on some ROC curve. Given that an enrollment station can do only a fixed and finite amount of work, there is a trade-off between efforts in preventing double enrolls (low FNMR_E) and efforts in preventing fake enrolls (low FMR_E).

This brings us to the enrolled population in database $\mathbf{M}$ as a system "variable" that influences the accuracy of a biometric authentication system. Database $\mathbf{M}$ contains the m records

$$T_i \;=\; (ID_i, K_i, B_i), \;\; i = 1, ..., m,$$

where ID_i is a unique machine identifier for subject d_i, K_i is knowledge like passwords selected by subject d_i and B_i is the biometric template of subject d_i.

What now is the integrity/quality of the data in this database? For the entries of T_i we have—

ID numbers: The integrity of the IDs, $\{ID_1, ID_2, ...\}$ as determined by the P_D and P_F, probability of a duplicate and probability of a fake (created or stolen) enrollment, respectively.

These variables describe the quality of the enrollment system. Factors like circumvention, collusion, coercion, impersonation (see Chapter 12), or just sloppy enrollment all contribute to the probability of fake enrolls, or FMR_E of enrollment. The probability of duplicate enrolls FNMR_E will be largely determined by the FNMR of the one-to-many matcher.

The FMR and FNMR of enrollment are elusive concepts. They are determined by the ability to authenticate seed documents and the ability to match biometrics on the seed documents and subjects d.

Knowledge: The integrity of the knowledge K_i, $i = 1, ..., m$ is another fuzzy concept. It refers, in some way to the "secrecy" and accuracy of passwords and other knowledge held by subjects d_i.

Biometrics: This is the quality of the biometric samples and templates that are on file. How well does biometric sample B for subject d represent the true biometric $\mathcal{B}$ as a function of time? It is this question that we try to answer in the next section.

A fourth aspect to data integrity can be included, if you will:

Other data: Traditional measures of data integrity of information that is held on file about subjects d_i that is not used for authentication purposes. This refers to measures of correctness of entries like birthdays and addresses, etc.

9.5.2 Probabilistic enrollment

The enrollment process has as goal to build an access control list of subjects $d_i, i = 1, ..., m$ of some subset **M** of population **W** authorized to use (or barred from using) some service. The access control list indicates whether the association between an enrolled identity d_i and the corresponding biometric $\mathcal{B}_i$ is to be trusted or not. This is an unrealistic goal, and it is best to assign some probability $p_i \geq 0$ to each subject $d_i, i = 1, ..., m$.

After all, upon enrollment a real world subject d_i becomes just a machine representation or token T_i, i.e., (ID_i, K_i, B_i), an identifier along with a password K_i and a biometric template B_i. And there is only a probabilistic link between this biometric template B_i and real world subject d_i. It is therefore desirable that the enrollment process also computes and stores likelihoods

$$Prob\,(d_i|B_i), \quad i = 1, ..., m. \tag{9.5}$$

This is the likelihood of d_i given the stored template; it expresses how well a subject's biometric $\mathcal{B}_i$ matches his or her template B_i.

The probabilities (9.5) can only be computed if there exist some machine representation of the real world biometrics $\mathcal{B}_i$. Let these representations be another set of templates $\tilde{B}_i$, $i = 1, ..., m$ and write $Prob\,(d_i|B_i)$ as $Prob\,(d_i|B_i; \tilde{B}_i)$. Then we get

$$\begin{aligned} Prob\,(\text{corrrect match}|\mathcal{B}_i \equiv \tilde{\mathcal{B}}_i) &= Prob(s > T|\mathcal{B}_i \equiv \tilde{\mathcal{B}}_i) \\ Prob\,(d_i|B_i; \tilde{B}_i) &= 1 - \text{FNMR}_i \approx 1 - (1 - s_i) = s_i. \end{aligned} \tag{9.6}$$

Where, for simplicity, we assume that the match score $s_i = s(B_i, \tilde{B}_i)$ is the likelihood that d_i is the true subject, given B_i. Here FNMR_i is derived from the match score: $\text{FNMR}_i = 1 - s_i$. This is only true if the $\text{FMR}_i \ll \text{FNMR}_i$ for that particular subject d_i; in effect, it will only be true if $s(B_i, \tilde{B}_k) \ll s_i$ for all $k \neq i$.

9.5.3 Modeling the world

As noted before, the likelihood $Prob\,(d_i|B_i)$ can be approximated by match scores s_i only under very unrealistic circumstances. More realistic approximations will have to involve the modeling of other subjects d_k, $k \neq i$ that are enrolled in **M**. We are interested, more generally, in $Prob\,(d_i|O)$, the likelihood of subject d_i given the biometric data O collected at enrollment time. Using Bayes' theorem

$$Prob\,(d_i|O) = \frac{Prob\,(O|d_i)Prob\,(d_i)}{Prob\,(O)}.$$

$Prob\,(d_i)$ is the prior probability that subject d_i is present, and $Prob\,(O)$ is the prior probability that this particular observation will occur (which cannot be computed exactly). Assume $Prob\,(d_i) = P_d$ is constant, for $i = 1, ..., m$, the set of enrolled subjects $\mathbf{M}$ and we get

$$Prob\,(d_i|O) = \frac{Prob\,(O|d_i)Prob\,(d_i)}{\sum_{j=1}^{m} Prob\,(O|d_j)Prob\,(d_j)} \approx \frac{Prob\,(O|d_i)}{\sum_{j=1}^{m} Prob\,(O|d_j)}. \quad (9.7)$$

Evaluating $Prob\,(O|d_i)$ is a matter of fitting model d_i to the data O and determining how well this can be done. Evaluating the rest of this expression is impractical and even impossible. It is computationally expensive and upon enrollment of some subject d_j, the samples of d_k, $k = j+1, ..., m$ are not yet available.

The most difficult issue in training a biometric authentication system is the modeling of data from unknown people. In general, a system has to model genuine subjects but *also* model imposters, forgers, and impersonators, i.e., modeling probabilities $Prob(\mathcal{B}' \neq \mathcal{B})$ that real-world biometric $\mathcal{B}'$ can, deliberately or unintentionally, impersonate some previously recorded biometric $\mathcal{B}$. An interesting idea to try to achieve this is "anti-speaker modeling," or "cohort modeling" which has been developed by the speaker recognition community [118,191].

9.5.4 Modeling the rest of the world—cohorts

Voice verification methods not only use a model describing the speaker's biometric machine representation, but also a model describing all other speakers. The reason for this has been that the probability of a match cannot be evaluated directly in many of the hidden Markov model voice recognition algorithms. The match probability can only be calculated indirectly through Bayes' rule that requires the likelihood of observing the voice data (independent of the speaker) to be calculated. Determining this likelihood of the voice data by modeling all possible speakers $\mathbf{W}$ is not possible and needs to be approximated.

Following the speaker recognition literature [190], there exist two techniques to approximate the denominator of (9.7).

1. *Cohort modeling* approximates the set $\mathbf{M}$ by a subset $\mathcal{M}_i$ that resemble subject d_i. For each subject d_i, a set of approximate forgeries (subject $d_i^1, d_i^2, \cdots$) is computed and stored. Note that this will increase the template size. We denote this set by D_i—the set is called the set of cohorts of speaker i.

2. *World modeling* reduces the set $\mathbf{M}$ to one fictitious model subject D, trained on a pool of data from many different speakers, who represent the "world" $\mathbf{W}$ of possible speakers.

   The great advantage is that no sets of cohorts have to be stored, but it may be unclear how to construct a world D. For applications where small template size is important, this may be the only solution.

Both these techniques, of course, mean that samples B need to be used to train the classifiers. An immediate problem here is that the developer will need to collect a representative training set of data, the training set has to represent the ultimate user population in some sense.

World modeling approximates (9.7) as

$$Prob\,(d_i|O) \approx \frac{Prob\,(O|d_i)}{\alpha \times Prob\,(O|D) + Prob\,(O|d_i)}. \tag{9.8}$$

Here we added the probability $Prob\,(O|d_i)$ and the factor α to the denominator to ensure that d_i is part of the world model and to normalize to database size m. A good choice for $\alpha = (m-1)$, so that the denominator reflects the whole population $D + d_i$. Here assume that D is some sort of model of the "average" person (as related to subject d_i), so that a database D of $m-1$ subjects behaves like database of $m-1$ subjects d_i. It could, for instance be a collection of goats, sheep, wolves, and lambs as discussed in Section 9.2. The speaker recognition community, in this case, trains the world model D using a pool of data from many different speakers.

For cohort modeling, we get

$$\begin{aligned} Prob\,(d_i|O) &\approx \frac{Prob\,(O|d_i)}{\alpha \times \sum_{j \in \mathcal{M}_i} Prob\,(O|d_j) + Prob\,(O|d_i)} \\ &= \frac{Prob\,(O|d_i)}{\alpha_i \times Prob\,(O|D_i) + Prob\,(O|d_i)}. \end{aligned} \tag{9.9}$$

Here we added again the probability $Prob\,(O|d_i)$ and the factor α for the same reason. Let the number of cohorts for d_i be c_i; the factor α is $\alpha_i = c_i$, assuming that subjects not in d_i's cohort pose negligible threats. With D_i the cohort for subject d_i, the model D_i behaves like a database of c_i subjects that try to "look" like subject d_i. For that reason, one may want to lower α_i because D_i is not a good representation of the population $\mathbf{M}$.

Note that up till now we have treated the problem as a *closed-world* authentication problem, in the sense that it is expected that only subjects $d_i \in \mathbf{M}$ are matched. In reality, models such as D and D_i are constructed from many databases of labeled biometric samples. The probabilities $Prob\,(O|d_i)$ in the denominators of (9.8) and (9.9) are often omitted.

Modeling techniques like this have resulted in great improvements in accuracy. We need to be able to somehow model the sample formation process, given the subject d_i; or, somewhat equivalently, we need to be able to model the feature formation process, given the subject d_i. We need to be able to somehow compute realistic *a posteriori* probabilities $Prob\,(O|d_i)$.

No conclusive difference in performance between the two techniques in text dependent speaker verification has been found, but world modeling has some computational and algorithmic advantages over cohort modeling:

- Less storage is required for the model parameters.
- It is computationally more efficient during recognition, since only one anti-speaker likelihood needs to be evaluated.

- It does not require the selection of cohort models $\mathcal{M}_i$ for each speaker i.

The world model approach is the commonly adopted one. This world model is computed in different ways:

- A completely speaker-independent implementation. For each speaker the same algorithms are used to model the anti-speaker.
- Gender- or handset-dependent implementations, where the choice for the world model depends on the gender of the claimed identity or the type of handset used by the actual speaker [186].

A promising avenue of research is cohort modeling for other biometric identifiers. Similar techniques are of course already in use, for example, training neural networks by giving negative examples, whereas cohort modeling creates an explicit probabilistic model of the negative examples.

9.5.5 Updating the probabilities

Denote $Prob\,(d_i|O)$ with P_i. From a biometrics point of view, after enrollment of a set of subjects, the enrollment database **M** contains $\{(B_i, P_i), i = 1, ..., m\}$. That is, for each d_i, we have a P_i that expresses the likelihood of d_i given the data O.

The probability P_i expresses the amount of faith we have in the fact that data O is obtained from subject d_i. During operation of the authentication system, data from subjects is collected and the likelihood P_i could be updated, perhaps by using cohort and world modeling. Upon authentication of subject d_i, a biometric sample is acquired that we denote here as ΔO. We are interested in computing $Prob(d_i|O, \Delta O)$; this can be written as (again using Bayes' rule)

$$\begin{aligned} Prob\,(d_i|\Delta O, O) &= \frac{Prob\,(\Delta O, O|d_i) Prob\,(d_i)}{Prob\,(\Delta O, O)} \\ &\approx \frac{Prob\,(\Delta O|d_i)}{Prob\,(\Delta O)} \times \frac{Prob\,(O|d_i) Prob\,(d_i)}{Prob\,(O)} \\ &= \frac{Prob\,(\Delta O|d_i)}{Prob\,(\Delta O)} \cdot Prob\,(d_i|O) \\ &= \frac{Prob\,(\Delta O|d_i)}{Prob\,(\Delta O)} \cdot P_i, \end{aligned} \tag{9.10}$$

where the approximation holds good when the data ΔO and O are independent. The second term in the multiplication of (9.10) is the likelihood of d_i given the data O, also denoted as P_i, and depending on whether we use world modeling or cohort modeling; we use corresponding approximations (9.8) or (9.9), respectively.

What needs to be evaluated in (9.10) is the denominator $Prob\,(\Delta O)$. We can write this as

$$Prob\,(\Delta O) = Prob\,(\Delta O|d_i)\, Prob\,(d_i) + Prob\,(\Delta O|D_i)\, Prob\,(D_i). \tag{9.11}$$

This expression can be used both for world modeling and for cohort modeling: for world modeling D_i is a constant "average" person d; for cohort modeling, D_i is a set of subjects $\mathcal{M}_i = \{d_i^1, ...\}$ that resemble subject d_i. Again, as above we have the problem that priors, $Prob\,(d_i)$ and $Prob\,(D_i)$, appear. The question then is how we approximate these priors. We could, for instance, set $Prob\,(d_i) = P_i$ in (9.11) and (9.10) becomes

$$\begin{aligned} Prob\,(d_i|\Delta O, O) &\approx \frac{Prob\,(\Delta O|d_i)}{Prob\,(\Delta O)} \cdot P_i \qquad (9.12) \\ &= \frac{Prob\,(\Delta O|d_i)\,P_i}{Prob\,(\Delta O|d_i)P_i + Prob\,(\Delta O|D_i)(1-P_i)}. \end{aligned}$$

In effect, this expression with $Prob\,(d_i) = P_i$ ensures that $0 \le Prob\,(d_i|\Delta O, O) \le 1$.

9.5.6 Use of the probabilities

Probabilistic modeling of the "world" $\mathbf{W}$ of subjects $\{d_1, d_2, ..., d_m\}$ has some appealing uses:

- As mentioned before, good results have been obtained. One can expect accuracy improvements, given that the probability $Prob\,(O|d_i)$ is well-formulated. Practical models here are multi-variate Gaussian for obvious reasons.

- It will be possible to define measures of biometric integrity of the enrollment database, an example is database integrity $\mathcal{I}$:

$$\mathcal{I}(\mathbf{M}) = \frac{\sum_{i=1}^{m} P_i}{m}. \qquad (9.13)$$

  Requiring high integrity involves optimizing the likelihoods P_i. This can be achieved through the control of biometric sample quality as discussed in Section 9.3. Real-time model building and computation of P_i allows for better-defined sample quality measures and interactive sample enhancement.

- The "integrity" P_i of the biometric data, i.e., the template B_i in the database $\mathbf{M}$, might be combined with the integrity of other database entries. An overall estimate of the integrity of subject d_i is expressed as $\mathcal{P}_i$, a measure that includes P_i and other factors that might affect the integrity of subject d_i. In the travel application a travel rating for each subject can be constructed and updated according to rules like (9.12). Ratings like "trustworthiness" can be defined just like today's credit rating.

10

Large-Scale Applications

Biometric identifiers, systems, and databases are really put to the test when 1:*many* searches of large biometric databases are part of the enrollment policy or authentication protocol. In such identification problems, not only are low error rates desired, high 1:1 match rates are quite often also required.

In the following sections we look at what happens with system requirements when the size of the database becomes large. Here system requirements include allowable error rates, computational resources, manual clerical labor, expert labor, etc.

Binning and filtering are two common meta-methods for alleviating some of this burden. That is, they help minimize the number of required biometric 1:1 matches:

Filtering: Narrowing down the search by using non-biometric parameters like sex, age group, nationality, etc.

Binning: Cutting down the search by using an additional biometric, e.g., fingerprint ridge flow pattern (whorl, arch, etc.).

Accuracy analysis then involves multi-biometrics and assumptions about the correctness of the parametric data.

While it is good to keep these techniques in mind, even when they are applied there is still a large residual problem that must be solved. In the rest of this chapter, we examine the design, construction, and application of large biometric systems where 1:*many* searching is part of the enrollment policy or authentication protocol.

10.1 Example systems

There is no general agreement on what database size m would be called "*large-scale*." Phillips [164] gives some approximate definitions for the scale of biometric identification systems, which are shown in Table 10.1. Of course, there is a theoretical (if non-constant) upper bound on the scale of the ultimate large-scale system, the *world scale*—a system which could identify every human being, perhaps by any finger, so the largest human identification system would have around 60 billion enrolled biometrics.

Scale	**# of individuals**	**# of stored samples**
Small	≈ 330	$\approx 1,000$
Medium	≈ 3300	$\approx 10,000$
Large	$\approx 33,000$	$\approx 100,000$
Very large	$\approx 330,000$	$\approx 1,000,000$
Extremely large	$\approx 3,300,000$	$\approx 10,000,000$

Table 10.1: The size, or scale, of identification systems.

We loosely use the term for problems that are of the size of a moderately large country. A number of large-scale systems are in operation, or have been in operation in various countries. Table 10.2 gives some examples of these systems.

The systems listed in Table 10.2 are really two *different* types of biometric systems due to the difference in timetables (see Figure 10.1):

Voter Registration: *Fixed and sequential registration and authentication periods.*

Transforming legacy databases to biometric databases in a *fixed* enrollment period of T_e days, followed by positive authentication of the entire population in a fixed period of T_a days at a later stage. This problem is addressed in Section 10.5.

Country	Program	Database Size (est.)	Biometric used
Nigeria	Voter Registration	60M	
Philippines	Social Security	40M	
Thailand	National ID	15–40M	fingerprint
Peru	Voter Registration	13M	
Hong Kong	National ID	7M	
Malaysia	National ID	22M	fingerprint and face

Table 10.2: A number of countries have implemented, or are implementing, large-scale biometric identification systems.

National ID: *An open-ended simultaneous enrollment and authentication.*

Creating or automating a biometric database during an *open-ended* registration or conversion period T_e. This can be the creation from "*scratch*" or the conversion of a biometric database, for example, converting police fingerprint paper records to a digitized system. This is the open-ended national ID registration problem described in Section 10.6. Such systems are used for authentication from early on in the registration period T_e; hence T_a overlaps T_e.

Many questions can be asked about such large 1:*many* systems. What are the requirements in terms of 1:1 match throughput and how do these requirements depend on m?

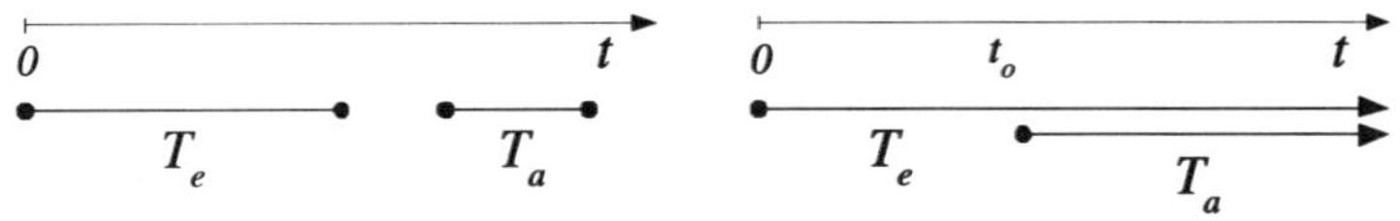

Figure 10.1: A voter registration period T_e is followed by an election T_a (left) during which authentication takes place. A national ID database is collected with concurrent enrollments and authentications (right).

	FAR(1)	FRR
1 finger	10^{-5}	10^{-3}
2 fingers	10^{-10}	10^{-3}

Table 10.3: Best approximate one- and two-finger error rates as derived from the numbers in Section 7.3.

What are the required accuracy numbers and can they be achieved? What is the cost of such a system and how does the cost depend on m? In general, what are the throughput and accuracy requirements as a function of m and what is the associated cost $\text{Cost}(m)$?

The central question is how does such a system scale with the size of the problem m ? That is, how hard is it to get from an m of, say, 1,000 to a database size of m in the millions of subjects. A second important question is what the logistical problems are in first enrolling a large number m subjects in a short time frame T_e and then authenticating these m subjects in time T_a.

In the following examples we will use fingerprints because this allows scaling of the biometric system: The more of a person's fingers that are used, the more subjects can be discriminated and, potentially, the larger one can make the size of biometric database. This should not necessarily be interpreted as if fingerprint is the preferred biometric to solve such large-scale applications.

10.2 Required accuracies

During enrollment screening, two types of errors can be made—the usual False Positive and False Negative errors. That is, erroneous matches that indicate possible double enrollments may be found in the database, and true existing matches (actual double enrollments) in the database may be missed.

10.2.1 Enrollment False Positives

As shown in Figure 10.2, the *total* number of False Alarms for double enrolls grows rapidly when m becomes large. Every single one of these False Positives needs to be resolved through some exception handling process during the enrollment period.

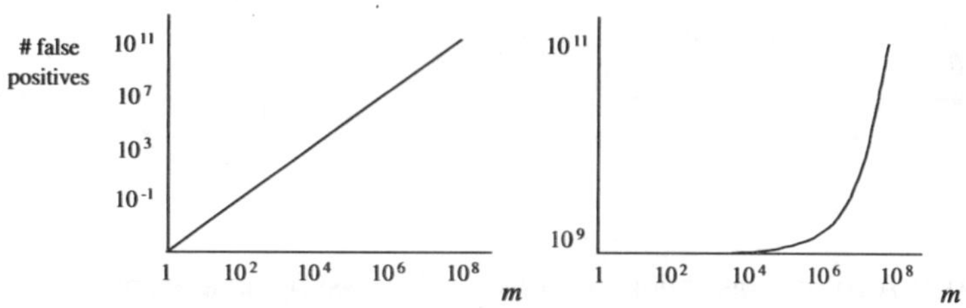

Figure 10.2: The total number of False Positives during enrollment depends quadratically on problem size m.

Let FAR(1) be the False Positive Rate and FRR be the False Negative Rate of the 1:1 matcher. Table 10.3 gives the rough state-of-the-art 1:1 error rates when using 1 or 2 fingers for biometric matching. This translates to a total number of False Positives during enrollment period as follows:

1. *One finger:* Then $\text{FAR}(1) \approx 10^{-5}$ is a potential operating point. Figure 10.2 shows that just as the number of required 1:1 matches is quadratic in m, the number of False Positives depends quadratically on m. The graph on the right on a linear scale shows that this is quite dramatic for even small m. That is, when $m = 10^8$, the total number of False Positives will be on the order of 10^{11}, when using one finger for matching $(\text{FAR}(1) = 10^{-5})$.

   Using Expression (10.4), the *total* number of *False Positives* during the entire enrollment period is

   $$\#\text{ of False Positives} \approx [m \times \text{FAR}(1)] \times m = 10^{-5} \times (10^8)^2 = 10^{11}.$$

   Clearly, handling on the order of 10^{11} False Positives, or exceptions, is unrealistic because that means that for every enrollment, there are on average $1,000$ False Alarms, or falsely associated identities!

2. *Two fingers:* The False Positive Rate is $\text{FAR}(1) \approx 10^{-10}$ (see Table 10.3). At this rate, the total number of False Positives is

   $$\#\text{ of False Positives} = [m \times \text{FAR}(1)] \times m = 10^{-10} \times (10^8)^2 = 10^6.$$

   On average, there is *one* False Positive per 100 enrollees. Stated differently, there will only be a problem with 1 percent of the enrollments. Assume the time it takes to (manually) resolve a False Positive is on the same order of magnitude as manually enrolling a subject. The total time needed to resolve the False Positives is then approximately 1 percent of the time needed to enroll to entire $m = 10^8$ subjects.

Table 10.4 summarizes 1:1 False Positive Rates, the total number of False Positives during the enrollment days, and the percentage of time devoted to False Positive resolution (assuming that False Positive resolution requires about the same amount of time as an enrollment).

FAR(1)	False Positives	Exception handling
10^{-5}	$\approx 10^{11}$	$1000\times$
10^{-10}	$\approx 10^{6}$	$1\,\%$
10^{-12}	$\approx 10^{4}$	$0.01\,\%$

Table 10.4: The 1:1 False Positive Rate during enrollment has to be much less than 10^{-5} in order to keep exception handling low.

10.2.2 Enrollment False Negatives

As can be seen in Table 10.3, when operating at a False Positive Rate of $\mathrm{FAR}(1) = 10^{-5}$ or 10^{-10}, we are at a *False Negative Rate* of $\mathrm{FRR} \approx 10^{-3}$. This could mean several things:

1. On the order of 0.1 percent of the people in the database will be enrolled more than once. This means that there are up to $10^{-3} \times 10^8 = 10^5$ double enrolls.

2. If a particular subject is enrolled in the database and then maliciously attempts to re-enroll, there is a probability of $\approx 10^{-3}$ that he will succeed at this attempt. Note that this probability is independent of the size of the database or the time during the enrollment period at which the attempt occurs.

3. If a random subject is enrolled in the database and is then re-enrolled, there is a probability of order $\approx 10^{-3}$ that this mistake will not be caught.

This last statement makes it very hard to *a priori* predict the integrity of the resulting database. Database integrity also depends on the integrity of the enrollment process, which is a matter of human integrity and defenses against security attacks (because of negligence, incompetence, or lack of human integrity). In practice, this source of "error" may actually outweigh any system limitations due to a high FRR rate.

The most dangerous double enrolls (False Negatives) are the ones created by undesirable subjects for some particular reason. For this application, the False Negative Rate should of course be as low as possible. It is unclear what the deterrence against double enrollment (double dipping) of the False Negative Rate really is, though. A determined subject will try to double enroll through security loopholes and it is unclear what the benefit of $\mathrm{FRR} = 10^{-6}$ compared to, for instance, $\mathrm{FNR} = 10^{-3}$ really is.

10.2.3 Authentication accuracy

The matcher operating point $(\mathrm{FPR}(1), \mathrm{FNR}) = (10^{-5}, 10^{-3})$ using *one* finger translates directly to $\mathrm{FAR}(1) = 10^{-5}$ and $\mathrm{FRR} = 10^{-3}$ when used for verification. Thus it appears on the surface that only one finger needs to be used for accurate verification of subjects at the voting booth.

False Rejects: For authentication during the election, $m = 10^8$ subjects need to be verified. Hence, the number of False Rejects is

$$\#\text{ of False Rejects} = \mathrm{FRR} \times m = 10^{-3} \times 10^8 = 10^5.$$

Naturally, one would like to operate at a lower False Reject Rate, e.g., FRR $= 10^{-5}$, which results only in about $1,000$ False Rejects over the entire election period T_a.

False Accepts: At FAR$(1) = 10^{-5}$ the probability that a subject claiming a random identity will be accepted is 10^{-5}. This usually corresponds to a FRR $\approx 10^{-3}$ (wrongful denial of voting rights). When operating instead at a lower False Reject Rate of FRR $= 10^{-5}$, the probability of a False Accept is often much higher, say, FAR $\approx 10^{-2}$, or on the order of 1 percent!

Thus, for instance, a double-voter claiming a false identity then has a 1 percent chance of succeeding. This is likely too high, demanding instead that authentication of the voters during the election be done with two fingers also.

10.3 Matcher sizing

In most large-scale applications the size of the database grows over time. One major ramification of this is that it becomes harder and harder to enroll new subjects. That is, to enroll a new person they must be checked against all the other m individuals already in the database at time t. To study this in more detail, we first need to introduce some terminology:

- $m(t)$ = **Database size,** the number of individuals enrolled in the biometric databases at time t. This typically grows over time.
- $J(t)$ = **Match throughput,** the number of 1:1 biometric judgements (matches) that the match engine has to perform per unit of time. This is related to the number of subjects that need to be "processed" per time unit (per second).
- E_{avg} = **Enrollment rate,** the average number of subjects that need to be enrolled per time unit (per second).
- A_{avg} = **Authentication rate,** the average number of subjects that need to be authenticated per time unit (per second), be it for verification, identification, or screening.

The quantities E_{avg} and A_{avg} are expressed as constants but in reality are functions of time, like $m(t)$ and $J(t)$, over the lifespan of a biometric system.

10.3.1 Match throughput

During enrollment the system will use 1:*many* matching. The difficulty of this depends on the size of the biometric database at time t:

$$m(t) = \int_{x=0}^{t} E(x)\,\mathrm{d}x = E_{avg} \times t.$$

Here we have replaced $E(t)$ by the average rate E_{avg}. Using this same substitution, during enrollment we find that

$$J_e(t) = E(t)\,m(t) = E_{avg}^2\,t. \tag{10.1}$$

By contrast, for authentication we assume that only 1:1 verification matching is needed. Thus the required match throughput rate is a constant, simply,

$$J_a(t) \;=\; A_{avg}. \tag{10.2}$$

The above equations are for a two-phase system, such as voter registration. For an on-going continuous system, enrollment and authentication proceed concurrently. Let us assume that at some time t_s the database size stabilizes at size m. That is, the same number of people are deleted from the database per day as are enrolled per day. We then find that

$$J(t) \;=\; J_a(t) \,+\, J_e(t) \;=\; \begin{cases} A_{avg} \,+\, {E_{avg}}^2\, t & \text{for } t < t_s \\ A_{avg} \,+\, m\, E_{avg} & \text{for } t \geq t_s. \end{cases} \tag{10.3}$$

This time variance of the match rate is shown graphically in Figure 10.3. For National ID type applications (left) the required match rate rises until the database reaches its final size. For Voter ID type systems (right) the match rate rises during the enrollment phase. During the following authentication phase there is a typically much smaller match rate required since only 1:1 verification is performed.

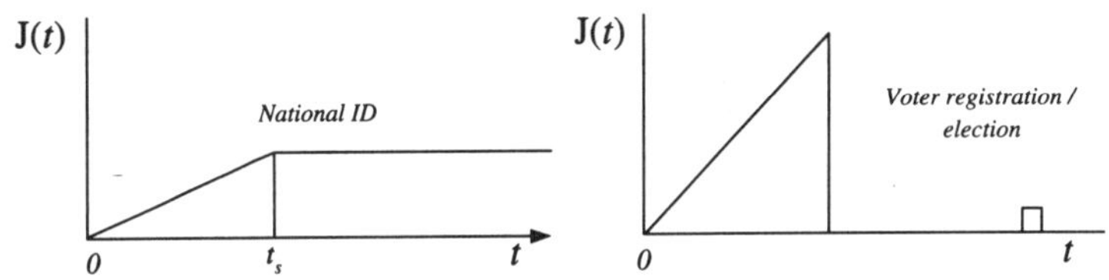

Figure 10.3: The required matching rate changes over time and depends on the specific type of application.

10.3.2 Total number of matches

Taking as an example the voter registration scenario, by integrating (10.2) the *total* number of biometric matches needed for verification on election day is

$$\int_{t=0}^{t_a} J(t)\,\mathrm{d}t = J_{avg} \int_{t=0}^{t_a} \mathrm{d}t = J_{avg}\, t_a = m.$$

The number of required matches is linear in m since only *one* 1:1 match is needed for each authentication.

By integrating (10.1), on the other hand, it is seen that the total number of 1:1 matches that need to be performed during the enrollment period T_e is quadratic in m, the problem size,

$$\int_{t=0}^{t_e} J(t)\,\mathrm{d}t = E_{avg}^2 \int_{t=0}^{t_e} t\,\mathrm{d}t = \frac{(E_{avg} t_e)^2}{2} = \frac{m^2}{2} = O(m^2). \tag{10.4}$$

Here $O()$ means "order of" and generally denotes the dominant terms (input variables and their exponents) that need to be considered when scaling up a problem. The number of

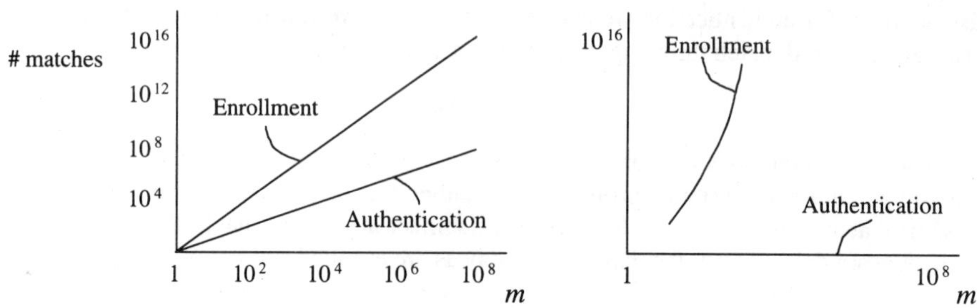

Figure 10.4: The total number of 1:1 matches required for enrolling m subjects is quadratic in m, while for authentication it is linear (logarithmic scales, left); number of enrollment matches with linear y-axis (right). Note that in the right illustration authentication plot is almost aligned with x-axis.

biometric matches is order m^2. As shown in Figure 10.4, on a log-log scale the slope of the enrollment curve is twice that of the authentication curve. On a linear scale (right), it is seen that the total number of matches needed for authentication is dwarfed by the number of matches needed for enrollment.

This is a nasty property of this type of enrollment. The number of required 1:1 matches during the enrollment period of T_e days is quadratic in m. The computing power needed to complete such an enrollment project within a given time frame is also quadratic in m. If the problem size is multiplied by 10, the computational requirements multiply by 100.

10.4 Exception handling

We already know that the total number of required 1:1 matches for enrolling m subjects is on the order of m^2 and therefore the computational cost is also on the order of m^2. What is of interest is the behavior of other costs $\text{Cost}(m)$ as function of m. Let this $\text{Cost}(m)$ only include manual labor, materials, overhead, and things like that. It has two components:

- **Incremental cost,** the cost of enrolling each of the m subjects, i.e., sampling the biometric, printing the ID cards, etc., or put differently, the cost aspects that are linear in m.

- **Exception handling,** the cost of dealing with a problem when enrolling these m subjects, in particular, the cost of resolving False Positives. We are interested in how this component of the cost depends on m.

Let the cost of handling *one* False Positive be h times the cost of an enrollment without False Positives. This h is the cost of manually comparing two pairs of fingerprints by fingerprint experts (when using two fingers). How does this h depend on m? Is the cost of comparing look-alike fingers higher when these are found in a larger database? This is unclear and therefore we use a first-order approximation,

$$h(m) \approx h_{avg} + i\,m.$$

This expression assumes there is some fixed cost per exception, like calling in a supervisor, plus some other cost that depends on the size of the database, such as manually examining the top 0.000,1 percent of the best-matching records.

Then the *total* cost of enrollment is a function of m :

$$\begin{aligned} \text{Cost}(m) &\approx m \times 1 + m \times [\text{FAR}(1) \times m] \times h(m) \\ &= m + \text{FAR}(1) \times (h_{avg} + i\,m) \times m^2 \\ &= m + h_{avg}\,\text{FAR}(1)\,m^2 + i\,\text{FAR}(1)\,m^3. \end{aligned} \tag{10.5}$$

Note that this cost is quadratic for large m, no matter whether $i = 0$ or FAR(1) is very small.

Let us first set $i = 0$ then from (10.5) we see that

$$\text{Cost}(m) = \begin{cases} m & \text{if FAR}(1) \ll (h_{avg}\,m)^{-1} \\ h_{avg}\,\text{FAR}(1)\,m^2 & \text{if FAR}(1) \gg (h_{avg}\,m)^{-1}. \end{cases}$$

Clearly for large m, (10.5) will be dominated by the second term and the total cost will explode since the cost increases quadratically with the number m subjects in the database.

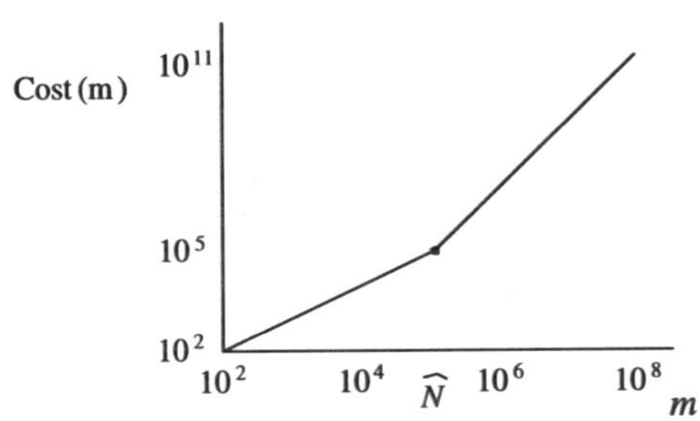

Figure 10.5: At a point $\hat{m}$ during enrollment, the enrollment cost begins to increase dramatically as a function of m (log-log scale).

This is shown in Figure 10.5, at point t in time where $\hat{m} = m(t)$ subjects are biometrically enrolled, the total cost of enrollment Cost(m) becomes quadratic in m. This crossover point occurs when the two terms are about equal:

$$\hat{m} \approx \frac{1}{h_{avg}\,\text{FAR}(1)}$$

If $\text{FAR}(1) = 10^{-5}$, the False Positive Rate using one finger, and $h_{avg} = 1$ (or any small constant), we have $\hat{m} \approx \text{FAR}^{-1}(1) = 10^5$. Clearly the use of two fingers with $\text{FAR}(1) = 10^{-10}$ is required because then $\hat{m} \approx \text{FAR}^{-1}(1) = 10^{10}$; otherwise the exception handling cost is prohibitive.

This particular $\hat{m}$ seems to be sufficiently high compared to the problem size $m = 10^8$, but we need to examine what happens when $i > 0$. Following (10.5) we have

$$\text{Cost}(m) = m + h_{avg}\,\text{FAR}(1)\,m^2 + i\,\text{FAR}(1)\,m^3.$$

Here the constant i may be ever so small, there *will be* a point $\hat{m}_{i>0} < \hat{m}_{i=0}$ where the cost of the enrollment process begins to explode. It is very hard to estimate i or $\hat{m}_{i>0}$ and therefore it is hard to predict exactly the cost of the registration phase of the election problem. Obviously, quality control of the biometric signal acquisition, and quality control of the enrollment process in general, is key to keeping h_{avg} and i low. Given the current state of the art, it may be very hard to achieve FAR(1), h_{avg}, and i that are small enough to avoid cost overruns and failures of such large scale endeavors.

In summary, the cost of the registration phase is hard to predict. What is the cost of manual resolution of False Positives as a function of m when relying solely on fingerprint? A step toward alleviating this problem is the use of a second biometric for human False Positive resolution. For instance, face recognition might be used for this purpose.

10.5 Voter registration

The problem is mandatory voter registration of 100 million voters in a biometric database during an enrollment period T_e, followed by a non-overlapping election period T_a. Specifically, the problem is the following:

1. $m = 10^8$ voters are to be biometrically enrolled during a registration period $T_e = 100$ days. During enrollment a subject is biometrically matched against the intermediate database of size $m(t) < 10^8$ at time t of enrollment. The purpose is to eliminate duplicates in legacy data and to prevent duplicates in the newly created biometric database.

2. The m enrolled voters need to be verified during an election period $T_a = 1$ day. Verification requires one biometric match per authentication.

Let us establish the 1:1 matching throughput requirements both for the enrollment and for the authentication phase of this biometric authentication system. A total of m voters need to be enrolled in a time frame T_e of t_e seconds, and hence the average enrollment rate requirement is

$$E_{avg} = m/t_e. \tag{10.6}$$

Equivalently, the average authentication rate requirement is

$$A_{avg} = m/t_a, \tag{10.7}$$

in time frame T_a of t_a seconds. For the moment, assume that it is indeed possible to ensure a sustained subject arrival rate of $E(t) = E_{avg}$ for t_e seconds to enroll m subjects at that rate. Similarly, we assume a constant subject arrival rate A_{avg} during the election.

Table 10.5 shows the required enrollment rate E_{avg} and maximum 1:1 match throughput, max J(t), for different registration periods (using 12-hour [720-minute] days). Remember, the required 1:1 match throughput is *not* constant but depends linearly on the enrollment time t or, perhaps more precisely, depends on intermediate database size $m(t)$. The maximum match throughput, $\max \mathrm{J_e} = \mathrm{J(t_e)} = \mathrm{E_{avg}} \times \mathrm{m}$, is needed at time t_e, the end of the enrollment period T_e. This maximum rate is on the order of 10^9 per second when the registration period is 50 or 100 days! In fact, 2.3×10^9 matches per second are required at the end of $T_e = 100$ days.

	$T_e = 50$ days	$T_e = 100$ days
E_{avg}	167K / hr = $2,778$ / minute = 46 / second	83K / hr = $1,389$ / minute = 23 / second
$\max \mathrm{J_e}$	4.6×10^9 / second	2.3×10^9 / second

Table 10.5: The enrollment rate and corresponding maximum 1:1 throughput needed for different enrollment periods T_e.

Also, using the enrollment rate E_{avg} from Table 10.5, we can estimate that the number of False Positives per minute as function of $m(t)$ is

$$\mathrm{FAR}(1) \times E_{avg} \times m(t) = 10^{-10} \times 1389 \times m(t) = 1.4 \times 10^{-7} \times m(t).$$

This means that at the end of the enrollment period, when $m(t_e) = 10^8$, there are 14 False Positives per minute.

If logistics is not an issue, the voter registration problem can be accomplished with a centralized database and all enrollment stations at the same geographic location, or with a distributed database and a distributed enrollment infrastructure.

1. One central database allows for straightforward $1{:}\,m$ searches against the database.
2. A distributed database, by nature, makes database updating (enrollments) and $1{:}\,m$ searching harder to implement and computationally more complex.

Even though a high enrollment (update) rate and an even higher search rate, or 1:1 match throughput, is needed, these are not the fundamental bottlenecks for scaling biometric systems. In principle, with enough computing power and bandwidth, such a large-scale system can be designed. Nevertheless, a biometric matching rate over 10^9 per second is quite a computational burden.

The election itself poses much less of a demand on the match engine. Authenticating 10^8 subjects in one day requires 1:1 match throughput,

$$J_u(t) = A_{avg} \approx 83\,\mathrm{M/hour} \approx 140\,\mathrm{K/minute} \approx 2,300/\mathrm{second},$$

when the polls are open for 12 hours (720 minutes) on election day.

Now let us look at the physical aspects of the problem of enrolling $m = 10^8$ subjects in 100 days, where each day the enrollment offices are open for 12 hours. Assume that the enrollment time of one subject is 6 minutes. At full capacity, therefore, any one enroll station can only process 120 subjects per day. Since a total of 10^6 subjects are to be registered per day, this means that 8,333 enrollment stations are needed.

During the election up to $m = 10^8$ subjects must be verified in one day (given a full turnout). Again, assuming that the polls are open for 12 hours and that each voter spends 6 minutes at a station, each station can handle only 120 voters. This means a total of over 833 *thousand* biometric verification stations will be needed on election day!

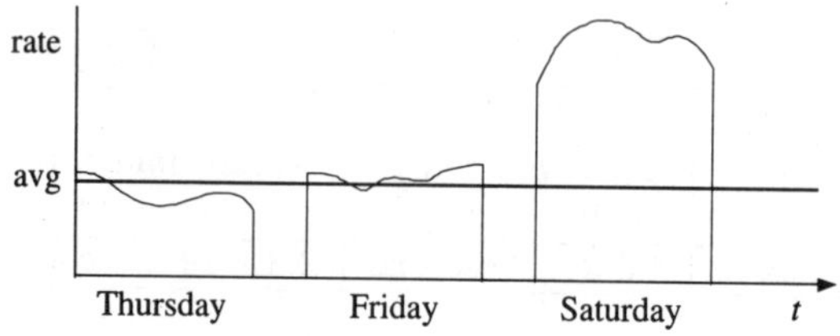

Figure 10.6: The processing rate for a system is generally not a constant. Depending on the time of day and day of week the demand can be significantly greater than the average.

When developing these estimates it is important to note that demand for such a system will not be constant. A typical demand profile is illustrated in Figure 10.6. Given the modern US lifestyle, people are more likely to try to handle the chore of enrolling either before or after work hours, or possibly during lunch. So the daily demand is likely to have two or three peaks separated by lulls. By the same reasoning, if the enrollment center is open on the weekend, the demand may be quite a bit higher than on the weekdays. A peaking factor of 2 or 3 times the average rate might be expected. If the system is not sized to handle this maximum demand, the response time per transaction will obviously increase. A similar effect is likely to occur for verification as well.

Table 10.6 summarizes the requirements developed in this section with a moderate amount of peaking taken into account. It is unclear whether these logistics can actually be achieved at anything like a reasonable cost. The enrollment aspect is particularly daunting.

However, it should be noted that during the election phase, controlling the arrival and processing of voters is no more complex than legacy voting infrastructures. This assumes the election only involves verification with a simple 1:1 match per voter and that False Rejects can be easily handled, for example, with a secondary biometric.

10.6 National ID systems

The second system to analyze is one where a country uses biometrics to uniquely identify all of its citizens. This is the easier of the two problems. The important properties here are that people are continually being enrolled and verified using the system. This is similar to a number of other scenarios such as registering for social benefits (e.g., unemployment subsidies) or commercial systems where a credit card transaction is authorized with a biometric (e.g., a thumbprint).

For a enrollment authentication, the system is used as a positive identification system based solely on a biometric identifier. The use of one finger is typically not enough to

Voter registration	Value	Explanation
m	10^8	Moderately large country
T_e	$1,200$ hrs	= 12 hrs / day x 100 days
E_{avg}	$83,000$ / hr	= m / T_e
Enroll time	0.1 hr	= 6 minute wait
T_a	120 hrs	= 12 hrs / day x 1 day
A_{avg}	$833,000$ / hr	= m / T_a
Verify time	0.1 hr	= 6 minutes in booth
max J_e	10^{13} / hr	Includes 20% peaking (278K processors x 10K / sec ?)
Enrollment stations	11K	Includes 20% peaking plus 10% spares
max J_a	10^6 / hr	Includes 20% peaking
Verification stations	$108,000$	Includes 20% peaking plus 10% spares

Table 10.6: A summary of the logistics for the voter registration scenario. The required matcher throughput during enrollment is phenomenally high.

handle a problem this size since it has a *matcher* operating point of only

$$(\text{FAR}(1), \text{FRR}) \approx (10^{-5}, 10^{-3}).$$

Therefore, we assume the combined use of two fingers as before to give

$$(\text{FAR}(1), \text{FRR}) \approx (10^{-10}, 10^{-3}).$$

This allows a large-scale biometric identification system of size $m = 10^8$ to run at a *system* operating point of

$$(\text{FAR}(10^8), \text{FRR}) \approx (10^{-2}, 10^{-3}).$$

Using these rates yields the following frequency of errors:

Enrollment False Accepts: When viewing identification as repeated verification, the number of False Accepts depends on the number of records an individual is matched against. Roughly one in every 100 subjects is falsely flagged as some identity already in the database.

Enrollment False Rejects: Since there is at most one record in the database that matches any individual, the False Reject rate does not depend on the database size. So for 1 in 1,000 identifications a subject is erroneously classified as absent from the database, whereas in reality that individual is properly enrolled.

When verifying enrolled individuals, the error rates are much better. From the chosen operating point it can be seen that only 0.1 percent of the applicants will be denied service. Moreover, only 1 in 10 *billion* false claims of identity will be accepted (essentially none).

This suggests that the system might be able to operate with only a single finger for the verification process. Assuming a single finger FAR of 10^{-5} means 1 in every $100,000$ fraudulent requests might succeed.

When installing such a system one must also be concerned about the logistics. Given the proper set of numbers describing the system performance parameters we can figure out the number of enrollment and verification stations needed as well as the peak match rates for each type of transaction. A set of such calculations is summarized in Table 10.7.

Suppose that the system is used to check the passports of travelers, specifically for immigration control at airports. There are several observations which can be made. First, the database is likely to be constantly changing—new people will sign up while others will have their passports expire and fall out of the program. It is not unreasonable to assume a 10 percent turnover in database population per year. Second, the number of verifications per year might also be a significant fraction of the database size, like 40 percent. This does not necessarily mean that 40 percent of all passport holders will travel once per year. Instead, it might be that 10 percent travel once per year, 5 percent travel twice a year, and 5 percent travel four times a year. The bulk of the population, 80 percent, might not travel at all in any given year.

The required enrollment and verification rates are calculated from the total number of transactions (enrollments or verifications) divided by the time available (the hours that the facility is open). The enrollment component of the matching rate is then simply the enrollment rate times the database size, while the verification component is governed by the authentication rate alone (see Expression (10.3)).

Notice the incredibly high value for the match rate due to enrollment. On a system basis this is only 6 ns per equivalent match! In general this is not feasible with a single processor, so some sort of parallel solution must be developed. Since the problem can be easily split up, one might imagine a central computing core with on the order of $17,000$ processors, each of which is capable of performing a more reasonable $10,000$ matches per second (i.e., 100 μs each). Note that although the individual processor clock speeds may eventually become very fast, there may still be a significant I/O limitation. That is, the query pattern must be passed in to each processor, then the resulting scores must be extracted and all merged together to form a decision. A possible alternative solution would be to use some sort of direct indexing matcher, such as the geometric hashing scheme described in [75].

If we set targets for the completion time for each sort of transaction, we can multiply the required rate by this time interval to determine the number of biometric acquisition stations needed. This calculation will just give a lower bound, as typically a number of spares might be desired to cover the inevitable hardware failures in the field. In the case of airport control, the computed number (7) was unreasonably low since in the US there are 16 major international airports. Each needs at least one, if not several, stations. So this overriding consideration actually governs the number of verifications stations deployed.

Once again the variation in demand needs to be taken into account. Remember the system is installed at international airports to check passports of citizens. The traffic has a natural cyclicity: many flights arrive during the day, far fewer late at night. Moreover, when a plane arrives hundreds of passengers disembark simultaneously and head for the border security terminals. Yet, between flights these terminals might all sit idle.

In Table 10.7 we have used relatively small peaking factors and applied them only to the signal acquisition hardware requirements. A substantially higher value, like 300 percent,

National ID	Value	Explanation
m	10^8	Moderately large country
T_e	2,000 hrs *continuous*	= 8 hrs / day x 250 days / yr
E_{avg}	5,000 / hr	= 10^7 / yr / T_e (US Passport statistics)
Enroll time	0.1 hr	= 6 minute wait
T_a	8,000 hrs *continuous*	= 22 hrs / day x 365 days / yr
A_{avg}	500 / hr	= 4×10^7 / yr / T_a (US Immigration: air stats)
Verify time	0.01 hr	= 36 second wait
max J_e	6×10^{11} / hr	Includes 20% peaking (17K processors x 10K / sec ?)
Enrollment stations	660	Includes 20% peaking plus 10% spares
max J_a	600 / hr	Includes 20% peaking
Verification stations	48	= 3 x 16 international airports (above the minimum of 7)

Table 10.7: A summary of the logistics of the national ID scenario. The system requirements are heavily dominated by the enrollment process.

might be more appropriate to satisfy user expectations of convenience. This same factor would also have to be applied to the matching rates, but luckily it is only a linear correction.

10.7 How feasible is large scale?

In general, national ID systems are more feasible that voter registration systems. This is because the logistical problems are not as severe as those encountered when there is a hard enrollment deadline. The big difference between a national identification system and a voter registration system is that the average transaction rate for national ID can be made much lower than the enrollment rate needed for voter registration. The transaction rate essentially can be set as low as needed by choosing E_{avg}, the enrollment rate, to be low and controlling the authentication rate, A_{avg}, the number of subjects per second that are authenticated. This has a number of ramifications:

- The average 1:1 match rate J_{avg} does not need to be as high as 2.3×10^9 matches per second, the rate for the voter registration problem at the end of a *short* enrollment period. There is no requirement that the 1:1 matcher need operate at that rate within a short time t_e.

- There is no need for such a large number $\approx 10^4$ of dedicated enrollment stations. In fact, the nature of the application probably means some enrollment infrastructure is already in place (e.g., the law enforcement infrastructure).

- The lurking problem of the explosion of the cost of manual False Positive resolution for larger database size m can be avoided, or at least minimized. This does not mean that the cost of (manually) resolving False Positives might not explode, i.e., become $O(m^2)$. However, when this happens, the rate of enrollment can be slowed down and there is no threat of a total system halt because of congestion and work backlog.

Regardless of time scales, to successfully implement any large-scale system both the $\text{FAR}(1)$ and the FRR need to be extremely low. This is requirement is hard to verify beforehand because it only expresses itself when m becomes large (see [110]). It does not necessarily express itself for prototypes of small problem size, and its behavior (sometimes called its *performance projection*) is hard to predict even with sophisticated techniques like *extreme value statistics* [119]. Ways to characterize and measure the accuracy of large-scale systems are described in Chapter 6. However, in general very little is known about the various biometric parameters for systems of large size m.

Since False Match Rates $\text{FAR}(1) \leq 10^{-10}$ are needed for large-scale problems, clearly, using only one finger for a problem of size $m = 10^8$ with the current state of the art is not adequate. Yet a serious potential problem is that $\text{FRR} \approx 10^{-3}$, even when using two fingers. This rate means—

1. During enrollment 1:1,000 subjects in the database might be enrolled more than once.

2. During subsequent verification 1:1,000 subjects will not be identified, even though the subject is in the database.

To a certain extent, it might be possible to design a large-scale national identification system "*while growing.*" That is, it can start at an operating point which achieves a lower False Negative Rate by giving up some of the False Positive Rate.

For the voter registration problem, the consequences of $\text{FRR} \approx 10^{-3}$, namely, double enrolls during registration and False Rejects during the election, appear to be less serious. However, as noted, the total cost (including labor) of the registration phase is hard to predict. Therefore, voter registration at this scale is a risky endeavor, again simply because of time constraints.

11

Integrating Information

For many applications there are additional non-biometric sources of information that can be used in person authentication. In other applications, the use of a single biometric is not sufficiently secure or does not provide sufficient coverage of the user population in that using *only one* biometric results in too many Failures to Acquire (high FTA rate), where it is simply impossible to sample the biometric of a subject. In both cases, the question becomes, How do we integrate multiple biometric (and non-biometric) sources of information (as in Figure 11.1) to make the overall application more accurate and hence more secure? In this chapter we describe a variety of techniques for doing so.

Note that the integration of information and the integration of biometric identifiers is an *integral part* of authentication protocols $\mathcal{A}_p$ described in Chapter 2.

11.1 Integration methods

Information contained in multiple biometrics can be integrated using a number of different methods, at various levels, and in different contexts. Figure 11.2 gives an example using the finger biometric, which is an interesting one because people have ten of them.

Each of the methods in the figure combines a first finger impression (the center of

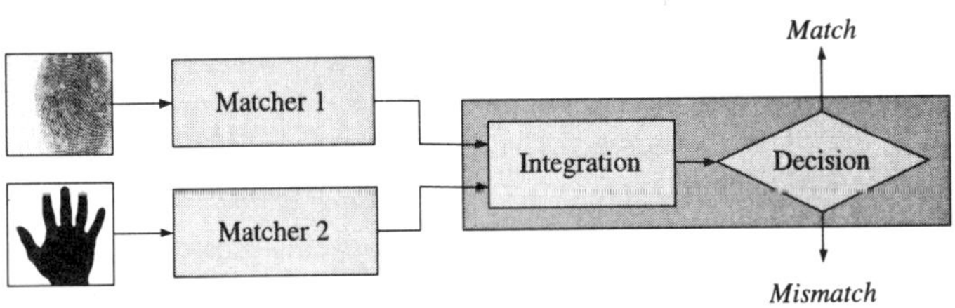

Figure 11.1: Multiple biometric authentication can be analyzed probabilistically and is intimately related to authentication protocols.

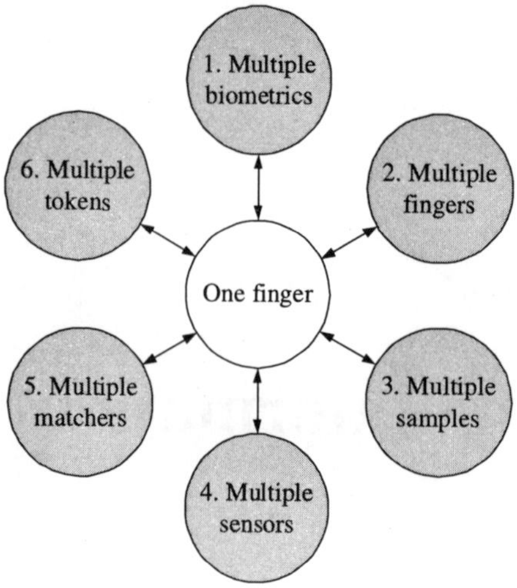

Figure 11.2: There are many different sources of information that can be used to expand a biometric system.

Figure 11.2) with a second biometric sample. Although discussed here in the context of fingerprints, the same methods can be used with other biometrics.

1. *Multiple different biometrics:* This is what typically is thought of as integrated biometric authentication. More than one, disparate biometric, such as face and voice, or fingerprint and hand geometry are used for authentication.

   The use of multiple fingers in this authentication protocol is a special case.

2. *Multiple fingers:* This authentication protocol is only possible when a number of the same biometric identifiers are present on subjects. It has been in use in law enforcement applications for a long time in the form of ten-print cards for positive identification.

   An example is to require that *both* the left index finger *and* the right index finger match, and so on.

3. *Multiple samples:* This includes, for example, the well-known authentication protocol "*three tries and you're out.*" That is, a subject is allowed to offer three samples of the biometric and has three chances to match this sample with the stored template.

   There are plenty of examples here. A variant of this protocol is to allow a subject an upper time limit Δt for authentication. This biometric integration of multiple samples over time is found, e.g., in voice and face recognition techniques.

4. *Multiple sensors:* Two or more different sensors can be used to acquire the same biometric identifier, e.g., a fingerprint sensor based on frustrated total internal reflec-

tion and a sensor based on CMOS technologies. Such protocols include the use of multiple cameras from different viewpoints for, say, face recognition.

5. *Multiple matchers:* Here different matching technologies are used on the same biometric sample acquired from the subject. These are the combinations of feature versus appearance representations (templates) for face images and fingerprint image, where the features would be eyes, nose and mouth and for fingerprints the features are minutiae.

6. *Multiple tokens:* For completeness we have added the integration of a biometric with a possession or knowledge token. This is the typical authentication with biometric verification protocol as discussed in Chapter 2 (Section 2.1).

Examples of these six protocols are given in Table 11.1. The dominant approach to integration is the application of the *AND* rule; more subtle integration rules are discussed in Section 11.1

		Example authentication protocol
1.	Multiple biometrics	Face image and voiceprint
2.	Multiple locations	Left and right iris
3.	Multiple sensing	Three tries of index finger
4.	Multiple sensors	Ultrasonic and optical sensing
5.	Multiple matchers	Minutiae and correlation fingerprint matcher
6.	Multiple tokens	Adding possession and/or knowledge

Table 11.1: Biometrics, and fingerprints in particular, lend themselves to many different types of integration.

Regardless of the method, there are two basic approaches to combining information from different sources:

1. *Tightly coupled integration:* The output from multiple biometric sensors could be used to create a more reliable and/or extensive (spatially, temporally, or both) input acquisition [27] (see Figure 11.3). The representations extracted from many biometric sensors could be collated and the decisions could be made based on the joint feature vectors. The integration at sensor or representation level assumes a strong interaction among the input measurements and the integration schemes [39].

2. *Loosely coupled integration:* This assumes very little or no interaction among the inputs (e.g., face and finger) and integration occurs at the output of relatively autonomous matching engines, each engine independently assessing the acquired biometric samples as shown in Figure 11.3 on the right.

We only discuss loosely coupled integration systems because they are not only simpler to implement, they are more feasible in commonly confronted integration scenarios. A typical scenario for integration is a system with two biometric subsystems (often proprietary) independently acquiring inputs and making an autonomous assessment of the

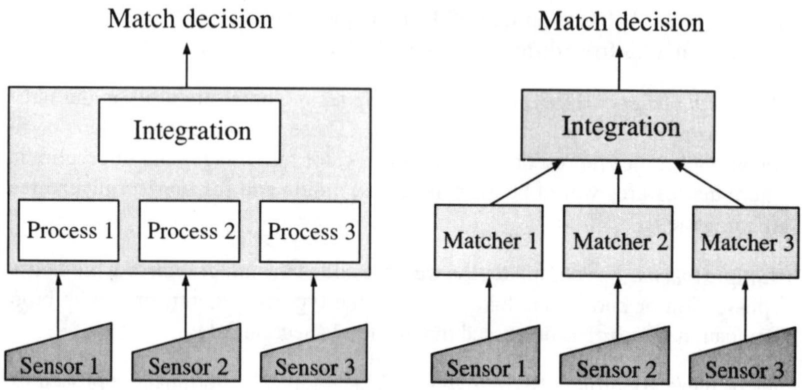

Figure 11.3: A tightly coupled system (left); a loosely coupled system (right).

"match" based on their respective biometrics. The features and templates extracted by one biometric system are *not* accessible to the other biometric system, while the decisions or scores of individual biometric systems are available for integration. Logically, one could try to integrate the biometrics either at the *decision* or at the *score* level.

11.2 Decision level integration

Decision level integration, or "fusion" as it is sometimes called, is typically concerned with method 5 from the previous list. That is, there are two or more matchers, either for the same or different biometrics (e.g., face and iris) and the task is to combine the decisions the matchers generate. Here the matchers are considered as black boxes—there is no access to their internal mechanisms or intermediate scores, each outputs simply a "Yes/No" verdict.

11.2.1 Boolean combination

By far the most prevalent rules for multiple biometrics combination in practical systems are the *AND* rule and the *OR* rule. Multiple biometrics are combined in that form in both identification and in verification protocols:

Identification protocol: An additional biometric is used to narrow down on a smaller and smaller set of biometric templates. Hence, the authentication protocol calls for first matching biometric a with a stored template and then matching biometric b with a stored template; hence only if both biometrics a *AND* b (see Figure 11.4) match, will a subject be considered a match for the candidate list.

In many practical situations, these techniques are applied both in positive and negative identification and essentially amount to sequential matching of the different

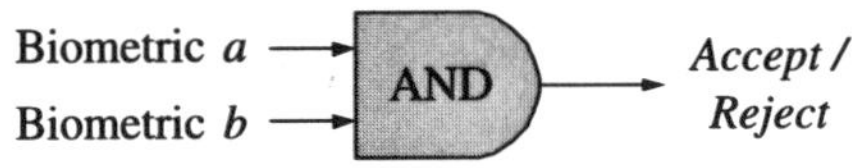

Figure 11.4: A subject is authenticated if *both* biometric a and biometric b match. This is an example of decision level fusion.

biometrics of a subject.[1] This is discussed in the next section. A discussion of biometrics integration and an example of a system that uses two biometrics can be found in [87]

Verification protocol: Here next to the *AND* rule, the *OR* (see Figure 11.5) is used to improve and encourage participation in biometric authentication systems.

That is, a second biometric is used to alleviate problems, or adverse properties, of an individual biometric as described in Chapter 8, i.e., the properties like universality, collectability, and acceptability. For example, an additional biometric beyond finger accommodates those subjects without fingers or with hard-to-acquire fingerprints and accommodates those who object to enrolling with a fingerprint.

The OR protocol can be implemented in various ways, in terms of the order of acquiring biometrics a and b.

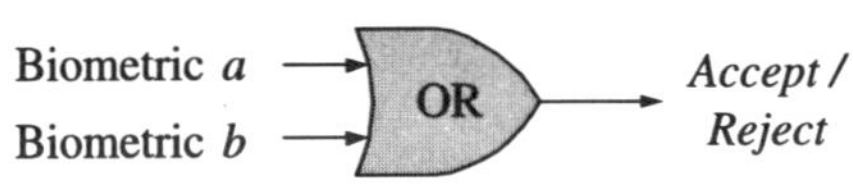

Figure 11.5: A subject is authenticated if *either* biometric a matches or if biometric b matches, the OR rule.

Let the False Accept and False Reject Rates for biometric a be FAR_a and FRR_a; and for biometric b denote these errors by FAR_b and FRR_b. Integration of multiple biometrics has an effect on the overall FAR and FRR of the integrated system. Combining the biometrics through an OR rule improves the FRR at the expense of FAR; combining the biometrics through an AND improves the FAR at the expense of FRR. This can be seen as follows:

- a *OR* b: To generate a False Reject, both biometrics a and b have to be falsely rejected; hence,

$$\text{FRR} = \text{FRR}_a\, \text{FRR}_b.$$

To generate a False Accept either biometric a or b needs to be falsely accepted; hence,

$$\begin{aligned} \text{FAR} &= \text{FAR}_a + \text{FAR}_b\,(1 - \text{FAR}_a) \\ &= \text{FAR}_a + \text{FAR}_b - \text{FAR}_a\, \text{FAR}_b, \end{aligned}$$

[1] When biometric measurements do not entail user cooperation the combination rules can also be applied in parallel.

assuming biometric a and b are independent pieces of information.

- *a AND b*: To generate a False Accept, both biometrics a and b have to be falsely accepted; hence,

$$\text{FAR} = \text{FAR}_a \times \text{FAR}_b.$$

To generate a False Reject either biometric a or biometric b needs to be falsely rejected; hence,

$$\begin{aligned}\text{FRR} &= \text{FRR}_a + \text{FRR}_b\,(1 - \text{FRR}_a) \\ &= \text{FRR}_a + \text{FRR}_b - \text{FRR}_a\,\text{FRR}_b,\end{aligned}$$

assuming a and b are independent pieces of information. Here the product is subtracted to avoid counting False Reject FRR_b when biometric a is falsely rejected.

Again, this is a trade-off between security and convenience. Table 11.2 summarizes the error rates, assuming that

$$\begin{aligned}\text{FRR}_a\,\text{FRR}_b &\ll \text{FRR}_a + \text{FRR}_b \text{ and} \\ \text{FAR}_a\,\text{FAR}_b &\ll \text{FAR}_a + \text{FAR}_b.\end{aligned}$$

	FAR	FRR
OR	$\text{FAR}_a + \text{FAR}_b$	$\text{FRR}_a \times \text{FRR}_b$
AND	$\text{FAR}_a \times \text{FAR}_b$	$\text{FRR}_a + \text{FRR}_b$

Table 11.2: Biometrics can be integrated to improve convenience, lower the FRR (OR rule); or to improve security, lower the FAR (AND rule).

11.2.2 Binning and filtering

For both positive and negative biometric identification, the capability of a biometric 1: *many* search of a database is a prerequisite. Information integration techniques have had a longstanding tradition in performing such biometric searches.

Let us look at a screening application here. Remember, by screening we mean negative identification, where the biometric system checks the claim "*I am not who I claim not to be*." It amounts to searching a biometric database **N** with N enrolled subjects d_n, denoted as set $\{d_n, n = 1, ..., N\}$. One way to implement such a search is by performing N biometric 1:1 matches, either sequential or in parallel, to ensure that no database subject d_n is missed. There are at least two problems associated with these biometric searches when N, the size of the database **N**, becomes large:

1. The computational requirements (real-time response!) become too high to implement a true N times 1:1 biometric search strategy.

2. The False Positive Rate of the 1:1 matcher is high, and the candidate lists, of expected length FPR $\times N$, become too large (when humans are expected to visually inspect the candidate lists).

Therefore, such searches are often constrained by a second biometric to reduce the search to smaller portions of the database.

Let the machine representation for subject d_n be credentials (P_n, B_n), with possessions P_n (that include identifiers such as name, etc.) and the biometric template B_n. For databases **N** holding a large number (N) of such biometric templates, search efficiencies can be improved by partitioning **N** into smaller groups based both on information contained within (endogenous to) the biometric templates B_n themselves and upon additional (exogenous) information such as the subject's name, obtained at the time of enrollment [229]. Constraining the search with parametric (e.g., non-biometric, demographic) data is called "filtering" and constraining the search with additional biometric data is referred to as "binning," using terms introduced by Wayman [229].

Filtering: Examples here of course are plenty. Filtering down the search of **N** by, say, a subject's surname, could partition the biometric template database into many sets, which can be overlapping when fuzzy, e.g., phonetic, comparison between name matches is used (some sets will be very large, e.g., "J. Smith"). Note that filtering then is an authentication protocol $\mathcal{A}_P$ that operates on authentication method, or credentials,

$$(possession, biometric) = (P, B) = (name, B).$$

The protocol $\mathcal{A}_P$ simply says—

1. Search the database **N** of enrolled subjects by surname (possibly with all possible spelling variations). In general, parametric data P_n is used for exact or approximate (fuzzy) matching.
2. Search the subjects with matching surnames P_n by matching input sample biometric B with the corresponding biometric templates B_n in **N**.

This returns a "candidate list" or "hit list" of subjects $\{d_n^1, d_n^2, ...\}$ from database **N** with the name *Smith* and a template that matches biometric sample B.

The extreme here is narrowing down the search using a unique (or almost unique) parametric identifier, i.e., social security number or some other national identifier. Then the credentials are

$$(possession, \text{biometric}) = (P, B) = (social\ security\ number, B)$$

and the problem of identification (1:N matching) is reduced to verification (1:1 matching).

Binning: This is filtering based on information contained within the biometric templates B_n in database **N** themselves, or more generally, based on other biometric identifiers B'. Such a search is performed by an authentication protocol operating with credentials $(B) = (\{B, B'\})$. Perhaps the best known instance of this technique is to *first* classify the type of the fingerprint and *then* to match the minutiae of the fingerprint. Here the fingerprint types are global print classes like "loop," "whorl," "arch," etc., and the minutiae are much finer-resolution features [66].

Binning could also be based on a different biometric of an individual, i.e., $(B) = (\{B, B'\})$. Here B could be a fingerprint and B' could face type, which may be obtained from a face image. If the search space (the number of matches) in **N** is to be reduced with this authentication method the sequential protocol is:

1. Select those subjects d_n in the database **N** whose biometric template matches biometric (sample) B', i.e., those subjects with face type B'.
2. Match the input biometric template B with the templates of those remaining subjects to find those subjects in **N** with both matching B' and B.

This ideally returns one unique enrolled subject d_n from database **N** or at most returns a small list.

This type of binning is commonly referred to as "*multiple biometrics*" and creates the same problems of biometric matcher integration as will be discussed later.

When using the filtering or binning authentication methods (P, B) or $(\{B, B'\})$, the space to which a subject biometric $\mathcal{B}$ is matched is limited to a smaller portion of the complete database of templates **N**, i.e., the protocol is directed at limiting the number of biometric 1:1 matches. In a sense, the database is divided up into multiple partitions or *bins;* these partitions may overlap in case of uncertainties in templates; a template may even be in all partitions.

Of course there is a trade-off between two large-scale system parameters when using these approaches. The fraction P_{pr} of the database **N** that is actually on average being matched to the biometric template B is called the *penetration rate:*

$$\text{P}_{pr} = E(\# \; of \; times \; B \; is \; matched)/N.$$

This is also sometimes called the *filter rate* ("binning" is not yet a commonly excepted term and is often also called filtering). Low penetration rates, however, come with high binning error rates, which result in missed matches. The binning error rate P_{be} is the percentage of subjects b_n in database **N** that are misclassified, e.g., a loop finger pattern misclassified as a whorl pattern. Obviously, we have a trade-off between penetration rate and binning error rate P_{be}.

Filtering is another type of integration, that of authentication protocols based on integrating possession type P_n of credentials such as *name*, *home state*, etc., to be used to assist and speed up the biometric search. Possession tokens can be added to an existing authentication protocol:

1. Adding P_n for subject d_n to *negative identification* prescribes narrowing down on a smaller and smaller set of biometric templates. This has the desirable effect that the chances of False Positives go down. The probability of a False Negative will go up dramatically; after all, biometric templates in the database **N** that are not associated with the additional possession identifiers P_n will *always* produce a False Negative.

2. Adding possession P_m or knowledge (passwords) K_m to *positive identification* (with positive database **M**) entails asking for additional possession or knowledge beyond the identifier ID_m, when d_m request authentication. Note that this type of integration does not decrease the *biometric* verification error rates and, therefore, does not

> increase the probability that the subject d_m is who he claims to be. The probability that a person is who he says he is equals the probability of a match between a stored biometric template and a newly acquired biometric sample. This probability does not increase when the subject supplies additional possession and knowledge, it simply means that the subject has the possession (bank account numbers, etc.) and knows the passwords.

Probabilistic analysis of filtering based on exact matching is much harder. False Matches and False Non-Matches will occur but these occurrences are due to data integrity problems. For password integrity of $\mathbf{M}$, one would have to estimate $p(K_m|d'_m)$, the probability that someone other than subject d_m has the knowledge K_m or the password of subject d_m.

There is confusion surrounding accuracy claims of biometric *identification* and *screening* systems. Only using False Negative and False Positive rates for negative identification or False Accept and False Reject Rates for positive identification may not be very informative. In Chapter 5, additional system parameters of large-scale systems have been discussed.

11.2.3 Dynamic authentication protocols

Protocols do not have to be static, indeed many dynamic protocols have been introduced, and of course manual/visual authentication by humans is always dynamic. Interactive input sample quality control implemented through communication between human and computer (see Chapter 9) is a dynamic protocol. For fingerprints the user might be requested to place his right index finger on the scanner, then his left, then his right middle finger. To obtain access, two out of three of these "questions" (fingerprint images) would have to be "answered" (matched) correctly. This, again, is a dynamic protocol.

One dynamic protocol for speaker verification is the idea of *conversational biometrics* where pre-specified knowledge K_m of the enrolled subjects d_m plays an important role. Quoting from [17]:

> ... A conversational biometric systems engages in a natural language dialog with the user, presents randomly generated questions, and collects the user's responses. When appropriate, the conversation may be embedded into the natural transaction dialog. The questions to be asked can be randomly generated from a large collection of questions and answers that the user provides during an explicit knowledge enrollment stage (e.g., "what is your favorite color?"), or may be generated from user-specific information available from the application (e.g., "what is the balance in your last statement?"). A protocol management module generates a protocol dynamically for each transaction, based on the application requirements. The dynamic protocol specifies the maximum number of questions to be asked, and the minimum number of correct answers needed, or alternatively, the protocol may specify the minimum score needed for the knowledge match. The protocol also specifies the minimum score required for the acoustic voiceprint match, and may adaptively modify the maximum number of questions to be asked, based on the voiceprint match

Conversational biometrics	
Place of birth	Virginia
Name of pet	Tiger
Bank balance	\$ 43.39
.	.
.	.
Q_c	A_c

Table 11.3: A list of questions and answers that represents the knowledge K_m of subject d_m in conversational biometric authentication.

scores. By generating appropriate protocols, one can provide higher accuracy or greater flexibility, or both [127,146].

Note that conversational biometrics does a biometric match between the speech sample B and the voiceprint B_m (biometric template) on file and a "knowledge match" between the collected responses through speech recognition and the knowledge K_m on file. Verifying the correctness of the answers, however, does not increase the probabilistic certainty about a subject d_m at the other end of the phone line; it may increase *the possibility* that the subject is authentic but it does not increase the probability that the subject is authentic. This is because the answers are not inherently associated with *only* the subject and *not* provably unknown to others, i.e., known to possible impersonators and imposters. Repudiation may be a problem then since the match probability of input biometric template B and the one on file B_m can be low, and the authentication is based on the subject's knowledge

Refer back to Chapter 2, where the three methods of authentication, *possession, knowledge,* and *biometrics* are defined. The answers to questions like, *What is your mother's maiden name?* and *How many miles do you have accrued?* are knowledge tokens K_m of a subject d_m. We can again view conversational biometrics as an authentication protocol $\mathcal{A}_P$:

$$\mathcal{A}_P(P, K, B).$$

This can be viewed as protocol $\mathcal{A}_P$ operating on the input credentials (P, K, B), the subject claiming identity d_m.

Consider the list of questions and the answers for the conversational biometric of Table 11.3. These questions have to be somehow agreed upon during an enrollment phase as described in enrollment policies E_M (see Chapter 9) there are two possibilities here:

1. Questions/answers about private or personal issues but about issues that are not really secret, like the name of a pet.

2. Questions/answers about issues that are somehow secret, or considered hard to obtain, such as the balance of a bank account.

In both cases, the conversational answers are knowledge tokens K_m of some subject d_m. Conversational knowledge may appear to be a possession P_m (the accrued mileage) and is

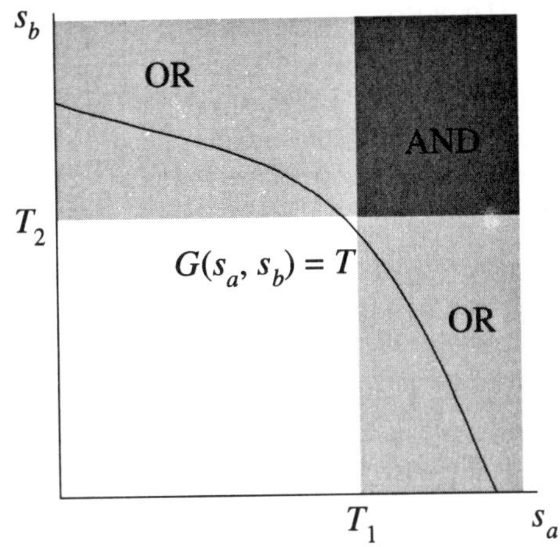

Figure 11.6: One decision region is admissible under the OR-rule; another region, contained in the first region, is admissible under the AND-rule.

often confused with biometric information B_m, for instance, a subject's *mother's maiden name* is sometimes mistakenly called a biometric identifier. Conversational biometrics has some of the properties of a biometric, in that it might not be changeable. Items of conversational biometric knowledge K are like passwords and phrases, however, because the knowledge K can be shared. Either way, the authentication protocol $\mathcal{A}_P(P, K, B)$ calls for simultaneously recognizing a subject d_m based on input voice B and enrolled voiceprint B_m while comparing the answers K to stored answers K_m on file.

Conversational biometrics could use other forms of man-machine communication, for example, using handwriting recognition or keystroke characteristics [154] (which is a behavioral biometric). This could be signature verification generalized to writer identification [88] with the same categories (as for speaker verification), i.e., *text-dependent* and *text-independent* writer verification. Authentication protocols where the subjects are asked to provide written or typed answers to questions can be imaged.

Of course, one can always use a question bank as in Table 11.3 with any biometric. In effect, conversational biometrics is an extension of machine-to-machine secure authentication protocols, called challenge and response protocols (see Chapter 12, Section 12.8). Here, electronic inquiries that can only be answered by an authentication system are dynamically formulated.

11.3 Score level integration

We now return to our two Boolean cases discussed in Section 11.2.1 (see Figures 11.4 and 11.5). The decisions made by the individual biometric matchers are integrated through the *OR* protocol or the *AND* protocol as in Figure 11.6. We have—

1. *The* OR *integration protocol:* A subject is asked to offer a *first* biometric identifier a and *then* identifier b. The subject is authenticated if $(s_a > T_1)$ *OR* $(s_b > T_2)$, where the advantage is that the second biometric only needs to be acquired if $s_a \leq T_1$. In Figure 11.6, a subject is authenticated if the pair of scores, (s_a, s_a), lies in the "OR" region.

2. *The* AND *integration protocol:* A subject is required to offer *both* the biometric identifier a and identifier b to the biometric authentication system. The subject is authenticated if $(s_a > T_1)$ *AND* $(s_b > T_2)$. In Figure 11.6 this means that a subject is authenticated if the score pair (s_a, s_b) lies in region labeled "AND."

Note, however, that the curved dark line in Figure 11.6 may be a better decision boundary (in terms of errors rates) than either the AND or OR solution. Thus, instead of limiting ourselves to a rectangular partitioning of the decision space, in this section will look at combining the underlying scores in a more intelligent fashion.

Let us examine loosely coupled integration at the score level. We start by assuming that the scores are normalized $s_a,\ s_b \in [0, 1]$, and denote the measurements of biometric a and b as B. There are then two principal cases:

1. The scores s_a and s_b are (monotonically) related to the likelihood $Prob\,(d_m|B)$;

2. The scores s_a and s_b are related to the likelihood $Prob\,(d_m|B)$ in more complex fashion.

We will briefly consider the latter situation first. Given ground-truth-marked data, it is possible to determine a function which relates (s_a, s_b) to $Prob\,(d_m|B)$. More specifically, using ground truth data one could estimate joint class conditional probability density functions $Prob(s_a, s_b|H_o)$, $Prob(s_a, s_b|H_a)$, where H_o and H_a represent hypotheses corresponding to match and mismatch scores, respectively. The likelihood $Prob\,(d_m|B)$ can then be inferred, and based on the relative values of the conditional densities, the combined decision can be reached. For example, Prabhakar and Jain [172] estimate the conditional densities using a non-parametric estimation method (e.g., Parzen window) and they design the combined matcher based on Neyman-Pearson likelihood rule:

$$Prob\,(d_m|B) \sim Prob(s_a, s_b|H_o)/Prob(s_a, s_b|H_a).$$

While this method of combining scores is most general and optimal, in practice, paucity of training data may restrict the utility of this approach.

Let us now focus on the situation where s_a and s_b are (monotonically) related to the likelihood $Prob\,(d_m|B)$; the better the model d_m fits the data, the higher the scores. In this case, the decision can be reduced to authenticating the subject when $G(s_a, s_b) > T$, i.e., if the score pair (s_a, s_b) lies above the curve $G(s_a, s_b) = T$ as in Figure 11.6.

In this situation, optimal integration amounts to finding a curve $G(s_a, s_b)$ such that, for example, the FA rate is below some desired minimum level and the FR is as low as possible. Hence, the authentication system is operating at a curve, rather than at an operating point on the ROC (see quantitative parameters listed in Chapter 5). This integrated operating curve could be found by using test databases of multiple biometric samples that are realistically acquired from *randomly* selected subjects.

11.3.1 Normal distributions

First, we present a glimpse toward determining an operating curve $\hat{G}(s_a, s_b)$ that satisfies the specified requirements of the False Accept and False Reject Rates, where we mean the requirements of the multi-biometric system. Hence, typically, the FAR is specified to be below some small number δ.

The question is how we might compute the curve $G(s_a, s_b)$? Let us first look at the case where we approximate the curve G with a linear function of the random variables s_a and s_b, i.e., we have decision rule

$$G(s_a, s_b) = \alpha\, s_a + (1 - \alpha)\, s_b > T, \tag{11.1}$$

as shown in Figure 11.7. In this case a score pair s_a, s_b is rejected if it lies within the gray area. The only case where such an approximation is correct is when both s_a and s_b are normally distributed $\mathcal{N}(\overline{s}_a, \sigma_a^2)$ and $\mathcal{N}(\overline{s}_b, \sigma_b^2)$ [56]. Additionally, when we assume that s_a and s_b are independent, (11.1) becomes

$$G(s_a, s_b) = \frac{s_a}{\sigma_a^2} + \frac{s_b}{\sigma_b^2} = \frac{\sigma_a^2\, s_a + \sigma_b^2\, s_b}{\sigma_a^2\, \sigma_b^2} > T. \tag{11.2}$$

This simply states that the scores are weighted by the variances, $1/\sigma_a^2$ and $1/\sigma_b^2$. The higher the variance, the less influence the corresponding biometric should have. Problems

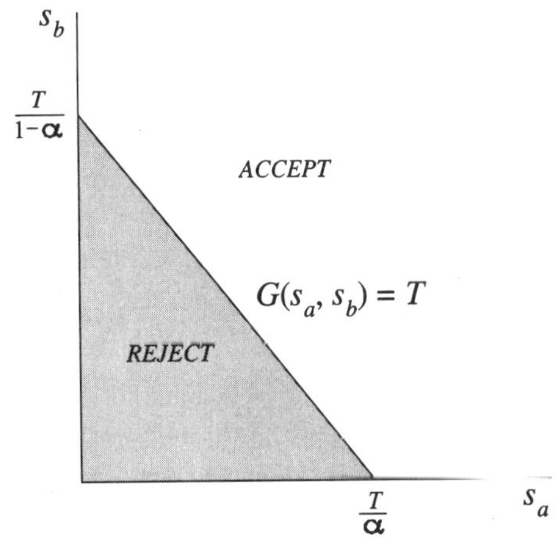

Figure 11.7: The use of linearly weighted decision scores to integrate biometric a and biometric b.

with (11.2) are twofold:

1. The covariance matrix $\Sigma(s_a, s_b)$ is assumed to be diagonal. This is probably a good assumption if disparate biometrics are used but is not true when, for instance, two fingers are used.

This is easily rectified by modeling s_a and s_b with a bivariate Gaussian.

2. However, modeling *match scores* with Gaussians is not realistic, because

$$Prob(s_a < 0) = Prob(s_b < 0) = Prob(s_a > 1) = Prob(s_b > 1) = 0.$$

The use of a Gaussian model for the probability distribution of the distance between two biometric templates B and B_m is a better assumption, such as modeling the distance between two sets of minutiae points. Hence we need to somehow define a distance measure between a query biometric sample (or template) B and a stored model biometric sample B_m (or template). A simple example would be

$$Dist\,(B, B_m) = 1 - s(B, B_m),$$

assuming the distance $Dist\,(B, B_m)$ is inversely proportional to score $s(B, B_m)$. And we have

$$Prob(d) = \frac{1}{\sqrt{2\pi}\,\sigma} \exp\left[-\frac{1}{2}\left(\frac{d - E_m}{\sigma_m}\right)^2\right], \tag{11.3}$$

with $E_m = E(Dist\,(B, B_m))$ the expected distance when B and B_m are templates from the same biometric and $\sigma_m = \sigma(Dist\,(B, B_m))$ is the variance of this distance. Depending on the quality of enrollment and the quality of the print σ_m could be high and E_m might be non-zero.

Now, as argued in Expression (9.6) in Chapter 9, *if* a match score means anything, it is somehow related to $Prob\,(d_m|B)$, the likelihood that subject d_m is present, given the data O, or the derived template B. A better expression for this likelihood is given by Expression (9.12) in Chapter 9:

$$Prob\,(d_m|B) \approx \frac{Prob\,(B|d_m)\,P_m}{Prob\,(B|d_m)P_m + Prob\,(B|D_m)(1 - P_m)}.$$

Computing a distance measure $Dist\,(B, B_m)$ now amounts to evaluating this expression, where $Prob\,(B|D_m)$ is the likelihood of b_m given some world model when $D_m = D$ or some cohort set with $D_m = \{d_m^1, d_m^2, ...\}$ for subject d_m (see Chapter 9). The probability P_m is the prior probability that subject d_m is present, which may be computed using a uniform distribution $1/M$ over the M enrolled subjects.

On a related note, integration research has also explored the issue of combining matchers based on weighed product of the individual matcher scores rather than their weighed sum. Note that under the assumption of matchers operating under independent information, like the weighted sum, the weighted product of the scores has a sound probabilistic basis. However, in practice, it was found that weighted product combination is more brittle and typically performs worse than the weighted sum combination [4].

It is also important to note that although the weighted sum integration has a number of restrictive underlying assumptions, it often outperforms or performs as well as many other sophisticated integration techniques [115].

11.3.2 Degenerate cases

Note that Expression (11.2), the use of a weighted sum of scores to integrate decisions, includes two degenerate cases.

1. $\sigma_a \ll \sigma_b$: The variance of score s_a is much smaller than the variance of score s_b; i.e., we will have a much better estimate of s_a than of s_b. Then decision rule (11.2) is approximated as

$$G(s_a, 0) = \frac{s_a}{\sigma_a^2} > T.$$

   This means that biometric b has relatively little class separation, or separation of subjects, to contribute to the random variable $G(s_a, s_b)$ and might as well be excluded.

2. $\sigma_b \ll \sigma_b$: Here the reverse is true, and it may very well be the case that biometric a is irrelevant in the decision making.

Hence, it is *not necessarily true* that integrating multiple biometrics gives better error rates.

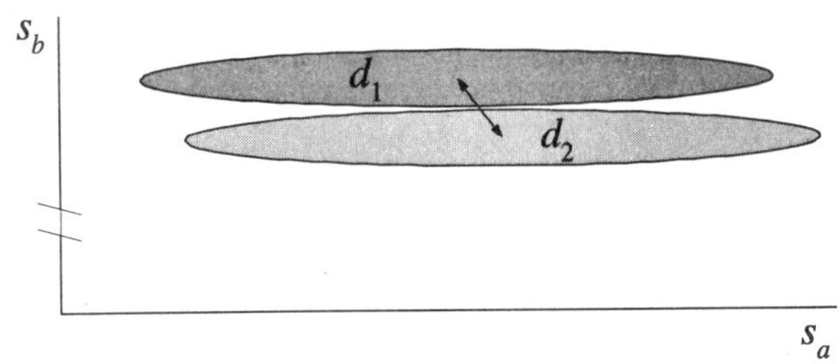

Figure 11.8: The (s_a, s_b) clusters, indicated with ellipses, of subject d_1 and subject d_2, with $\sigma_a \gg \sigma_b$.

Figure 11.8 shows possible score clusters (s_a, s_b) of two subjects d_1 and d_2, where we have $\sigma_a \gg \sigma_b$. The clusters are represented with ellipses. The subjects d_1 and d_2 have approximately the same distribution, but have different means. The major axis of the ellipses is proportional to σ_a and the minor is proportional to σ_b. Assuming that the cluster centers are nearby, it is likely that biometric a does not contribute much to class separation, and if so, it is perhaps better to ignore biometric a all together. There are no research results available as to whether this may actually occur when trying to integrate a set of disparate biometrics, like face and finger.

The biometric a in Figure 11.8 could become important if the two clusters are moved away from each other along the x-direction, i.e., when the inter-class distance is relatively large. Of course, with scores ranging between a small finite interval (i.e., $[0, 1]$), separating the clusters is only going to be possible to a limited extent. When a cohort method or a world model is used, the inter-subject variation (or intra-cluster distance, see Figure 11.8) can be taken into account more explicitly. For subject d_m not only the probability $Prob\,(B|D_m)$ is evaluated, which models the inter-cluster variation but also the

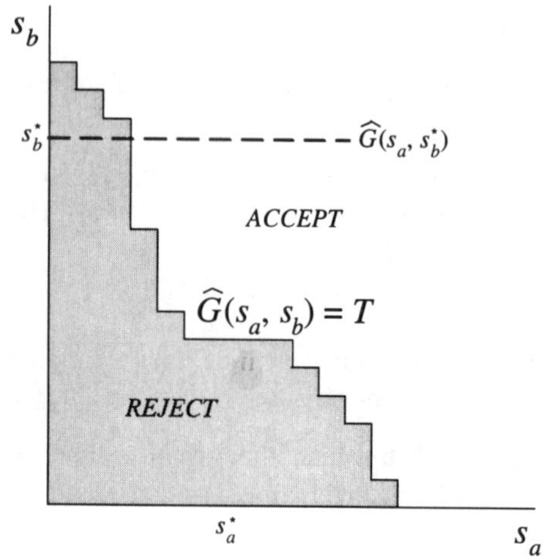

Figure 11.9: Computing a decision boundary based on an estimate of the correct accept rate $\hat{G}(s_a, s_b)$.

$Prob\,(B|d_m)$, which models the intra-cluster variation. Better yet, with good models for the probability of false association, as for example the models discussed in Chapter 14.

11.3.3 From thresholds to boundaries

So we have seen that under the assumption that the match scores, or match probabilities, of the biometrics s_a and s_b are normally distributed, the optimal decision threshold T is a line (Figure 11.7) in the s_a-s_b plane. The Gaussian assumption may be too much to ask for and the optimal decision boundary $G(s_a, s_b)$ might be nonlinear; or we may simply want to test how well our weighted scheme holds. The question then becomes, Is there a way to estimate some decision boundary $\hat{G}(s_a, s_b)$?

One approach is the extension of the way we looked at the False Accept Rate (FAR) and the False Reject Rates (FRR) in Chapter 5: Derive estimates of the match and mismatch score cumulative distributions from training data and determine the operating point T that satisfies the design criteria. Except now we have decision regions that are determined by "operating curves" in the support region $s_a,\ s_b \in [0, 1]$ for the joint score probability density $p(s_a, s_b)$.

Assume that for the score pairs (s_a, s_b) we have the cumulative match score distribution $F(s_a, s_b)$ and the cumulative mismatch score distributions $G(s_a, s_b)$, which we need to determine the FAR and the FRR. Remember, the match score distribution is equal to the FRR, $F(s_a, s_b) = \text{FRR}(s_a, s_b)$; and that the mismatch score distribution is the complement of the FAR, $G(s_a, s_b) = 1 - \text{FAR}(s_a, s_b)$, because the FAR is the portion of the mismatch distribution in the *ACCEPT* decision region. Now we want to accept only score

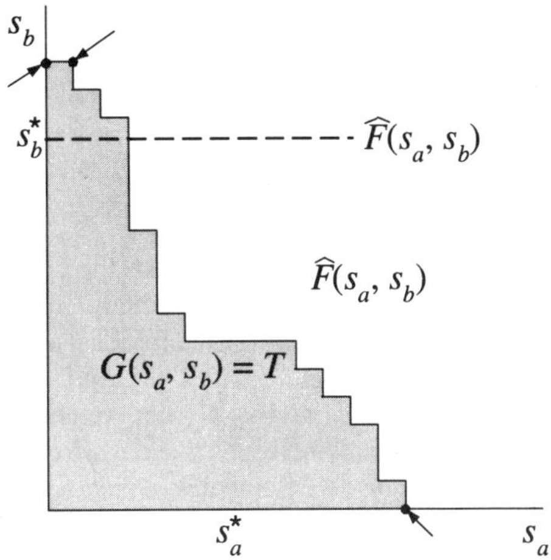

Figure 11.10: Computing a sequence of False Reject Rates at points $(s_a^1, 0)$, (s_a^2, s_b^2) to $(0, s_b^K)$ on the decision boundary $\hat{G}(s_a, s_b) = T$.

pairs (s_a, s_b) that fall in the region with low FAR, i.e., where $\text{FAR}(s_a, s_b) \leq \delta$ with small δ. Then we have

$$G(s_a, s_b) = \begin{cases} \text{FAR}(s_a, s_b) & \leq \delta \\ 1 - \text{FAR}(s_a, s_b) & > 1 - \delta = T, \end{cases}$$

and we get back a decision rule of the form of Expressions (11.1) and (11.2) and we get our curve of Figure 11.6. Requiring that $\text{FAR}(s_a, s_b)$ be small is equivalent to requiring that $G(s_a, s_b)$, the correct accept rate at (s_a, s_b), be high $G(s_a, s_b) > T$. So we need to get an estimate $\hat{G}(s_a, s_b)$ and determine for which region in the s_a-s_b plane this function is greater than T, with $T = 1 - \delta$ and δ the specified FA rate.

Looking at a quantized curve in Figure 11.9 it becomes clear how to estimate the curve $G(s_a, s_b) = T$. For each fixed s_b^*, the distribution $G(s_a, s_b^*)$ is one-dimensional and estimates can be obtained as described in Chapter 15. Hence the s_b-axis can be quantized into numbers $\{s_b^*\}$, and a set of cumulative distribution estimates $\{\hat{G}(s_a, s_b^*)\}$ can be estimated; equivalently, by quantizing the s_a-axis, a set of $\{\hat{G}(s_a, s_b^*)\}$ is obtained. These two sets of cumulative distributions allow for the construction of the decision boundary $\hat{G}(s_a, s_b) = T$, as shown in Figure 11.9.

One remaining question that we would like to address here is the False Reject Rate of this multi-biometric system; remember, the FAR is fixed at δ. The FRR, which equals the match score cumulative distribution, $\text{FRR}(s_a, s_b) = F(s_a, s_b)$, can be estimated by estimating $\hat{F}(a_s, b_s)$, as described above for the estimate of $\hat{G}(a_s, b_s)$. The False Reject estimates are now the value of $\hat{F}(s_a, s_b)$ along the curve $\hat{G}(s_a, s_b) = T$. With the curve specified as a sequence of K points $(s_a^1, 0)$, (s_a^2, s_b^2), ..., $(0, s_b^K)$ (see Figure 11.10), the

False Reject Rates are given by $\hat{F}(s_a^k, s_b^k), k = 1, ..., K$. It is interesting to note that a multi-biometric system should not just be associated with *one* FAR and *one* FRR but with *one* FAR and a sequence of $\text{FRR}_k, k = 1, ..., K$ for some K.

11.4 Alternatives

The OR and AND rules for combining *two* decisions are the only ways to combine single *"YES/NO"* match decisions about two different biometric samples from two different matchers. Typically, the OR rule models sequential, conditional authentication. Even when the match scores s_a and s_b are independent variables, the AND rule is a provable suboptimal solution, since the match scores s_a and s_b are not weighted. That is, irrespective of the match score variances σ_a and σ_b, biometrics a and b are considered equally important. Other types of integration will involve the scores s_a and s_b of the two matchers. So, certainly, biometric software interfaces should expose the match scores of the biometric matcher to the system developer.

When *more than two* decisions are needed to be combined, the choice of fusion is not merely limited to applying an overall (global) OR and AND rules to all the decisions. For instance, it is possible to group the individual decisions and to apply the choice of operators (e.g., OR or AND) iteratively to each group of decisions until a final decision is obtained. A second, obvious method of decision combination is voting. Perhaps each of the matcher's votes can be weighted (e.g., by the corresponding matcher reliability/confidence value). This situation is similar to score combination schemes discussed in Section 11.3. Referring to Figure 11.6, it may be noted that it is typically suboptimal first to arrive at decisions and then combine the decisions using sum/voting schemes; it makes more sense directly to involve metrics (e.g., similarity/distance) used for the decision-making in the individual matchers in a score combination (Section 11.3) strategy. Kittler and Alkoot [114], however, have shown that score combination strategies are superior when the individual matcher scores follow Gaussian distributions and voting is superior when the scores distributions are heavy tailed.

So far, we assumed that the decision combinations involved only authentication systems. In an identification system, *"YES/NO"* decisions are replaced by a single absolute identity decisions (e.g., "this is John"), lists of candidate identities (e.g., "either this subject is John or Barry"), ranked/weighted lists of candidate identities (e.g., "my first choice for this subject's identity is John and my second choice for his identity is Barry"). One way to handle decision fusion in such cases is to view the problem as a decision to include a (candidate) identity in the overall decision (i.e., to include or not to include John Smith in the combined decision) and the problem then is reduced to fusion of *"YES/NO"* decisions. In case of ranked/weighted lists, the ranks/weights can be considered as the weights of the individuals decisions, as discussed before.

12

Thwarting Attacks

Automated biometric authentication systems help to alleviate the problems associated with existing methods of user authentication. Biometrics can improve convenience or security, or ideally both. However, security weak points will exist or will be introduced in any biometric installation, unintended by the designers of the system. These weak points will be discovered during operation of a system when the system is attacked, and the system will be attacked most successfully at the security weak points.

Unlike password systems, which are prone to brute-force dictionary attacks, biometric systems require substantially more effort to attack successfully. Although standard encryption techniques are useful in many ways to prevent a breach of security, there are several new attack points possible in a biometric system.

In remote, unattended applications, such as web-based e-commerce applications, attackers may have enough time to make numerous attempts, at a safe distance from the server, before being noticed, or they may be able to attack the client physically. At first glance, supervised biometric installations, such as those at airports, may not be that vulnerable to brute-force attacks. But such installations can certainly be the victim of replay attacks. Below, in Section 12.1, we develop a generic pattern recognition model that enables the study of security weak points. Such understanding is needed when designing biometric systems, while still keeping in mind the security *versus* convenience trade-off.

A biometric authentication system looks like the system in Figure 12.1. An input device measures a biometric sample from a human being. This sample is converted into a machine representation of the sample (which can be the sample itself) by the "feature extractor." This machine representation is matched against a reference representation, previously stored during enrollment. In Section 12.1, we describe the various using a model like this.

12.1 Pattern recognition model

A generic biometric system can be cast in the framework of a pattern recognition system [19,179]. The stages of such a generic pattern recognition system are shown in Figure 12.1, indicated by A, B, C, D; enrollment is in two stages: E and F. Excellent introductions to such automated biometric systems can be found in, e.g., [99,136]. More recent descriptions can be found in reports like the *Biometric Device Protection Profile (DBPP)* [49,117].

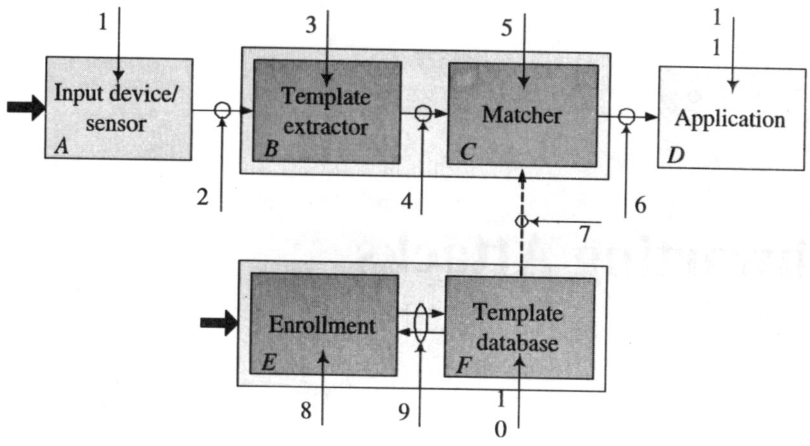

Figure 12.1: Stages of authentication system and enrollment system and points of attack in a generic biometric authentication system.

Any biometric system can be described as a four-stage system, as shown in Figure 12.1. More specifically, these stages are—

A. The first stage is the actual acquisition of the input biometric sample. This process can be as simple as recording telephonic speech, or as intrusive as drawing a blood sample for DNA testing.

The actual process and logistics of this biometric acquisition process is perhaps the most important factor in the convenience of a particular biometric installation. On the other hand, the amount of control one has over the signal acquisition greatly influences the quality of the sample and therefore the accuracy of the system. Often the difficulty here is proper control of signal acquisition without overly inconveniencing the user.

B. The second stage is the processing of the biometric sample to obtain some digital machine representation that can be matched to other such representations. This stage can be as simple as just storing the biometric sample, or it can be as complicated as computing a complex index structure, as in [75].

The problem of designing representations for biometrics is a complex, difficult task and the subject of much research.

C. This stage is the computation of match score between two or more representations: one representation of the input sample and one or more stored (enrolled) representations (from database F). There are several factors that determine how precisely this score can be estimated for some input sample. These factors are the quality of the input, the precision with which the representation can be computed from a biometric sample, and the effectiveness with which the matcher can compare two representations.

Another factor here is the quality of the training data that is used to optimize the matching process. Intricacies of this optimization process are discussed in Chapter 9.

This stage also makes the actual *match* versus *no match* decision. This decision is best made with the particular application in mind.

D. This is the application, which is protected by the biometric authentication system. This might be immigration control, a financial institution's ATM, or a computer account.

At this stage, the ultimate desired security of the decision, in terms of allowable False Accept Rate (FAR), is chosen as the operating point, which gives the False Reject Rate (FRR). This is a *compromise* between security and convenience, based on either static and or dynamic policies (see Chapter 2).

Besides the day-to-day work flow of a biometric authentication or identification application, the implementation of the enrollment process and the organization of the database of biometric samples are important aspects of the application. Specifically, referring to Figure 12.1, while enrollment E is more complicated than in password systems, a biometric sample database F can be considered similar to password storage.

E. The enrollment procedure and logistics are often-underemphasized aspects of a biometric system. Unlike in password systems, it is not easy to re-enroll with a new biometric sample and therefore the enrollment process needs careful consideration. Chapter 9 discusses the intricacies of this part of a biometric authentication system.

F. The database of biometric samples F is either distributed (e.g., recording the information on individual smartcards [178]) or centrally organized. This is a design choice in a biometric system that can have many ramifications.

Within this model (Figure 12.1) we identify 11 basic points of attack that plague biometric authentication systems. We subdivide these various attack points into several classes and expand on them below.

12.2 Attacking biometric identifiers

Many attacks on biometric authentication applications are based on somehow ensuring that the biometric data the system processes is not the correct data for the attacker. This is achieved either by mimicking another person, distorting one's own appearance, or changing the data stored inside the biometric system.

1. Threat 1 is an attack involving the presentation biometric data at the sensor A. This can be implemented as a *coercive attack*, where the true biometric is presented but in some unauthorized manner; an *impersonation attack*, where an unauthorized individual changes his or her biometric (particularly face or voice) to appear like an authorized individual; or a *replay attack*, where a recording of true data is presented to the sensor.

A coercive attack involves the presentation of a legitimate user's biometric data in an illegitimate scenario. The most obvious case is when a genuine user is forced by an attacker to identify him or herself to an authentication system. The authentication means is forcibly extracted from a genuine user to gain access to the system with concomitant privileges, in just the same way that an ATM user could be forced to give away his or her ATM card and PIN at gunpoint. It is desirable to detect instances of coercion reliably without endangering the lives of genuine users so that an appropriate action can be taken when such coercion is detected. Such detection could take the form of a stress analysis of the user, or having a guard supervise the system. Such attacks might also be deterred, for instance, by recording video of every biometric transaction.

The correct biometric could also be presented, in the scenario beloved of action movies, after physical removal from the rightful owner. This leads to one of the popular fears about biometrics—that fraud can result in physical threat, with eyes or fingers being removed by criminals for access to biometric devices. While this is conceivable, biometric systems designers have sought to counter such attacks by "liveness" detection—measuring the movements of the iris; electrical activity, temperature, or pulse in fingers [200]; facial expression change in faces; or challenge-response through a retina-recognition system. With computing power becoming more abundant, the technology for detecting fake biometrics will likely keep improving, but so too does the ability to spoof biometrics.

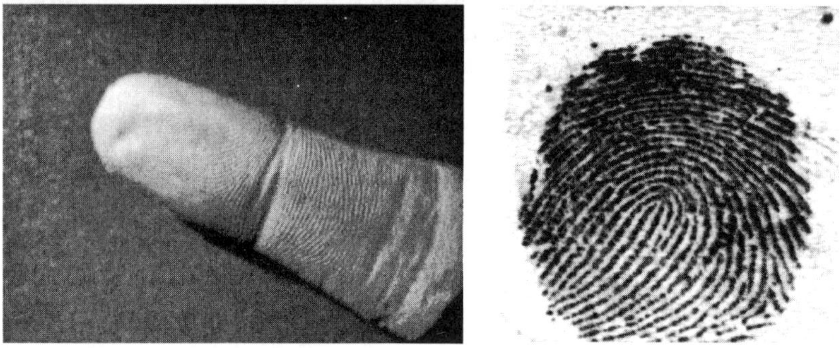

Figure 12.2: Fake rubber fingerprints can be easily fabricated and scanned.

An impersonation attack involves changing one's appearance so that the measured biometric matches an authorized individual (or fails to match in screening scenarios). Face and voice are particularly subject to attack, since some people are very skilled at sounding or looking like other individuals, and would be able to cause a system falsely to accept them. Even for such people, though, there are individuals that they would not be able to impersonate effectively. However, in the screening scenario, it is sufficient to change one's appearance sufficiently not to be recognized as your true self, which is much easier to achieve with simple disguises or plastic surgery. In this case the attacker is seeking to cause a False Negative error. For other biometrics, impersonation is harder—though several groups [133] have reported success in defeating fingerprint systems with rubber casts of fingerprints. An image of fake finger and the corresponding fingerprint is shown in Figure 12.2. It is also possible to have surgery to destroy one's fingerprints.

The combination of multiple biometrics also reduces the exposure to impersonation attacks since it means more attributes of an individual have to be copied with good fidelity. This is particularly true with joint face, speech, and lip-movement recognition—acquiring (or synthesizing) all three biometrics in a form suitable for replay will be harder than doing any one, particularly if the system is checking for consistency (particularly synchronization) between the three.

When cost is not an issue, all biometrics can be, and probably will be, subject to impersonation. The ease with which successful impersonation can be achieved varies from very easy to close to impossible and of course very much depends on the implementation of the specific system and on the intrinsic error rates (Chapter 14) of the particular biometric.

Replay attacks involve the re-presentation of previously recorded biometric information. This is most commonly thought of in the scenario of recording someone speaking a fixed-text password to a speaker ID system and playing back the recording to the system to gain entry. Such simple attacks can be prevented with prompted text systems—see Chapter 3. Simply holding up a photograph was enough to defeat early face-recognition systems, though these can now detect three-dimensionality and prompt for a change in facial expression (Section 12.8)—requiring a more sophisticated attack—perhaps holding up an LCD screen in front of the camera.

The inclusion of biometrics in the authentication protocol may inadvertently introduce points of attacks or security vulnerabilities. It is *these* points of attack that we study in the next section, and as we will see, biometric systems have many points of attack in common with password systems.

12.3 Front-end attacks

The front end of the system is where the bulk of the authentication activity occurs. The front end is responsible for turning the sensed biometric signal into some sort of invariant representation and then matching this against a retrieved reference template for the individual. This opens up several possibilities for attack:

2. Threat 2 is an attack on the channel between the sensor and the biometric system. This can again be a replay attack—resubmission of previously stored biometric signal—or an electronic impersonation. This would include circumventing the sensor by playing back video of a fingerprint or a face into the "video in" of a system, or electrically injecting an audio signal at a microphone output. If physical access to this attack point is available, it is simpler to attack than attacking the sensor (point 1 above)—the spoof signal can be both acquired and then replayed at this point—but techniques such as digital encryption and time-stamping can protect against these attacks. Further, a system might detect perfect matches against past data. Electronic impersonation might include the injection at this stage of a fingerprint image that had been created artificially from information about the minutiae locations derived from

a minutia record read from a card.

3. Threat 3 is attack by "Trojan horse" on the feature extractor B. The feature extractor could be attacked so that it will produce a pre-selected feature set at some given time or under some specific condition. That is, after the features have been extracted from the input signal they are replaced with a different synthesized feature set (assuming the representation is known).

4. Attack point 4 is the communication channel between the feature extractor and matcher. In the case of fingerprints, if minutiae are transmitted to a remote matcher (which can be the case when using smartcards [178] to store the template) then this threat is very real.

5. Attack 5 is again a Trojan horse: the matcher is attacked directly to produce an artificially high or low match score, thereby manipulating the match decisions. For instance, a hacker could replace the biometric library on a computer with a library that always declared a true match (for a particular user).

12.4 Circumvention

An often overlooked vulnerability of an authentication system is—

6. Threat 6, a very important threat, the overriding of the output of the matching module C. The output of the matching module could either be a hard *match* versus *no match* decision, or it could be just the probability of a match where the final decision is left up to the application. The attack point 6 is the same in both cases.

Some problems plague all authentication technologies (based on possession, knowledge, or biometrics) alike. Fraud in an authentication system is possible in different forms. Some forms of fraud are characterized as loopholes in the system: possibilities of illegitimate access to a system not envisioned by its designers. Other forms involve using intentionally incorporated mechanisms to transcend the authentication used by the system (super-user) and hence, in principle, cannot be eliminated using any strategies embedded inside the system (intra-system). The types of fraud could be categorized as follows:

- *Collusion:* In any application, some operators of the system will have a super-user status, which allows them to bypass the authentication component of the processing and to overrule the decision made by the system. This facility is incorporated in the system work flow to permit handling of exceptional situations, e.g., processing of individuals with no fingers in fingerprint authentication systems.

- *Covert acquisition:* It is possible that the means of identification could be compromised without the knowledge of a legitimate user and be subsequently abused. For instance, there is a significant amount of fraud in covertly observing personal identification numbers (PINs) at automated teller machines and at public telephones.

  This could be called an *impersonation attack* like the ones discussed in Section 12.2; however, only a user's parametric data are used—the biometric is not impersonated.

Yet it is unclear how one could realistically steal someone's fingerprint template by mere physical observation, much less supply it to the authentication system. It is more likely that the attacker will duplicate some possession or knowledge, such as a smartcard or PIN, which must be used in conjunction with the biometric (which has also been impersonated).

- *Denial:* It is possible that a genuine user may identify himself to the system using legitimate means of identification, through, say, a smartcard, to gain access to the privileges and be subsequently denied such an access, i.e., evoke a False Reject because of compromising biometric authentication templates. While this is not exactly fraud—no unauthorized access was granted to the protected resource—it disrupts the functioning of the system without explicitly breaking any of its components.

Many of these problems may not be fully eliminated. Currently, attempts to reduce fraud in authentication systems are process-oriented and *ad hoc*. There is a need to focus research efforts on systematic, technology-intensive approaches to combat fraud in the system. This is especially true in biometric authentication systems where the captured biometric measurements and context may have sufficient information to deter some forms of fraud. (One is literally leaving ones fingerprints at the scene of the crime). In particular, multiple biometrics show promise in approaching solutions to many of the above-mentioned problems.

12.5 Back-end attacks

The database of enrolled individuals is available locally or remotely, possibly distributed over several servers. Unauthorized modification of one or more machine representations in the database, which could result in authorization of a fraudulent individual, or at least in denial of service to the person associated with the corrupted (or inserted/excised) template (again, it is assumed that the representation is known).

7. Threat 7 is another channel attack but on the communication between the central or distributed database and the authentication system. The (biometric) representations from the stored database F are sent to the matcher through a channel, which is attacked to change the representations before they reach the matcher.

Processes E and F perform an extremely important function in a biometric authentication system—the enrollment of the eligible subjects, or the access control list. The "cleanness" of that database F is of extreme importance because the final authentication system is only as secure as its enrollment database itself (see Chapter 9). Three points of attack can be identified:

8. Attack point 8 is the enrollment center or application (E in Figure 12.1). The enrollment and authentication processes have similarities in the sense that they are both implementations of an authentication protocol, and therefore enrollment is vulnerable at attack points 1, ..., 6.

9. This point of attack is a channel (similar to attack point 10). Control of this channel allows an attacker to override the (biometric) representation that is sent from F, the biometric database, to C.

10. Attack on the database F itself. The database of enrolled (biometric) representations is available locally or remotely possibly distributed over several servers. This threat is the unauthorized modification of one or more representations in the database. This could result in authorization of a fraudulent individual, denial of service to the person associated with the corrupted template (again, it is assumed that the representation format is known), or removal of a known "wanted" person from a screening list.

This also opens up the possibility of privacy attacks—an attack on the confidentiality of the biometric authentication system, i.e., on the access control list or database of members. This attack is not aimed at the application but is aimed at the biometric authentication system database. Privacy aspects of a biometric installation, include the possibilities and consequences of such attacks.

With collusion between the hacker and the supervisor of the enrollment center, it is easy to enroll a newly created identity or stolen identity, the consequences of which could be severe. This threat is also very real in manual authentication systems. Purely from a biometrics point of view, the threat is ultimately related to the ease with which a biometric can be impersonated (see Section 12.2 above) and hence to the intrinsic FA. Enrollment needs to be more secure than authentication and is best performed under trusted and competent supervision.

Of course, what also needs to be remembered is the application or system that is being protected:

11. As noted in [107], the actual application D is a point of attack, too. This means that biometric authentication systems should take advantage of all the security services that are offered in a traditional authentication system, too.

In addition, Schneier describes many other types of abuses of biometrics in [197].

Overall, the greatest threat to biometric authentication systems is presenting, either physically or electronically, fake biometrics or previously acquired biometrics. This is a threat that needs to be addressed, especially for enrollment; that is, careful thought must go into how easy it is to enroll a newly created identity. In particular, threat 1 and threat 2, somehow tampering with the input device or the communication channel so that the sensed biometric resembles the biometric of a target, need to be examined. As we note below, electronic impersonation is becoming, or will become, feasible. However, the threat of an electronic replay of previously recorded biometric samples can be prevented using judicious combinations of challenge and response systems and data hiding (see Section 12.8 and [181]).

The fact that the biometric templates of the system in Figure 12.1 are stored either in a central database or distributed over clients or smartcards, does not necessarily affect the security of a biometric-based system since this portion of the system can be made secure with traditional technology. While a smartcard solution may provide a privacy advantage since the biometric database F is not centralized, it may also give attackers ample time to tamper with the smartcard. In many cases the biometric may be stored for practical purposes on the smartcard but also be stored in a central server, (e.g., for reissuing lost cards) opening such a system to more attacks.

The issue of privacy raises many concerns and needs to be solved by designing identity-encrypting technology. In that fashion, identity-encrypted templates can be matched in the "encrypted" domain without any possibility of ever tracing back the original identities. (A first attempt is described below, in Section 12.9.) The detection of fake biometrics is also a very important research topic that has not been addressed sufficiently yet.

12.6 Other attacks

Password systems are vulnerable to brute force attacks. Here the number of characters in the password is proportional to the bit-strength of the password, which expresses the amount of effort it takes (on average) to break into password authentication systems. Because people tend to choose simple passwords, within whatever system of rules is enforced, dictionary attacks make brute force attacks faster than implied by the theoretical maximum bit-strength of a password. There exists an "equivalent" notion of bit-strength of a biometric, also called *intrinsic error rate* (see Chapter 14). Such attacks would occur either at point 2 or point 4, but in general, the number of variations that would have to be tried is prohibitive.

In Chapter 14 we discuss some of the motivations for research in the bit strength of biometric identifiers. More on this subject can be found in [179,181].

There are further attacks that we have not mentioned some of which are mentioned in [117]:

- *Hill climbing attack:* This attack involves repeatedly submitting biometric data to an algorithm with slight modifications, and preserving modifications that result in an improved score. More sophisticated versions may actually try to model the scores returned to give faster improvements. Eventually, a score exceeding the match threshold might be achieved. This method is particularly suited to breaking in to an electronic system when the attacker has no knowledge of the legitimate user's biometric data. Such an attack can be prevented by disallowing repeated trials, and by giving out only yes/no matches, or at least quantizing match scores or adding a small amount of noise to the scores.

- *Swamping attack:* This would be similar to a brute force attack, exploiting weakness in the algorithm to obtain a match for incorrect data. For instance, for fingerprints the attacker might submit a print with hundreds of minutiae in the hope that at least the threshold number N of them will match the stored template (assuming the matcher does not reject such representations or normalize for this effect).

- *Piggy-back attack:* This type of malicious attack involves an unauthorized user gaining access to the protected assets through simultaneous entry with a legitimate user. This attack may involve physical threat, simple tailgating, or logical entry beyond the portal." This corresponds in some ways with coercion as mentioned in attack point 1.

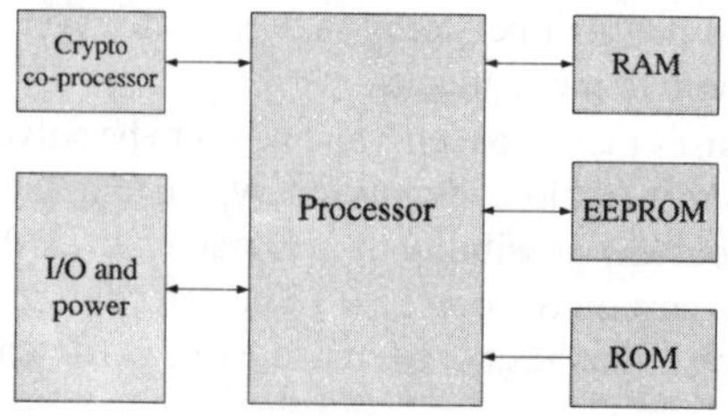

Figure 12.3: System architecture of a smartcard.

- *Illegitimate enrollment:* A threat that applies to any security system is simply allowing the enrollment of (or the assignment of inappropriate access rights to) attackers. For a door-entry system, if an attacker can once present credentials allowing enrollment, the security system will not prevent his subsequent access.

12.7 Combining smartcards and biometrics

Smartcards can play an important role in biometrics-based authentication systems. While biometrics are the pinnacle of reliable authentication and non-reputability, smartcards have the ability to store the biometrics and other relevant data, execute on-card functions, and interact with a card reader securely. Authentication systems that require high security can combine biometrics with a smartcard to get the best benefits from both these technologies.

A typical smartcard consists of an embedded processor and memory, as shown in Figure 12.3. Smartcards can be of two basic types: contact and contactless. In the contact type, the card has to be in physical contact with a reader. In the contactless variety the card can in the field of a reader without touching the reader and the contents can be read. A hybrid card can have both the interfaces. Over the years, smartcards have evolved from simple memory cards to more complex systems with 32-bit CPUs and hardware encryption engines to carry out secure computations such as encryption and digital signatures.

A number of measures make smartcards a strong candidate to support the privacy of stored data. Data stored in smartcard memory is protected by combined hardware and software mechanisms. Special on-board electronics protects data from unwanted alterations during memory read or write operations. Chemical and electrical corruption of memory contents are detected in hardware using special memory integrity checks and in software using checksums. Illegal tape-out reading of smartcard data is also protected using both hardware and software techniques. Inclusion of a hardware encryption engine makes the card a particularly reliable storage device.

Consider combining biometrics and smartcards into a two-factor authentication system. Such a system, consisting of a valid card and a valid enrolled biometrics, is always more secure than any one single factor system. If the user loses the card, access is prohibited until a new replacement card is obtained. The lost card, even when it falls into wrong hands, does not pose any threat to the system as one also needs the biometrics also to successfully pass the authentication system.

To implement this, during the enrollment of the subject the biometrics template would be stored in the smartcard along with other information. For authentication of the user, the smartcard and the live biometrics would be presented to the system. The system then generates the matching template from the presented biometrics, decrypts the stored template from the smartcard, and examines the similarity between the two templates. Having all the data locally available significantly cuts down on communication with the database server.

Storing biometric information in the smartcard has the added advantage of privacy. As pointed out earlier, with the central storage of a biometrics, there is the possibility of misuse of the biometrics for purposes unknown to the owner of the biometrics (attack point 10). Large collections of biometrics data can be sold or given away to unlicensed parties who can use the information however they choose. Yet smartcards provide a way of decentralizing the database into millions of pieces and thus giving the owners (enrollees) control. Furthermore, there is no central database which, if compromised for even one user, might invalidate the complete system for all users.

Once a secure storage is available, many other kinds of information can also recorded that can aid the biometrics application. For example, in a fingerprint-based authentication system, the fingerprint features along with, say, the threshold to be used during matching might be stored in the smartcard. Similarly, in a face-recognition system, the face template and the expected ranges of variation in its parameters can be stored. For speaker recognition systems, this method can provide extra benefit. In addition to storing a representation of the user's voiceprint, a generic voice model can also be saved. These parameters can then be restored by the service machine and used by a speech recognition driven menu system, thus resulting in improved interface performance.

The smartcards above are not assumed to have a significant amount of memory and processing power. A high-end smartcard can have up to 64 KB of ROM memory and run at 32 MHz clock rate [80]. However, the RAM space is quite limited—often only on the order of 1–2 Kilobytes. This, for the present, prohibits running any computing-intensive applications such as signal processing of biometrics input. However, by proper design, some or all of the *matcher* applications can be made to run on the smartcard even now. The applications running on a smartcard are virtually attack-free by design. When the matcher runs on the smartcard, the enrolled template never has to leave the card and hence further enhancing security. A secure protocol can be used to read the verdict computed by the card.

Several new applications in need of secure authentication use smartcards and biometrics, e.g., the US Department of Defence Common Access Card and the tamper-proof visa. Smartcard's tamper-proof design and low power consumption, even with 128-bit hardware encryption, makes them an attractive candidate for many such secure authentication applications.

12.8 Challenge and response

One of the prime vulnerabilities of biometric authentication systems is replay attacks, particularly at points 2 and 4. One way to guard against these is to use a *challenge-response* protocol. Challenge-response involves the security system's issuing a challenge to the user. The user must respond appropriately to be authorized. In this way the system can demand secret knowledge and also protect against replay or other attacks.

For example, prompted-text or text-independent verification (which are well established in the speaker identification literature [6]) can avoid a simple replay attack, but at the cost of a more complex system. Ironically, advances in trainable speech synthesis algorithms (both audio and video [53,77]) provide tools for attacking even these more sophisticated systems. Such challenge and response systems can be extended to other biometrics. Interactive fingerprint authentication systems can use video streams of fingerprints [54]. Of course, even these systems will eventually be attacked successfully. It is difficult to predict whether man will beat machine, or vice versa. However, by making impersonation more difficult, at least casual attacks (the majority) can be deterred. Hill [85, p.137] describes a retina-identification challenge-response system that projects information, say, a number, onto the retina being verified, which must then be typed by the user. This prevents "stolen" eyes from being used, and means that any mechanical forgery must be extremely sophisticated.

Another type of challenge and response to detect replay attacks in hardware is proposed in [181]. This approach is based on challenges to the sensor that is assumed to have enough intelligence to respond to the challenges. Many silicon fingerprint scanners [10,192] would be able to exploit the proposed method as a local secure processor can be integrated without much effort.

These type of challenge and response implementations, either between human and computer or between computer and computer, are naturally part of the authentication protocol $\mathcal{A}_P(P, K, B)$ and fall into the category of a dynamic authentication protocol as described in Chapter 2.

12.9 Cancellable biometrics

If not carefully implemented, biometric authentication may be fooled by presenting or acquiring and transmitting of biometric identifiers for the enrollment of fake identities (Chapter 9) or by accessing applications through the generation of a biometric identifiers (Section 12.6). Biometric authentication further presents the dilemma that once a biometric identifier is somehow compromised, the identifier is compromised forever, which leads to security and privacy concerns.

12.9.1 Privacy

Automated biometrics authentication methods are being developed, field tested, and installed for various larger-scale financial access and physical access control applications. Consequently, privacy issues (data confidentiality issues, in security terms) with biometric authentication have been a topic of concern for quite some time now in the security literature [18,181].

Here we really have two related issues [97]:

Privacy: Any biometric technology is traditionally perceived as dehumanizing and as a threat to privacy rights. As biometric technology becomes more and more foolproof, the process of getting authenticated itself leaves trails of undeniable private information, e.g., Where is an individual? What is the individual buying? etc.

In case of biometric authentication, this problem is perceived to be even more serious because biometric features might additionally inform others about the medical history or susceptibilities of a subject. For example, retinal vasculature may divulge information about diabetes or hypertension [134], but it should be realized that biometric systems in general are just not built to collect such information.

Nevertheless, there is a concern about privacy issues associated with biometric authentication. When it comes to our rights to privacy and biometrics, however, it may be good to remember Thomas Jefferson, who said, "If we cannot secure all our rights, let us secure what we can."

(There is a trend toward central registration: things that used to be carried and possessed "by the bearer" are now carefully registered, such as university degrees and transcripts, deeds to homes (now "recorded" by the town), passports, etc. Information about these things flows between trusted institutions, bypassing the individual who is the principal involved.)

Proscription: When a biometric measurement is offered to a given system, the information contained in it should not be used for any other purpose than its intended use. In any (networked) information processing system, it is difficult to ensure that the biometric measurement will only be used for intended purposes. This may be hard to enforce (see Figure 12.4), where it is shown how easily verification databases of individuals can be linked to criminal databases.

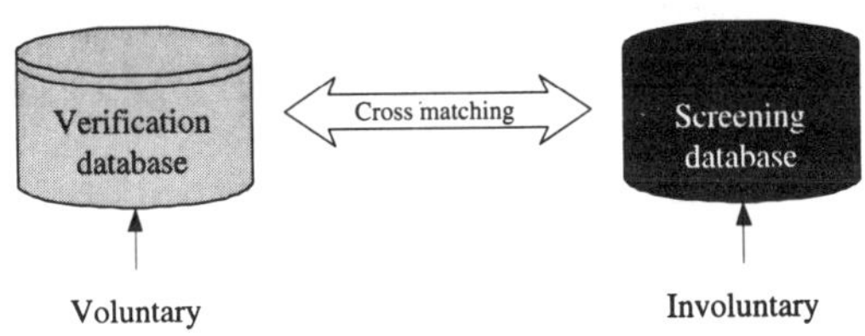

Figure 12.4: Cross-matching of large databases is a privacy concern.

Privacy and proscription concerns are summarized as follows:

1. Much data about people is already collected. There is concern about every bit of additional information that is stored about people, especially when it involves personal traits like biometrics.

2. Traditional security issues like *data integrity* and *data confidentiality* with respect to violations of personal information.

3. Biometrics databases can be used for cross-matching: for example, matching against law enforcement databases, such as the FBI or INS databases. This is a proscription issue and becomes a real problem when authentication databases are matched against legacy criminal databases (Figure 12.4).

These concerns are aggravated by the fact that a biometric cannot be changed. One of the properties that make biometrics so attractive for authentication purposes, their invariance over time (permanence), is also one of their liabilities. When a credit card number is somehow compromised, the issuing bank can just assign the customer a new credit card number. When a biometric is compromised, however, a new one cannot be issued. A person only has so many fingers, and only one face. A technique called cancellable biometrics, or identity encryption, that may alleviate privacy concerns is introduced in [181].

12.9.2 Intentional, repeatable transforms

The companion papers [19,182] offer a range of biometric security solutions, among them "cancellable biometrics." This is an intentional, repeatable distortion of a biometrics signal based on a chosen transform. The biometric signal is distorted in the same fashion at each presentation, that is, during enrollment and for every subsequent authentication. With this approach, every instance of enrollment can use a different transform, thus rendering cross-matching impossible. Furthermore, if one variant of the biometric is compromised, then the transformation can simply be changed to create a new variant (transformed representation) for re-enrollment as, essentially, a new person. In general, the distortion transforms are selected to be non-invertible. So even if the exact transform is known and the resulting transformed biometrics is known, the original (undistorted) biometrics cannot be recovered.

Cancellable transforms can be applied in the biometric signal domain or in the domain of features that are used to represent the biometrics. That is, the biometric signal can be transformed directly after acquisition, or the signal can be processed as usual and the extracted features can be transformed. We describe two particular transforms in the next sections, though many other types of transforms are of course possible.

12.9.3 Signal domain distortions

This category refers to the distortion (preferably non-invertible) of the raw biometric signal as it is acquired by the sensor, e.g., the original voice recording or fingerprint impression.

For face or fingerprint images, a morphed version of the image can be enrolled. Such morphing can be achieved in various ways. For example, a regular point pattern can be overlaid on the image. The morphed image is then obtained by randomly perturbing this point pattern in a structured fashion. Note that a subject can be enrolled in a legacy fingerprint or face authentication systems with such a morphed image. The legacy authentication system does not have to be aware of the fact that the image is morphed. Moreover, matching of this morphed image to any other existing fingerprint or face database will not identify the identity of the owner of the fingerprint or face. An example of transformed fingerprint images is shown in Figure 12.5. Other examples of transformed biometric images can be found in [20] and [181].

Note that in order to apply the same image morphing for each authentication, the fingerprint or face image needs to be transformed into a canonical position before the distortion. This can be done by aligning intrinsic points in the image, such as the intra-eye distance in a face or the core and delta in a fingerprint.

The iris, the colored area around the pupil of the eye, is another biometric (see Chapter 3 and [47]). This biometric is derived from an image of the user's iris as depicted in

Figure 12.5: Two prints of a finger (a, b). The same distortion applied to the original images gives (c, d). Note that (c) and (d) appear similar, yet do not match (a) and (b).

Figure 12.6. User authentication and identification through the iris image is achieved by developing a binary code, an iriscode, $c =$'0100101110....011' from a processed image of the iris. Identification can then be achieved very quickly even on large databases of enrolled users, since comparing these codes is very simple (Hamming distance).

Again there is the problem that if a person's iriscode is compromised, it is compromised forever. Thus, it is desirable to have a cancellable version of iriscodes. Figure 12.6 shows an example where the iris image on the right is an intentionally distorted version of the iriscode on the left.

A similar morphing technique can be applied to signals not usually considered images. Figure 12.7 shows a two-dimensional original voice biometrics signal $D(f,t)$. At each given time t_o, $D(f,t_o)$ gives the frequency content of the voice signal at that time point, like a spectrogram.

This fixed-text voiceprint in Figure 12.7 can be divided into time segments A, B, C, and D arranged in a time sequence (A,B,C,D). In this figure the time segments are of equal length but that is not a requirement. A scrambled voice signal is constructed as a sequence $(\underline{A},C,\underline{D},B)$. Here the underscore notation $\underline{A}$ means that the time segment A is played in reverse.

Note that for voice scrambling as in Figure 12.7, only minimal registration of the query

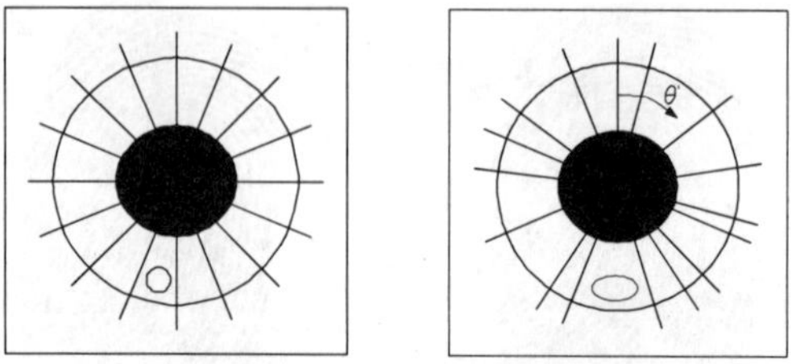

Figure 12.6: The original iris image can be scrambled in many ways.

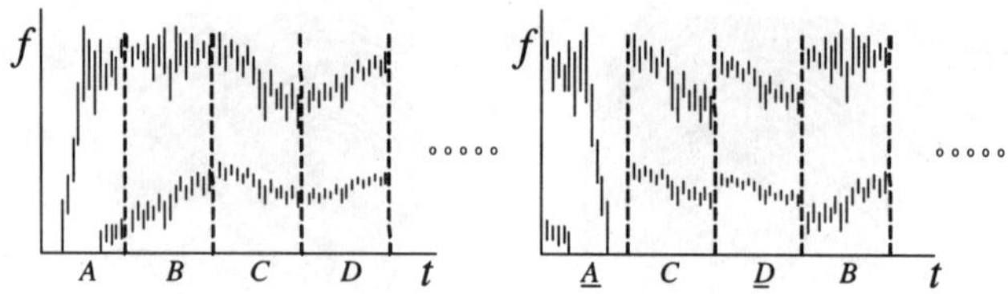

Figure 12.7: The original voice print (left) is scrambled by dividing the signal up in segments. The scrambled voice print (right) is constructed by randomly reordering and reversing the segments.

voiceprint with the enrolled voiceprint is needed, such as aligning the onset time.

12.9.4 Feature domain distortions

Processed biometrics signals (templates) can also be intentionally distorted. Here, we present an example of a non-invertible distortion of a point pattern. For example, consider a set of fingerprint minutiae

$$S = \{(x_i, y_i, \theta_i); i = 1, ..., M\}.$$

A non-invertible transform maps such a set S into a new set S' so that the original set S cannot be recovered from S', i.e.,

$$S = \{(x_i, y_i, \theta_i); i = 1, ..., M\} \mapsto S' = \{(X_i, Y_i, \Theta_i); i = 1, ..., M\}. \qquad (12.1)$$

Figure 12.8 shows how the x-coordinates of the point set S can be transformed through a mapping $x \mapsto X$, or $X = F(x)$.

Similar polynomial non-invertible transforms,

$$Y = G(y) \quad \text{and} \quad \Theta = H(\theta),$$

can be used for the remaining coordinates of the point set of (12.1).

12.9.5 Relation to compression and encryption

Cancellable biometrics differ significantly from compression of the signal using standard image-processing techniques. In signal compression, the signal temporarily loses its spa-

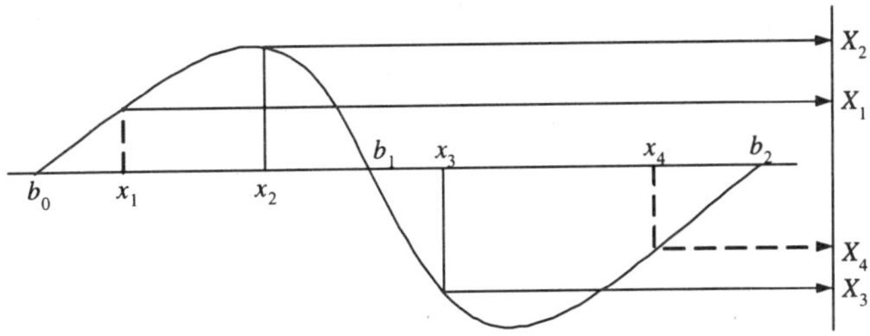

Figure 12.8: The mapping of one of the coordinates of a higher-dimensional point set into a new set of coordinates. A third-order polynomial is used.

tial domain characteristics such as geometric proximity. That is, two points in the original uncompressed signal are unlikely to remain at a comparable distance in the compressed domain. However, after decoding the original signal is either perfectly restored, or approximately restored if the compression is "lossy." In cancellable biometrics, by contrast, much of the local geometry is retained.

Cancellable biometrics are also quite different from encryption technologies. In encryption, the desire is to recover the original signal at the other end of the secure transmission. Whereas for the non-invertible distortions we use, the original signal is not restored, and in fact, it should be (close to) impossible to do this.

Furthermore, existing biometric systems cannot directly authenticate compressed or encrypted signals, whereas cancellable signals are intended to be processed by existing legacy software as if they were normal signals.

13

APIs, Standards, and Databases

For wide acceptance of biometrics, standards for interfaces and methods for performance evaluation are needed. Unfortunately, at the hardware level, the devices for biometrics still remain non-inter-operable except when sharing a common existing standard such as NTSC video. However, at the next level several standards are in the process of being developed and promoted. For instance, the BioAPI [11,215] is a standard for the application programming interface allowing the decoupling of biometrics-technologies from the applications that use them. In terms of standardized evaluation, NIST is playing an important role in designing several fingerprint databases [222–226] and conducting speaker verification tests. Similarly, the US Department of Defense runs the FERET and FRVT face recognition tests. This chapter will further describe the current inter-operability standards and common public databases.

13.1 Interface standards

A rapidly developing technology like biometrics needs to develop standards to address issues of inter-operability, plug-and-play compatibility, and building common Application Program Interfaces (APIs). In a typical biometrics recognition system there are three layers of interaction, shown in Figure 13.1. The lowest layer is the hardware layer, involved with interfacing with the biometrics-specific hardware. Also at the hardware level, standards like what ports and connectors the devices should use are being developed. The next level deals with the recognition of the basic biometrics signal with vendor-specific representation templates. The highest level is the layer involving the application. Standards are needed at every layer. In this section we briefly present the existing standards for these layers.

Inter-operability of biometric systems will be a necessity in large-scale deployments of biometric recognition systems. At the hardware level, as each sensor involves a different interface, the devices are incompatible even within the same class of biometrics. For example, two USB-ported fingerprint scanners are incompatible and the same driver cannot interface with them. However, when the devices involve video frame interface, software standards such as Video for Windows and Windows Driver Model and signal standards such as NTSC and RS-170 ensure that frame grabbing is unaffected.

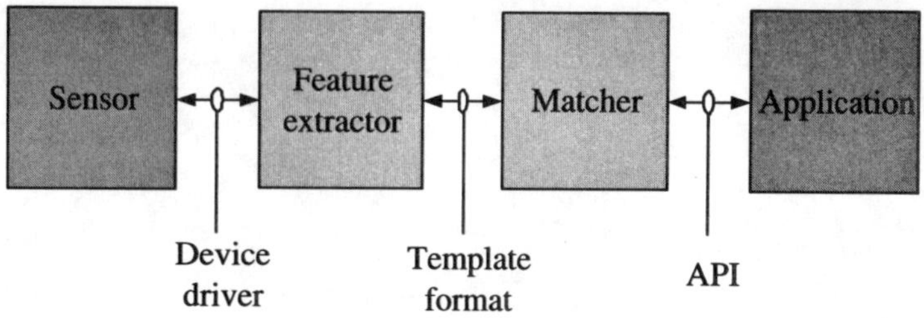

Figure 13.1: Layers of interaction with biometric authentication systems.

Once the signal is acquired, inter-operability at the template level is almost impossible for any technology as the individual representations vary significantly and quite often are proprietary and undisclosed. Hence the applications may have to re-enroll the users if there is a change in the vendor or hardware. The use of a known standard APIs make the change of the software easier; the industry is, however, very far removed from plug-and-play operation between sensors and vendors.

13.1.1 Application programming interfaces

There is much work ongoing in the area of standardization of biometric Application Programming Interfaces (APIs), at all levels of interaction (Figure 13.1). Given all the possible biometric protocols, in particular the dynamic conversational protocols, obviously this is quite a complicated task. We mention a few API standardization efforts:

- *BioAPI:*

  The BioAPI is the most popular API in the biometrics area. It is intended to provide a common interface for authentication based on any biometrics. The primitives included in BioAPI allow applications to manage enrollment, verification, and identification tasks on a client/server and the biometrics signal acquisition on a client.

  At the highest level, BioAPI defines a Biometrics Service Provider (BSP) which deals with all the aspects of biometrics signal processing. The three steps in the processing are capture, process, and match. Any biometric data that is returned to the application is referred to as Biometric Identification Record (BIR). The three main classes of functions supported by the API deal with enrollment, verification, and identification. BioAPI components post information about themselves in the BioAPI module registry during installation. The module registry can be used by applications to check the BSPs installed and their functionalities. Any device-specific parameters including the status of the device can also be contained in the registry to let the application decide the correct device and its usage.

  The specification has been released as the ANSI/INCITS standard also. More details and reference implementation code are available in [11].

- *SVAPI:*

  This standard was meant for an API for speaker verification. It merged with BioAPI and hence has been discontinued.

- *BioAPI for Java Card:*

  Building on the existing API for Java Cards for security and maximal functionality, the Java Card forum's biometrics task force defined a more generic biometric API in order to support secure match-on-card functionality and security of the templates. The main goal of the API is to enable on-card applets to support match-on-card for biometrics template matching. The API uses Java Card and its security mechanisms for storage on card and supports match-on-card. The API supports secure matching of the templates without exposing the reference template stored in the card, supports multiple biometrics templates used in the industry and protects the template by limiting the number of failed match attempts while staying compatible with BioAPI and CBEFF (Common Biometric Exchange File Format). More details are available in [151].

- *BAPI:*

  This is an API advocated by I/O Software Inc. [92] API that allows the programmer to develop applications for a broad range of Virtual Biometric Devices (VBDs) without knowing the specific capabilities of the device. The API is composed of three distinct levels of functionality from high device abstraction to low (device-specific) abstraction.

  BAPI is a standard software protocol and API for communication between software applications, the operating system, and biometrics device. Individual hardware and related modules can easily be swapped in or out. This API is device agnostic. There is no consortium for BAPI. Any software/hardware developer, integrator, or solutions provider wishing to integrate with BAPI and Microsoft Windows should contact I/O Software [92].

13.1.2 Data structure and security standards

Here we mention some of the other relevant standards known to the authors. Unlike the previously listed API standards which describe the processing functionalities available and how to invoke them, the standards listed here relate to the structure of declarative data passed between modules and policies for communicating and storing this information. We believe that this list is representative, but we do not guarantee that it is exhaustive.

- *CBEFF:*

  The common Biometrics Exchange File Format (CBEFF) is an emerging standard that deals with the issue of sharing biometrics in raw signal form or in template form. The standard is recommended for facilitating data interchange between different components of a system or even across systems. The data described by CBEFF includes security information in the form of digital signatures and data encryption; identification of biometric type and information about the biometric sample and the actual biometric data. More details about CBEFF are available in [35].

- *ANSI-NIST-ITL-2000:*

  This standard specifies a common format to be used to exchange fingerprint, facial, scar, mark, and tattoo identification data effectively across dissimilar systems made by different manufacturers. This standard [70] is a revised version of the ANSI-NIST-CSL-1993 standard and ANSI-NIST-ITL 1A-1997.

- *AAMVA DL / ID 2000:*

  The American Association for Motor Vehicle Administration (AAMVA) Driver's License and Identification (DL/ID) Standard [1] describes a uniform means to identify issuers and holders of driver's license cards. The standard specifies identification information on driver's license and ID card applications. It also specifies a format for fingerprint minutiae data that would be readable across state and province boundaries for driver's licenses.

- The *1996 FBI WSQ* standard:

  For compression and decompression of gray scale fingerprint images, the FBI in collaboration with LANL has developed a wavelet-based standard known as Wavelet Scalar Quantization (WSQ). The original standard was proposed in 1993 and revised in 1996. The significant advantage of the standard is that even at higher compression ratio, the minutiae features are less disturbed compared to other image compression standards. For law enforcement agencies where storage of the fingerprint images is required, WSQ plays a crucial role. More details about WSQ are available in [67].

- *Intel's CDSA:*

  Intel's Common Data Security Architecture (CDSA) standard provides a set of security building blocks for application developers. CDSA is designed as an overall infrastructure for data security on PCs, workstations, and servers. It is founded on two fundamental data security premises: digital certificates (a form of electronic identification that enables a hierarchy of trust, dependent on the identity of the user) and portable digital tokens, which store cryptographic keys and perform cryptographic operations. The CDSA 2.0 specification introduces Elective Module Managers (EMM) as a key CDSA component. An EMM can add new and compelling security features such as biometric authentication.

  Human Recognition Services (HRS) is an EMM under CDSA. Its basic purpose is to verify the identity of a person based on some combination of password knowledge and biometric measurement. The HRS EMM supports the integrity features of the CDSA stack which are currently not provided by the BioAPI. More details about CDSA can be found in [91].

- *ANSI X9.84:*

  For use of biometrics in the financial industry, X9.84 describes a standard for biometric information management and security. The X9.84 standard defines requirements for managing and securing biometric information such as customer identification and employee verification. X9.84-2000 specifies the minimum security requirements for effective management of biometrics data for the financial services industry and the security for the collection, distribution, and processing of biometrics data. The standard includes specification for the security of the physical hardware used, the management of the biometric data, the utilization of biometric technology for verification/identification of banking customers and employees, the application of biometric technology for physical and logical access controls, the encapsulation of biometric data and techniques for securely transmitting and storing biometric data. The biometric data object specified in X9.84 is compatible with CBEFF. More details about X9.84 are available in [5].

13.2 Databases

For impartial performance evaluation of biometric recognition systems, the industry needs to have access to large public databases. In this area, NIST has been a leader for a long time with a large variety of databases for fingerprints, mug-shots, and voice samples. With significant interest in the face recognition, there are now several large databases, many collected by academic institutions. For speaker recognition, there are also additional academic databases available publicly. However, for iris *no known public database* is available at this time.

13.2.1 Fingerprint databases

As mentioned NIST has been very active in the area of fingerprint database collection. These are databases with prints from digitized paper impressions, annotated fingerprint classes, and fingerprint image sequences. Information about the NIST databases can be found at http://www.nist.gov/srd/ but we give a brief description here:

- *NIST-4* (http://www.nist.gov/srd/nistsd4.htm):

  The NIST-4 fingerprint database is a very popular database consisting of 2,000 fingerprint pairs scanned at 500 dpi. Each image is 512 $\times$ 512 pixels and has a class label. The database consists of 400 image pairs from each of the five classes. Even though this database is more suitable for classification tests, researchers have been using it for testing matching performance also.

- *NIST-9* (http://www.nist.gov/srd/nistsd9.htm):

  The NIST-9 is one of the largest rolled fingerprint databases, consisting of five volumes. Each volume has 270 mated fingerprint cards resulting in 5,400 images. The cards have been selected to reflect the natural fingerprint distribution. Each image is 832 by 768 pixels scanned at 500 dpi resolution and the National Crime Information Center (NCIC) class marked by FBI is available.

- *NIST-10* (http://www.nist.gov/srd/nistsd10.htm):

  The NIST-10 database is a supplemental database to NIST-9 consisting of a larger sample of fingerprint patterns that have a low natural frequency of occurrence and transitional fingerprint classes. The database consists of 5,520 images. Each segmented image is 832 by 768 pixels scanned at 500 dpi resolution. The NCIC class provided by the FBI is also available. This database is more suitable for algorithm testing involving rolled fingerprints.

- *NIST-14* (http://www.nist.gov/srd/nistsd14.htm):

  NIST Special Database 14 is suitable for development and testing of automated fingerprint classification and matching systems on a set of images which approximate a natural distribution of the fingerprint classes. Each segmented image is 832 by 768 pixels, scanned at 500 dpi and classified using the NCIC classes given by the FBI and WSQ compressed. The full database consists of 27,000 fingerprint pairs.

- *NIST-24* (http://www.nist.gov/srd/nistsd24.htm):

  There are two components in the NIST-24 livescan fingerprint video database. The first component consists of MPEG-2 compressed video of fingerprint images acquired over 10 seconds when the user intentionally distorts the impression. Each frame is 720×480 pixels. All ten fingers of ten subjects are available. This database is useful for studying the impact of distortion on matcher performance as well as distortion detection. The second component of the database deals with image acquisition at various rotations. Similar to the first component, it consists of sequences of all ten fingers of ten subjects. The MPEG-2 compression for both the databases have been carried out at 5 Mbits/second.

- *NIST-27* (http://www.nist.gov/srd/nistsd27.htm):

  In the area of latent scene-of-crime fingerprint analysis, NIST-27 is a very useful database. It consists of 258 latent fingerprints from crime scenes and their matching rolled fingerprint mates with the minutiae marked on both the latent and the rolled fingerprints. All the paired minutiae between the latent and its corresponding rolled print are also identified. This database is useful for testing latent matching performance.

- *NIST-29* (http://www.nist.gov/srd/nistsd29.htm):

  The NIST-29 fingerprint database consists of both rolled and plain fingerprints of the same fingers that can be used for system performance testing. The images have been scanned at 500 dpi and compressed using WSQ. There are a total of 4,320 rolled fingers from 216 persons and all ten fingers are scanned. Two impressions of each finger is available in addition to the plain finger images. All the images of a person are formatted as a NIST record.

- *NIST-30* (http://www.nist.gov/srd/nistsd30.htm):

  NIST-30 are dual-resolution (500 dpi and 1,000 dpi) fingerprint images to test fingerprint compression algorithms. The data consists of 36 ten-print paired cards with both the rolled and plain images scanned at dual resolution. The cards are segmented into the ten rolled prints and the plain prints and stored in ANSI/NIST formatted files.

- *FVC 2000 and FVC 2002 Fingerprint Databases*:

  At the 2000 IEEE International Conference on Pattern Recognition, the first Fingerprint Verification Contest (FVC) was organized by the University of Bologna [130]. This competition and the follow-up one has resulted in the a large public database. The FVC 2000 databases were collected using livescan devices using three different scanners including low-resolution optical (256 dpi), CMOS (smaller scan area of 0.6" $\times$ 0.8"), and a full (1" $\times$ 1") gray scale optical FBI-certified scanner. In addition there is a database of synthetically generated fingerprints. Yet another difference of the FVC 2000 database is that there are eight samples per finger, which can enable algorithms to understand the variations across matching fingers. However, the database consists of 100 fingers only and the same persons have not been imaged for all the scanners. Each database has a total of 800 images.

  A newer version of the database is being planned in FVC 2002. More details about FVC 2000 and FVC 2002 are available at [13].

13.2.2 Face databases

Because of forensic applications of face recognition, interest in face databases has been in the mission of NIST all along; more recently, because of the tremendous interest in face recognition in academia, a plethora of university face databases is emerging:

- *The Olivetti (ORL, now AT&T) database* (http://www.uk.research.att.com/facedatabase.html):

  This database [195] consists of ten 92 $\times$ 112 pixel gray scale images of 40 subjects with a variety of lighting and facial expressions. This database has often been used in the research community for testing new algorithms, but since recognition results on it are very high, it is no longer very useful for testing state-of-the-art algorithms.

- *FERET* (http://www.dodcounterdrug.com/facialrecognition):

  The Facial Recognition Technology (FERET) database was collected over several sessions spanning over three years. The database contains 1,564 sets of images for a total of 14,126 images that includes 1,199 individuals and 365 duplicate sets of images. A duplicate set is a second set of images of a person already in the database and was usually taken on a different day.

  For some individuals, over two years has elapsed between their first and last sittings, with some subjects being photographed multiple times. More exhaustive face databases are now being made available through FRVT 2002. More details about how to get a copy of FERET database are available at [165].

- *FRVT 2002* (http://www.dodcounterdrug.com/facialrecognition):

  The main idea of the Facial Recognition Vendor Test (FRVT) 2000 was to evaluate face recognition technology performance in the real world. The FRVT 2000 used most of the FERET database. The FRVT 2002 extends the test to cover more difficult databases including a short video of faces. The largest face database to be used in the test involves 120,000 faces. More details about the data will be made available at http://www.frvt.org/FRVT2002/default.htm.

- *NIST 18 Mugshot Identification Database* (http://www.nist.gov/srd/nistsd18.htm):

  The NIST 18 database consists of 3,248 mugshot images for testing mugshot recognition systems. The database contains both front and side (profile) views when available. Separating front views and profiles, there are 131 cases with two or more front views and 1418 with only one front view. The images have been scanned at a resolution of 500 dpi.

- *The MIT database* (ftp://whitechapel.media.mit.edu/pub/images/):

  The MIT face database is one of the oldest public face databases. It consists of faces of 16 subjects and 27 images per subject with varying illumination, scale, and head orientation. The data set can be directly downloaded from the above site. For large scale system testing, this database may not be suitable.

- *The CMU database* (http://www.ri.cmu.edu/projects/project_418.html):

  There are several different face databases available from the CMU Robotics Institute. The CMU Pose, Illumination, and Expression (PIE) database consists of 41,368 images of 68 people. Each person face has been imaged under 13 different poses, 43 different illumination conditions, and with 4 different expressions. The other databases include facial expression analysis database and face detection database. More details about these databases are available at the web site mentioned above.

- *The Yale database* (http://cvc.yale.edu/projects/yalefacesB/yalefacesB.html):

  The Yale Face database B contains of 5,850 single light source images of ten subjects each imaged under 576 viewing conditions arising from nine poses and 64 illumination conditions. For every subject in a particular pose, an image with ambient (background) illumination is also captured. The acquired images are 640(w) $\times$ 480 (h) in size and have 256 gray scales. Even though a large number of poses are available, the database has only ten subjects. Hence, it is better suited for face modeling.

- *The Purdue database* (http://rvl1.ecn.purdue.edu/ aleix/aleix_face_DB.html):

  A relatively large face database is available from Purdue University. It contains over 4,000 color images corresponding to 126 subjects imaged with different facial expressions, illumination conditions, and occlusion. Each person participated in two sessions, separated by two weeks time. The images are 768 $\times$ 576 pixels with 24 bit color. The database is available freely for research purposes from the URL specified above.

- *M2VTS* (http://www.de.infowin.org/ACTS/RUS/PROJECTS/ac102.htm):

  Other than the government-initiated programs for face databases, M2VTS (deals with Multi-modal Biometric Person Authentication) has been an active player in face databases. In fact, it is the only **multi-modal** database that includes face and speech. The database consists of 37 persons and faces taken at five different orientations. Later this database has been extended to a large face video database and called *XM2VTSDB*. The new database consists of 295 persons and four sessions taken over several months. High-quality color images are available.

- *The Equinox database* (http://www.equinoxsensors.com/products/HID.html):

  Equinox is collecting an extensive database of face imagery, including **thermal** imagery. It contains the following modalities: broadband-visible images co-registered with body heat imagery (LWIR = 8–12 μs), medium infrared imagery (MWIR = 3–5 μs), and near infrared imagery (SWIR = 0.9–1.7 μs). This data collection is made available for experimentation and statistical performance evaluations. Initial performance evaluations will be made to assess the utility of thermal infrared images for face recognition, detection, and tracking.

13.2.3 Speaker recognition databases

As in other biometric databases, there is variation in the number of speakers, the time lapse between sessions, and the number of sessions recorded. However, there are also a number of other issues in designing a speaker recognition database. These include the type of speech (e.g., prompted, conversational, or fixed); the duration of the acquired speech signal; the acoustic environment; and the type of microphone and the transmission channel used. Hence, there end up being many more parameters to explore for voice databases than for finger or face databases.

There are several public speech databases, including the multi-modal database (i.e., M2VTS) described above that can be used for speaker recognition. For a good survey article on speaker databases, see [32]. Here, we list some of the more common databases, most of which were designed for telephone applications.

- *NIST* (http://www.nist.gov/speech/tests/spk/index.htm):

  The original NIST speaker database is derived from Switchboard I and Switchboard II data sets. The data sets were collected by the Linguistic Data Consortium (LDC) and consist of 2,728 five-minute conversations involving 640 college student subjects as speakers. The subjects are allowed to make one call per day. Every call by the same person involved use of a different handset, whereas the receiver always had the same handset.

- *TIMIT and NTIMIT* (http://www.ldc.upenn.edu/Catalog/LDC93S2.html):

  The DARPA Acoustic-Phonetic Continuous Speech Corpus (TIMIT) is one of the first databases in the speaker recognition area. The database consists of 630 speakers where the subjects read sentences in a acoustic sound booth. The NYNEX Speech Corpus CD-ROMs (NTIMIT) database has been recreated by an artificial mouth talking into a handset and recording after the signal was transmitted through long distance telephone lines.

- *The YOHO Database* (http://www.ldc.upenn.edu/Catalog/LDC94S16.html):

  The YOHO Speaker Verification data set is useful for text-dependent speaker recognition experiments. The data set consists of 128 subjects speaking the prompted digit phrases into high-quality handsets in an office environment. The data was collected from four enrollment sessions followed by ten verification sessions with at least several days intervening between the sessions.

- *OGI* (http://cslu.cse.ogi.edu/corpora/spkrec/):

  A dedicated speaker recognition database involving 100 subjects is available from the OGI School of Science & Engineering. The subjects called the system 12 times over a two-year period under different noise conditions and using different types of handsets and different telephone environments. The speakers discourse on a variety of topics to make the database vocabulary independent.

- *ELRA databases* (http://www.icp.grenet.fr/ELRA/):

  The European Language Resources Association (ELRA) has several speaker databases in non-English languages and non-native English speakers.

  SIVA - The Speaker Identification and Verification Archives is an Italian-speaker database consisting of 671 subjects. The subjects speak the prompted words and digits and read text into a telephone handset in an office or home environment. The data has been collected over several sessions with the sessions separated by several days.

  PolyVar - This is a speaker verification corpus comprising of native and non-native speakers of French, mainly from Switzerland. It consists of read and spontaneous speech in Swiss-German and French amounting to 160 hours of speech. Thirty-one speakers called from two to ten times, and 41 speakers made more than ten calls.

  POLYCOST - This database was collected under the European COST 250 project. Most of the speech is that of non-native English speakers with some speech in the speaker's native tongue covering 13 European countries. The speech was collected over ISDN phone lines. The impact of language in speaker recognition can be tested with this database.

13.2.4 Other databases

On-line handwriting recognition generally addresses the problem of recognizing handwriting from data collected with a sensitive pad which provides discretized pen trajectory information. In contrast to other pattern recognition fields, however, there are few large corpora of training and test data publicly available, and no open competitions have been organized.

- *UNIPEN database* (http://hwr.nici.kun.nl/unipen/):

  As of 1999, the international Unipen Foundation was installed to safeguard the distribution of a large database of on-line handwritten samples, collected by a consortium of 40 companies and institutes.

Recently there has been a lot of interest in gait-based identification—where a person is recognized by how they walk. This biometric, although fairly weak, is attractive because it can be performed on fairly low resolution imagery and hence can be employed for recognition at a distance. The bulk of the early exploratory work on this method has been done at universities and hence the publicly available databases are all from academia.

- *USF Gait Baseline database* (http://marathon.csee.usf.edu/GitBaseline/):

  This web site contains a large set of video sequences (about 300 GB of data related to 452 sequences from 74 subjects) acquired to investigate important dimensions such as variations due to viewpoint, footwear, and walking surface of the gait recognition problem. For each individual, video is collected in up to eight conditions: two surface conditions, two shoe types, and two camera views. Each subject walks a similar-sized elliptical course with major axis of about 15 m and minor axis of about 5 m. All data is collected outdoors with the adverse complications of shadows, motions in the background, and moving shadows due to cloud cover.

- *CMU "Moving Body" corpus* (http://hid.ri.cmu.edu/HidEval/index.html):

  The MoBo data set was collected at Carnegie Mellon University as part of the Human Identification at a Distance project consists of videos of subjects walking on a treadmill. Each sequence is filmed simultaneously from six different viewpoints. In total, 25 subjects are filmed walking with four gaits (slow walk, fast walk, on inclined treadmill and carrying a ball) for 11 seconds.

13.3 Certifications

Certified products and solutions in the biometrics area are not yet there. Once again, fingerprint-related solutions have been more advanced in this area because of the longer history. There are two FBI certification programs in the fingerprint area. The first one deals with certification of optical fingerprint scanners and the second one deals with the certification of WSQ implementation for accuracy. The International Computer Security Association (ICSA, formerly NCSA) also has a certification program in the biometrics area. *Common criteria* certification has become more popular in biometrics recently as vendors are busy developing the "protection profiles" (see below).

- *FBI* fingerprint scanners (http://www.fbi.gov/hq/cjisd/iafis/efts70/appendixf.htm):

  For the FBI's IAFIS system, there are strict guidelines for fingerprint image quality adherence during capture, display, and printing. The specifications are covered in the Appendix-F of Criminal Justice Information Services (CJIS) Electronic Fingerprint Transmission Specification (http://www.fbi.gov/hq/cjisd/iafis/efts70/cover.htm).

  Appendix G describes the requirements for accrediting current live- and card-scan equipment for fingerprint acquisition. A list of FBI certified scanners as having been tested and found to be in compliance with the FBI's Integrated Automated Fingerprint Identification System (IAFIS) Image Quality Specifications (IQS) is available at http://www.fbi.gov/hq/cjisd/iafis/cert.htm. Note that the list includes certified live scanners, card scanners and printers.

- *WSQ* fingerprint compression (http://www.itl.nist.gov/iad/894.03/fing/cert_gui.html):

  The Wavelet Scalar Quantization (WSQ) fingerprint image compression logarithm is the standard for the exchange of gray level fingerprint images. The WSQ Specification defines a class of encoders and a single decoder with sufficient generality to decode compressed image data produced by any compliant encoder. Part III of the WSQ specification contains the specific parameter values that must be implemented by encoders for certification. A WSQ decoder must implement the full range of functionality contained in the WSQ Specification including even and odd length filters. Compliance with the WSQ Specification is determined by comparing the output from the implementation under test with the output from a NIST double-precision reference implementation. The comparison criteria and accuracy requirements are contained in the WSQ Specification.

- *Common Criteria* (http://www.commoncriteria.org):

  Common Criteria is a set of generic security requirements to aid in the specification of products and functions. A number of assurances are related to system attributes involving security measures implemented in hardware, software, and firmware. The Common Criteria certification has been adopted by the ISO as an international standard ISO 15408.

  In the Common Criteria framework consumers use a *protection profile* to specify security parameters. The main objectives of such a profile are—

  - to describe the Target of Evaluation (TOE), i.e., identify the part of the system that is the subject of evaluation;
  - to describe the security environment of the TOE, including the assets to be protected and the specification of the threats to be countered;
  - to describe the security objectives of the TOE and its supporting environment;
  - to specify the security requirements, which include the TOE functional requirements and the TOE assurance requirements.

  A protection profile for medium robust environments involving biometrics for the US Department of Defense is described in [117]. Similar protection files are also found at http://www.cesg.gov.uk/technology/biometrics/media/bdpp082.pdf.

  Consumers and other parties can then independently select an evaluation assurance level from a defined set of seven Evaluation Assurance Levels (EALs) *EAL1-EAL7*. These assurance measures along with the TOE security threats, objectives, requirements, security function specifications are collectively known as the *Security Target* for the purpose of evaluation and certification.

  The scope of Common Criteria includes standards for acceptance of evaluations of security products performed by independent labs and addresses protection of information from unauthorized disclosure, modification, or loss of use. However, Common Criteria does not address accreditation, cryptographic algorithm selection, physical security related to the hardware, and legal implications of common criteria.

- *ICSA* certification:

  The ICSA (International Computer Security Association) used to certify biometric authentication products. However, recently they have discontinued this effort.

13.4 Legislation

Biometrics as a tool for identification and authentication is being incorporated into several new pieces of legislation (particularly since the terrorist events of September 11, 2001), and have been also proposed as additions to some older pieces of legislation as well. A number of government agencies are addressing the important role that biometrics will play in identifying and verifying the identity of individuals.

- *USA Patriot Act;* US Public Law 107-56 (http://frwebgate.access.gpo.gov/cgi-bin/getdoc.cgi?dbname=107_cong_public_laws&docid=f:publ056.107)

  This act specifically addresses the need for tools required for strengthening law enforcement agencies to intercept and obstruct terrorism. It requires use of fingerprints to obtain criminal history background of visa applicants to the USA. A special feasibility study is envisaged in this act to investigate the use of biometrics with special emphasis on fingerprints for providing access to the FBI IAFIS system at overseas consular posts and points of entry to the USA.

- *Improve Aviation Security using emerging technologies;* US Public Law 107-71 (http://frwebgate.access.gpo.gov/cgi-bin/getdoc.cgi?dbname=107_cong_public_laws&docid=f:publ071.107)

  This act requires use of emerging technologies such as biometrics to improve aviation security. Specifically, it proposes to investigate use of biometrics for (i) positive identification/verification of employees or law enforcement officers entitled to enter a secure area of the airport; (ii) use of biometrics to strengthen access control points in air traffic control operation areas, crew lounges, catering delivery areas; (iii) include biometrics in pilot licenses; (iv) use of biometrics (voice) in stress level analysis to prevent a person who may be considered a danger to air safety or security; and (v) analyze threat level of a person using biometrics.

- *Enhanced Border Security and Visa Reform Act;* US Public Law 107-173 (http://frwebgate.access.gpo.gov/cgi-bin/getdoc.cgi?dbname=107_cong_public_laws&docid=f:publ173.107)

  This act requires use of biometrics in visa and other travel-related documents. An attractive feature of this act is that it requires the documents to be machine readable while tamper resistant. Biometric readers are required to be installed at the port of entry into the US to enable biometric acquisition. It also requires that countries with no visa requirement for travel to the USA should include biometrics in their passports.

- *Government Paperwork Elimination Act* (GPEA)
  (http://www.whitehouse.gov/omb/circulars/a130/a130appendix_ii.html)

  The GPEA requires federal agencies to allow individuals or entities that deal with the agencies the option to submit information or transact with the agency electronically. The Act specifically states that electronic records and their related electronic signatures are not to be denied legal effect, validity, or enforceability and encourages Federal government use of a range of electronic signature alternatives including biometrics.

- *Health Insurance Portability and Accountability Act* (HIPAA)
  (http://cms.hhs.gov/hipaa/hipaa2/default.asp)

  The US Department of Health and Human Services (HHS) has proposed HIPAA to protect privacy of personal health information in 1996. The final Rule adopting HIPAA standards for the security of electronic health information as published in the Federal Register on February 20, 2003, specifies a series of administrative, technical, and physical security procedures for covered entities to use to assure the confidentiality of electronic protected health information. The HIPAA security standards final rule specifies use of biometric identification for person authentication.

Also, at the US state level several bills are being discussed, and some have even been passed, involving the protection of privacy for biometrics data, especially regarding the use of biometrics for driver's licenses.

- *Driver's license related:*

  Use of biometrics in driver's licenses has been considered in several states: Florida S 124, Missouri HB 178 and SB 0163, Washington SB 5412.

- *Privacy issues related to biometrics data:*

  Several states have declared it illegal to collect and distribute/sell biometrics data without the knowledge of the subject: California SB 186, California AB 48, Maryland SB 264, South Carolina S 43 and S 222, Virginia SB 979 and HB 2175. EU Data Directive (95/46/EC, Feb. 1995) also prohibits sharing of biometric data and personal information.

We have just highlighted some of the more important items here. For a more thorough (and continuously updated) survey of pertinent legislative actions, readers are referred to the International Biometric Industry Association newsletters published online at http://www.ibia.org/newslett.htm

Part IV

Mathematical Analyses

14

A Biometric's Individuality

There has been much interest in the *individuality* of some of the biometric identifiers. Loosely speaking, individuality has to do with comparing a biometric with a password [179]; it is related to how easy it is randomly to "guess" a biometric machine representation, given current sensing and matching capabilities. Reasons for the interest in individuality include the many recent challenges to fingerprint, signature, and writer authentication expert testimony in the courts of law. As well as the published empirical studies on the relatively new iris biometric (Section 14.2).

14.1 Approaches to individuality

In order to have a meaningful formulation of the individuality problem, we need to define—

1. representations of the biometric identifier;
2. a metric of similarity of two biometric identifiers; and
3. representations of the target population (or their representative samples).

For example, if two fingerprints originating from two different fingers are examined at a very high level of detail, it may be decided that the fingerprints are indeed different. The question is, however, How do we address the above five points of Justice Blackmun [105] and statistically justify such a decision?

Fingerprints in particular can be represented by a large number of features, including the overall ridge flow pattern and ridge frequency. Ordering the features from coarse to fine, we have, as shown in Figure 14.1—

1. location and position of singular points, core(s) and delta(s);
2. ridge counts between pairs of minutiae;
3. type, direction, and location of minutiae points; and even
4. locations of the sweat pores.

Figure 14.1: A fingerprint image of type "right loop" includes feature like the overall ridge structure, minutiae, and ridge counts; depending on sensing quality, sweat pores may be detectable.

All these features contribute in establishing fingerprint individuality. Hence, the exact representation of a biometric identifier such as fingerprints is important because only then can individuality issues be objectively studied. Note that sensing capabilities will directly affect a biometric's individuality; consider, for example, the detection of sweat pores on a finger flow pattern. However, when latent prints from a crime scene are examined, little influence can be exerted on the fingerprint acquisition process.

Now what does biometric individuality objectively mean? Here is *one* formulation of individuality problem from Pankanti et al. [158]: Given a biometric sample, determine the probability of finding an arbitrary biometric sample from the target population sufficiently similar to it. Pankanti et al. define this as the probability of a false association: Given two fingerprints from two different fingers, determine the probability that they are "sufficiently" similar. In other words, what are the theoretical lower bounds on the False Accept and False Reject Rates, often called the *intrinsic error rates.* Most experts and automatic fingerprint identification systems (AFIS) declare that the fingerprints originate from the same source if they are "sufficiently" similar. However, how much similarity is enough depends on typical (within-class *and* between-class) variations observed for multiple impressions of two or more fingers.

Given a representation scheme and a similarity metric, there are two approaches to determining the individuality of the fingerprints: theoretical and empirical [158]. In a theoretical approach, one models all realistic phenomenon affecting between-class and within-class pattern variations. Given this similarity metric, one can then theoretically estimate the probability of a false association. In an empirical approach (or technology evaluation), as discussed in Chapter 7, *representative* samples of biometric identifiers are collected, and using a *typical* matcher, empirical accuracy estimates are obtained. The accuracy estimates provide an indication of the uniqueness of the biometric identifier with respect to *that* matcher.

Both approaches have their advantages and limitations. In the theoretical approach, the problems with realistic modeling of all the parameters of the biometric matcher may become unsurmountable. But a good theoretical model would be a quick method of gaining insights into the fundamental accuracy bounds of a biometric identifier; more typically, these models may be relatively less effective in predicting performance of a practical system. In the empirical approach, there are known problems (and costs) associated with the collection of the *representative* samples and with the possibly large numbers of samples that are required. Many of these issues are discussed in Chapters 5, 7, and 15. The empirical method is *the* only way of measuring effective performance of a real system, but it does not readily lead us into any fundamental understanding of the underlying system design issues.

So far, only fingerprints have been theoretically investigated for individuality analysis. Below, we first briefly outline empirical individuality studies carried out for iris and handwriting biometrics.

14.2 Empirical individuality studies

In his iris person identification studies [47], Daugman proposed degrees of freedom of iris mismatch score distribution as a measure of individuality, or uniqueness of an iris pattern. Daugman studied iris individuality based on "iriscode" representation (see Figure 14.2). The iriscode is a 256–byte binary code representing an iris and the Hamming distance is used as the distance metric for comparing iriscodes. Daugman assumed that a chance match of independent bits of iriscode follows a binomial distribution and he then used the empirical distribution of the mismatch scores of the iriscode to estimate the number of independent bits in 256 byte iriscode to be 173 bits. This study used a sample size of 1,800 iris images from 323 persons. Daugman's analysis estimates that the probability of finding two arbitrary iriscodes that are sufficiently similar is $2^{-173} \approx 10^{-52}$. This effectively implies that the probability of a chance match of two unrelated iriscodes is 10^{-52}. The issues around these intrinsic error rates are very controversial; for example, there is much discussion about how precisely one can measure this problem a little.

Figure 14.2: The bit strength of the iris representation is defined by using the Hamming distance between iriscodes (courtesy of J. Daugman, University of Cambridge, UK).

Recently, Srihari et al. [207] published results on the individuality information contained in natural English handwriting. Their study involves 1,000 subjects representing the United States population demographics. Each subject provided three samples of handwriting by carefully copying a letter containing 636 words on a white piece of paper. Each document was then digitized; and the document, paragraph, word, and character level features were extracted from each sample. The features consisted of more than two dozen types, including gray level entropy of the foreground, thresholds used for segmentation, number of foreground pixels, lengths of words, numbers of interior/exterior contours, frequencies of different slants, average character heights, slants, and aspect ratios. Using these features, they compared the documents using an automatic document understanding system. Based on these handwriting samples, the individuality information was estimated for two different modes based on the matching accuracies.

It was determined by Srihari et al. [207] that writers of any two documents can be discriminated based solely on document level features with 95 percent accuracy on average. This means that the handwriting biometric has a False Accept Rate of about 5 percent. The probability of correctly identifying a writer from a database of 1,000 writers (i.e., identifying a single writer from 1,000 subjects) was estimated to be on average 80 percent.

The rest of this chapter is devoted to theoretical individuality models.

14.3 A partial iris model

The iris biometric [46,47] is promising yet mysterious and not well understood. What is clear is that an enrolled reference iris is represented by a unique N-bit iriscode.

The problem is simple: An iris is photographed, processed and represented as a reference N-bit iriscode R; an N-bit query bit string Q is similarly extracted from an unknown iris. The Hamming distance $h(Q, R)$ is used to identify *one* particular iriscode Q from among many with great accuracy. That is, a small Hamming distance h, meaning a large number $(N-h)$ of agreeing bits, between iriscodes Q and R, indicates strong evidence that real world irises $\mathcal{R}$ and $\mathcal{Q}$ are from the same individual. Extremely high accuracy numbers are claimed for iris, but what is really going on?

Let $\mathcal{Q}$ and $\mathcal{R}$ be the real world irises. We have that $Q = Q(\mathcal{Q})$ and $R = R(\mathcal{R})$, both Q and R are iriscodes, binary strings "01000100101" of length N. In terms of hypothesis testing, the hypotheses are

$$\begin{array}{lll} H_o: & \mathcal{Q} \equiv \mathcal{R}, & Q \text{ and } R \text{ are from the same iris;} \\ H_a: & \mathcal{Q} \not\equiv \mathcal{R}, & Q \text{ and } R \text{ are from different irises.} \end{array}$$

Let $h \in \{0, ..., N\}$ be the Hamming distance between the iriscodes and d_T a decision threshold. When distance $h(Q, R)$ is smaller than d_T, $\mathcal{Q}$ and $\mathcal{R}$ are likely to be the same; when $h(Q, R) > d_T$, $\mathcal{Q}$ and $\mathcal{R}$ are likely to be the different. This gives the following expressions for the FRR and FAR:

$$\begin{array}{lllll} \text{FRR}(d_T) & = & Prob\,(h > d_T | H_o) & = & Prob\,(h > d_T | \mathcal{Q} \equiv \mathcal{R}) \\ \text{FAR}(d_T) & = & Prob\,(h \leq d_T | H_a) & = & Prob\,(h \leq d_T | \mathcal{Q} \not\equiv \mathcal{R}). \end{array} \tag{14.1}$$

Can we estimate these error rates, just like the complex model for a fingerprint above? Can we develop an iris model and express *both* likelihoods in (14.1)? To accomplish this, just as in Section 14.4.3 for the finger biometric, there are three requirements:

1. A mathematical representation of the biometric identifier.
2. A measure of similarity between two biometric identifiers, or some measure of dissimilarity, like a distance metric, to compare representations.
3. Some formulation for the probability distribution of the iris population (or their representative samples), in particular, the distribution of the unenrolled irises.

The former two we have: iriscode and the Hamming distance between two codes. The latter is needed to express the probability of a False Reject FRR(d_T) and the probability of a False Accept FAR(d_T).

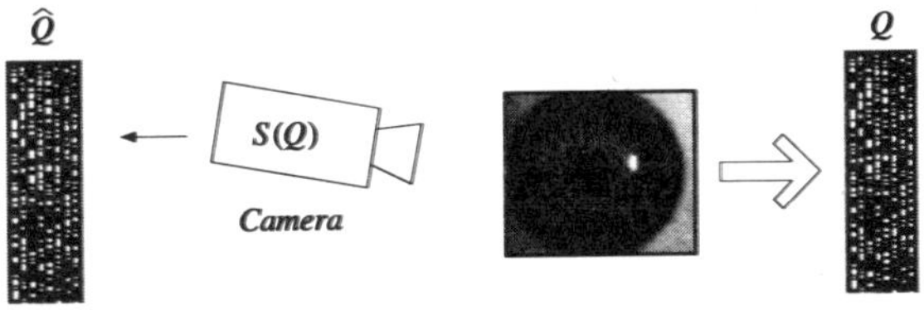

Figure 14.3: The subject's iris has a true, noise-free representation Q (right). It is assumed that this can be somehow extracted and used as the reference iriscode R. In the authentication system the camera only forms an approximation $\hat{Q}$ (left) to this representation.

Figure 14.3 is a simplified view of the iris imaging and iris recognition problem. Using a camera, the "true" underlying bit string Q is sensed, giving an estimated bit string $\hat{Q} = S(Q)$. The probability p of a bit flip is the probability that a bit read error occurs, we assume that the probability of missing a "1" and the probability of falsely detecting a "1" are the same. There is also a reference iriscode R, which was measured from iris $\mathcal{R}$ on which we impose no restriction. Essentially,

$$\hat{Q} = Q + noise.$$

14.3.1 FAR modeling

Let us first focus on False Accepts. Since the definition of an iriscode match is based on the Hamming distance and threshold d_T, the probability of a False Accept is

$$\text{FAR}(d_T) = Prob\,(h(\hat{Q}, R) \leq d_T),$$

where Q and R are non-matching iriscodes. We have to determine the probability that enough bits flip in the sensing process $\hat{Q} = S(Q)$ so that the resulting Hamming distance $h(\hat{Q}, R)$ is small. This distance is discrete and denoted as $h(Q, R) = n \geq 0$, where n can be 0, ..., N and n is the number of *disagreeing* bits between iriscodes Q and R. Hence, we can rewrite the above expression as a summation:

$$\text{FAR}(d_T) = \sum_{n=1}^{N} Prob\,[h(\hat{Q}, R) \leq d_T \,|\, h(Q, R) = n] \times Prob\,[h(Q, R) = n]$$

$$= \sum_{n=1}^{N} P_n(d_T) \times G_n. \qquad (14.2)$$

We have broken the complete formula into two components for convenience. The factor G_n is the probability of generating bit strings Q and R of length N that disagree in n bits, i.e., with $h(Q,R) = n$. The factor $P_n(d_T)$ is the probability that $h(\hat{Q}, R) \leq d_T$ simply because of bit read errors when estimating $\hat{Q}$ from the true Q that disagrees from the true Q in n bits. The summation gives the probability of False Accept, the probability that some unknown identity with true iriscode Q is accepted as the subject associated with R simply because of bit read errors in Figure 14.3.

Assume that we have a fixed $h(Q,R) = n$; i.e., there are $(N-n)$ bits that are the same in both code R and code Q. We need the probability that enough bits flip by chance in the sensing process $\hat{Q} = S(Q)$ so that $h(\hat{Q}, R) \leq d_T$; i.e.,

$$P_n(d_T) = Prob\,(h(\hat{Q}, R) \leq d_T \,|\, h(Q,R) = n).$$

In other words, we need to figure what the chances are that a False Accept is generated when estimating $\hat{Q}$ from Q.

A Hamming distance of $h(Q,R) = n$ means there are *exactly* $(N-n)$ matching bits between codes Q and R. Let us assume that i and j bits flip among non-matching (n) and matching $(N-n)$ bits, respectively. So in order to get a False Accept,

$$(n + j - i) \leq d_T.$$

The probability of i bits flipping in the non-matching n bits is

$$\binom{n}{i} p^i (1-p)^{n-i}. \qquad (14.3)$$

The probability of j bits flipping in the matching $(N-n)$ bits is

$$\binom{N-n}{j} p^j (1-p)^{N-n-j}. \qquad (14.4)$$

It follows that $P_n(d_T)$ is

$$\sum_{(n+j-i) \leq d_T} \binom{N-n}{j} p^j (1-p)^{N-n-j} \times \binom{n}{i} p^i (1-p)^{n-i}. \qquad (14.5)$$

We further need to determine $Prob\,(h(Q,R) = n) = G(n)$ in Expression (14.2). This is the probability that n bits between Q and R do not agree, or equivalently, the probability that $(N-n)$ bits do agree. Let the probability that an individual bit agrees be g, and the probability that an individual bit does not be $(1-g)$. The probability that $(N-n)$ bits agree and n bits disagree is

$$G_n = Prob\,(h(Q,R) = n) = \binom{N}{n} g^{N-n} (1-g)^n = \frac{1}{2^N} \binom{N}{n}. \qquad (14.6)$$

When we assume that $g = 1/2$, the product of probabilities simplifies to a fraction. What is left to do is to put (14.2), (14.5), and (14.6) together and we get

$$\text{FAR}(d_T) = \sum_{n=1}^{N} \left\{ \frac{1}{2^N} \binom{N}{n} \sum_{(n+j-i) \leq d_T} \left[\begin{array}{c} \binom{N-n}{j} p^j (1-p)^{N-n-j} \\ \times \binom{n}{i} p^i (1-p)^{n-i} \end{array} \right] \right\}. \tag{14.7}$$

14.3.2 FRR calculation

Let us now determine the False Reject Rate. For a False Reject, we need a large Hamming distance $h(\hat{Q}, R) > d_T$, while $h(Q, R) \leq d_T$. Therefore, the probability of a False reject is

$$\begin{aligned} \text{FRR}(d_T) &= Prob\,(h(\hat{Q}, R) > d_T \,|\, h(Q, R) = 0) \\ &= \sum_{i=d_T+1}^{N} Prob\,(h(\hat{Q}, R) = i \,|\, h(Q, R) = 0), \end{aligned}$$

which is the probability of i bits flip, with $i \in \{d_T + 1 \leq i \leq N\}$, when reading the true iriscode Q to obtain $\hat{Q}$. Assuming independence of the flip and non-flip events, this probability is

$$\text{FRR}(d_T) = \sum_{i=d_T+1}^{N} \binom{N}{i} p^i (1-p)^{N-i}, \tag{14.8}$$

where $(1 - p)$ is the probability of not flipping a bit.

14.3.3 Numeric evaluation

Usually, when dealing with iriscodes and Hamming distances, the relative distance $0 \leq h(Q, R)/N \leq 1$ is used. Consequently, a relative distance threshold $T = d_T/N$ is used in the decision rule. From here on, we will use this distance measure so that we can compare the above theoretical results to empirical results in the literature.

Given some N, we are now in a position to compute FAR(T) and FRR(T) as a function of $T \in [0, 1]$. We can compute the ROC of the "virtual" iris recognition system we have analyzed. We take $N = 173$, the number of independent bits in the iriscode, although 2,048 bits are in the surface representation. Daugman derived the 173 number from a fit of empirical data to a straight binomial model. Also, his system effectively compares seven rotated version of the input iriscode with each iriscode in the enrollment database. For this reason, we used $\text{FAR}_k(T) \approx k \times \text{FAR}(T)$, with $k = 7$ for these experiments.

Figure 14.4 shows the empirical ROC (dark arc) as reported on page 1,159 in [47], compared to the virtual ROCs for bit flip probabilities $p = 0.110$, $p = 0.150$, $p = 0.174$, $p = 0.200$, $p = 0.250$, and $p = 0.300$. Further, locii of constant threshold ($T = 0.250$, $T = 0.280$, $T = 0.300$, $T = 0.320$ and $T = 0.350$) are overlaid on the ROCs.

First, our model predicts that the iris FAR performance is relatively stable and is not affected by p. From this illustration, it can be seen that system parameter of p for a practical

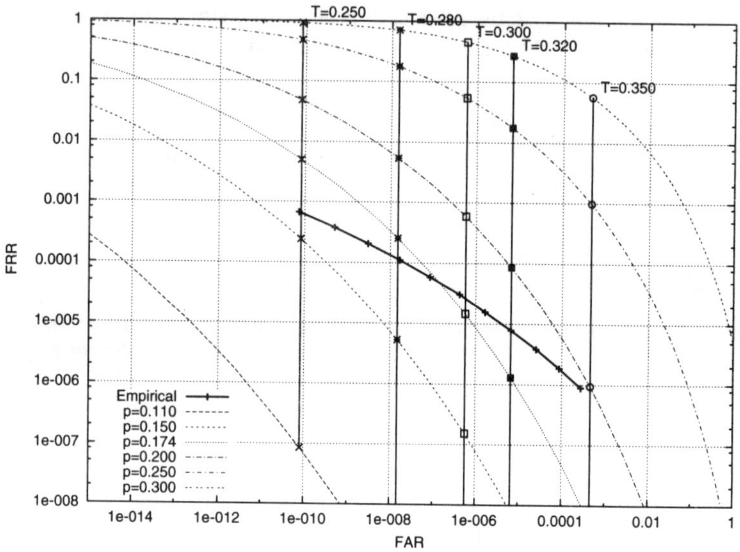

Figure 14.4: The empirical Receiver Operating Characteristic curve for iris recognition as reported by Daugman and theoretical ROCs for varying bit flip rates p, assuming $N = 173$ bits.

system cannot be reliably estimated from the empirical ROC alone. It can also be readily seen that FAR performance predicted by the foregoing analytical model is in excellent agreement with the empirical numbers reported by Daugman (the threshold lines closely match the values for his data).

Note, however, that the theoretical FRR accuracy performance degrades rapidly when the bit flip rate p increases. A surprising observation is that the reported empirical FRR performance degradation is significantly more stable with respect to the system threshold variation than predicted by the theory. This implies that the invariant bits in the iriscode representation are dramatically robust to the imaging noise.

This phenomenon is further illustrated in Figure 14.5. Here we find that the data in this region fit an exponential very well, namely,

$$\text{FRR}(T) \approx (1.0235)^{-T}$$

One way of explaining the empirical results is that perhaps not all bits are equally likely to flip, that there are some particularly "fragile" bits. Suppose the 173 bits are actually coefficients governing the decomposition of a particular iris into a basis set of "eigen-irises." You could then code each coefficient using 8 bits (or perhaps less for the high-order basis vectors). When adding noise to the input it is plausible that the low-order bits would be more likely to change than higher-order bits. It may also be the case that the bits for the higher-order basis vectors might be more susceptible because they in general account for smaller amounts of residual energy. Consequently, the number of bits flips in not linearly

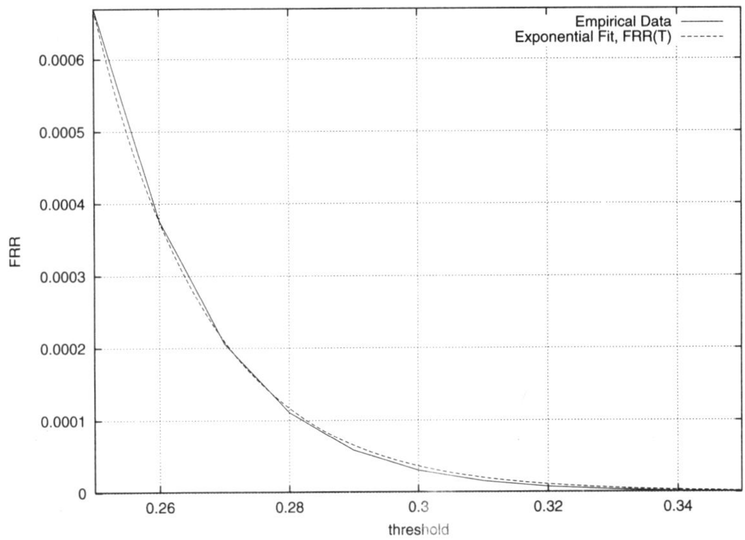

Figure 14.5: In the empirical results reported by Daugman, the FRR error rates are exponentially related to the decision threshold, as suggested by this exercise in curve fitting.

related to p but rather governed by some other function such as

$$p_{eff} = \frac{\log(kp+1)}{\log(2k+1)}.$$

Note that such a bias in bit flipping mechanism would not affect FAR performance (still dependent on normal p), but the FRR performance would be shifted based on this non-linear mapping.

As iris based identification is attempted in a wider range of applications, we will attain a better understanding of the issues related to the reliable large-scale identification and how to achieve the requisite feature invariance. For now, the iris individuality model, as formulated here, can fully explain the FAR performance, while a better model is still needed to explain the robust FRR performance.

However, the iris recognition problem is harder: it amounts to having to read the N bits as "wrapped" around a virtual pupil.

14.4 Fingerprint individuality

The two fundamental premises on which fingerprint identification is based are—

1. fingerprint details are permanent; and

2. fingerprints of an individual are unique.

The validity of the first premise has been established by empirical observations and is based on the anatomy and morphogenesis of friction ridge skin. It is the second premise that is being challenged in recent court cases. The notion of fingerprint individuality has been widely accepted based on manual inspection (by experts) of millions of fingerprints. However, the underlying scientific basis of fingerprint individuality has not been rigorously studied or tested. In March 2000, the US Department of Justice recognized that no such testing has been done and acknowledged its need [218].

In this section, we summarize two initial such studies. Both these studies use models of fingerprints constructed from minutiae; this, as seen above, has been demonstrated to be relatively stable, and it has been adopted by most of the Automatic Fingerprint Matching Systems (AFIS). The models define *a priori* both the representation of a fingerprint (template) and a metric for the similarity of two templates.

14.4.1 A simple model

To compute individuality, in [181] an estimate of the likelihood of randomly generating a minutiae template that matches the template of a fingerprint of an enrolled person is established. The enrolled, or reference, fingerprint has R minutiae; such a print is shown in Figure 14.6. Each minutiae has w possible directions and one of K possible locations. The number K is found by dividing the area of the overall image by the area of the "tolerance box" used to determine whether two minutiae match.

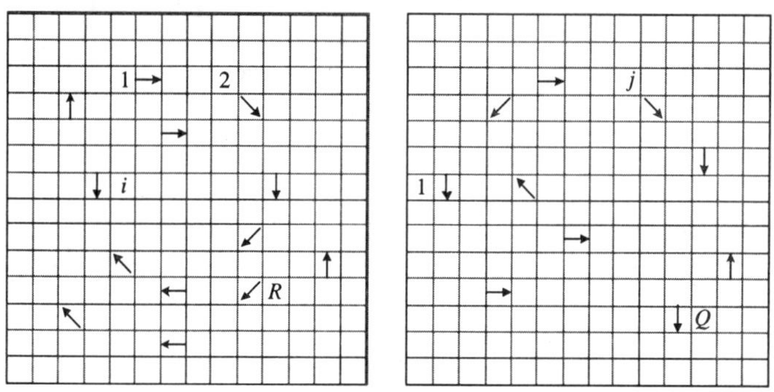

Figure 14.6: The enrolled template on the left with R minutiae is attacked by randomly generating templates with Q minutiae (right).

As a first step we need to estimate the probability that a randomly generated minutia will match one of the R minutiae in the reference print. This probability is approximated as

$$Prob\,(1) = \frac{R}{K\,w},$$

the probability that a first generated query minutia will match in both location and direction. When generating these random minutiae in Figure 14.6 it is not desirable to generate two minutiae at the same site. So after $(j-1)$ minutiae have been generated, the probability

$Prob\,(j)$ that the jth minutia will match could instead be as high as

$$Prob\,(j) \leq \frac{R}{(K-j+1)\,w}.$$

So to be conservative while generating Q random minutiae as in Figure 14.6 we assume each minutia has matching probability

$$P = \max_j Prob(j) = Prob\,(Q) = \frac{R}{(K-Q+1)\,w}. \tag{14.9}$$

This can break down for small K because the minutiae matching probability changes depending on how many minutiae have already been generated as well as on how many of those minutiae have matched. Also, the numerator might grow successively smaller than R if only one query minutia is allowed to match to each reference minutia.

Still, assuming that some constant value can be assigned to P of (14.9) is not unreasonable. Using this, we can derive that the chance of matching exactly t of the Q minutiae is

$$P_t = P^t\,(1-P)^{Q-t}.$$

There are a number of ways of selecting which t out of the Q minutiae in the query print are the ones that match. Thus the total match probability becomes

$$P_T(t) = \binom{Q}{t} P^t\,(1-P)^{Q-t}.$$

Matches of m or more minutiae typically count as a verification, so we get

$$Prob\,(K,R,Q,m) = \sum_{t=m}^{Q} \binom{Q}{t} P^t\,(1-P)^{Q-t}.$$

For convenience, let us introduce N and set $N = Q = R$, so this equation can be rewritten as

$$Prob\,(K,N,m) = \sum_{t=m}^{N} \binom{N}{t} P^t\,(1-P)^{N-t},$$

and since P is fairly small in our case, we can use a Poisson approximation to this binomial probability density function, and we get

$$Prob\,(K,N,m) = \sum_{t=m}^{N} \frac{(NP)^t\,e^{-NP}}{t!}.$$

This summation is usually heavily dominated by its first term; neglecting all but the first term results in

$$Prob\,(K,N,m) = \frac{(NP)^m\,e^{-NP}}{m!}. \tag{14.10}$$

Now because m is moderately large in practice (> 10), we can use Stirling's approximation and approximate (14.10) as

$$Prob\,(K,N,m) = \frac{(NP)^m\,e^{-NP}}{\sqrt{2\pi\,m}\;e^{-m}\,m^m}.$$

This equation can be further rewritten to emphasize its exponential nature

$$Prob\,(K, N, m) = \left(\frac{eNP}{m}\right)^{m} \frac{e^{-NP}}{\sqrt{2\pi\, m}}. \tag{14.11}$$

This is the probability of erroneously declaring a match because m minutiae between reference and query template match by chance; i.e., it is an estimate of the FAR.

Examining this expression for $Prob\,(K, N, m)$, we make several observations:

1. It can be seen that if we have other local characteristics that can be associated with minutiae, the probability of a successful chance match is much lower since w is larger and P of Equation (14.9) is lower.

2. If the extent of spatial domain is increased or the resolution is increased, K in Equation (14.9) increases so P is smaller. As in the above case, the strength of a fingerprint representation (template) also increases.

3. There is a strong dependence of (14.11) on N, the overall number of minutiae in the fingerprints. The probability of mismatching two prints based on mismatching m minutiae increases with N (remember, we assume $N = R = Q$).

4. When K, the number of possible minutiae locations, is fixed, the number N of minutiae that can be detected reliably is bounded $N \ll K$, and depends on the noise. Expression (14.11) is bounded from above, and is an estimate of the intrinsic strength of a fingerprint, under the assumptions (e.g., noise) of the model.

For best security, this means that N needs to be kept as low as possible, spurious minutiae from poor fingerprint images are particularly detrimental.

14.4.2 Probabilistic scoring

An interesting use for Equation (14.11) is as the basis for a symmetric scoring function. How should one measure the degree of match between two prints? One could simply count the number of minutiae that match, $S(m, Q, R) = m$, and use this as the match score (as in Table 14.1). However, matching 10 minutiae from a query print with 40 minutiae seems more meaningful than matching 10 from a print with 100. To counteract this, some have suggested weighting the score by the average number of minutiae in the reference and query prints, $S(m, Q, R) = m/(Q + R)$.

Perhaps a more cogent proposal is to construct a scoring function based on the relative "surprise" associated with matching a certain number of minutiae (as opposed to achieving this by chance alone). A standard way of measuring such unexpectedness is to take $1/Prob(event)$. Applying this to Equation (14.11) and then using the logarithm to compress the range of scores, we get

$$\begin{aligned} S(m, Q, R) &= -logP(K, N, m) \\ &= -mlog(\tfrac{eNp}{m}) + Np + \tfrac{1}{2}log(m) + \tfrac{1}{2}log(2\pi). \end{aligned} \tag{14.12}$$

To handle the normal case where R and Q are different, let us replace N with Q in the above expression, and use the approximation $p = \frac{R}{Kd}$ (from before). Also, since we are only ranking matches, the last (constant) term can be omitted. Doing this, we find

$$S(m,Q,R) = mlog(\frac{mKd}{eQR}) + \frac{QR}{Kd} - \frac{1}{2}log(m) \tag{14.13}$$

For most cases this expression is heavily dominated by its first term. Expanding just this term, regrouping, and multiplying through by $w = 1/log(\frac{e}{Kd})$ yields the final result:

$$S(m,Q,R) = m \times [1 - w\, log(m) + 2w\, log(\frac{m}{\sqrt{QR}})] \tag{14.14}$$

Overall the score is largely dependent on m, as expected, but deviates from linear somewhat as m gets large (the $log(m)$ term). Notice, more importantly, that S also depends on the ratio of the number of minutiae matched to the *geometric mean* of the number of total minutiae in each print. As mentioned before, just having a large number of minutiae match is not a sure indicator of identity. The scoring function here takes this into account in at least a semi-principled way.

14.4.3 A more complex model

Pankanti et al. [158] develop another fingerprint individuality model. As in Section 14.4.1, a fingerprint is modeled in terms of minutiae because of forensic tradition, though the analyses can be extended to other fingerprint features. Also, as in Section 14.4.1, uniform distributions of minutiae location and minutiae angle are assumed. More about that later.

Given a query fingerprint containing Q minutiae, the goal again is to compute the probability that any arbitrary reference fingerprint matches by chance. Each reference fingerprint is assumed to have R minutiae, and a match is declared if print $\mathcal{Q}$ has m or more corresponding (matching) minutiae with the print $\mathcal{R}$. That is, just like in the previous section, Pankanti et al. [158] develop a model to compute the intrinsic fingerprint error probabilities.

Fingerprint minutiae are defined by their location, (x_c, y_c), and the angle θ of the ridge on which they reside. The query and the reference fingerprints are defined as, $\mathcal{Q}$ and $\mathcal{R}$, respectively, with:

$$\begin{aligned} \mathcal{Q} &= \{(x_1, y_1, \theta_1), (x_2, y_2, \theta_2), ..., (x_Q, y_Q, \theta_Q)\}, \text{ and} \\ \mathcal{R} &= \{(x'_1, y'_1, \theta_1), (x'_2, y'_2, \theta'_2), ..., (x'_R, y'_R, \theta'_R)\}. \end{aligned}$$

Treat the minutiae coordinates (x_j, y_j) and (x'_i, y'_i) as if expressed in pixel locations, for the moment. A minutia j in the query fingerprint is considered as "corresponding" or "matching" to the minutia i in the reference template, if and only if

$$\begin{aligned} &\sqrt{(x'_i - x_j)^2 + (y'_i - y_j)^2} \leq r_0; \text{ and,} \\ &\min(|\theta'_i - \theta_j|, 360 \quad |\theta'_i \quad \theta_j|) \leq \theta_0/2, \end{aligned} \tag{14.15}$$

where r_0 is the tolerance distance and θ_0 is the tolerance angle, as shown in Figure 14.7. Note that Expression (14.15) implies that the minutiae are distributed according to a multivariate Gaussian in the half-sphere.

The model first concentrates on only the probabilities of spatial matching of minutiae. The total area of overlap between the query fingerprint and the reference fingerprint are,

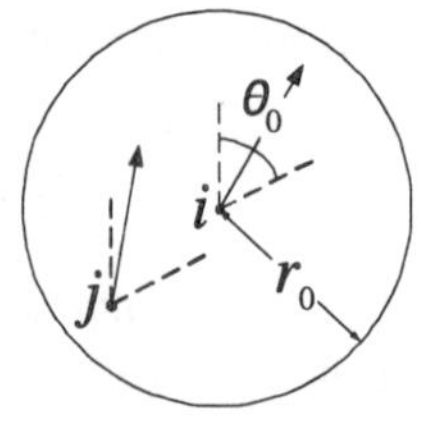

Figure 14.7: Minutia j in $\mathcal{Q}$ matches minutia i in $\mathcal{R}$ if the locations are within r_0 and the angles are within θ_0.

after somehow aligning the point patterns $\mathcal{R}$ and $\mathcal{Q}$, of an area of size A (in pixel^2 or # of pixels). The probability that there are exactly ρ minutiae correspondences based on *only the locations* of the Q query minutiae and R reference minutiae is given by

$$Prob\,(A, C, R, Q, \rho) = \binom{Q}{\rho}$$

$$\times \underbrace{\left(\frac{RC}{A}\right)\left(\frac{(R-1)C}{A-C}\right)\cdots\left(\frac{(R-\rho-1)C}{A-(\rho-1)C}\right)}_{\rho \text{ terms}}$$

$$\times \underbrace{\left(\frac{A-RC}{A-\rho C}\right)\left(\frac{A-(R-1)C}{A-(\rho+1)C}\right)\cdots\left(\frac{(A-(R-(Q-\rho+1))C}{A-(Q-1)C}\right)}_{Q-\rho \text{ terms}},$$

where $C = \pi\, r_0^2$, which is the circular tolerance region as in Figure 14.7.

1. The first component of this expression is simply the number of ways ρ minutiae can be chosen from Q query minutiae,

2. The second component, a product of ρ terms, denotes the probability of matching ρ minutiae between the reference template and the query template. The first term is the probability of matching the first query minutia to one of R reference minutiae; the second term is the probability of matching the second query minutia to the remaining $(R-1)$ minutiae of the reference minutiae $\mathcal{R}$; and so on.

   These probabilities are calculated as the fractions of fingerprint areas that can contain minutiae and the leftover overlapping area $(A - (i-1)C)$ of the query and the reference print.

3. The third component is a product of $(Q-\rho)$ terms expressing the probability that $(Q-\rho)$ minutiae in the query template do not match any remaining unmatched minutiae in the reference fingerprint. The first term denotes the probability that the $(\rho{+}1)$-st query minutia *does not match* any of the so far $(R{-}\rho)$ unmatched reference minutiae. This is estimated as $(A - RC)$, the left-over area with minutiae in $\mathcal{Q}$ divided by $(A-\rho C)$, the left-over area in A, $(A-\rho C)$ without minutiae. Here, we have $A = A \cap \mathcal{R}$ if one is willing to view $\mathcal{R}$ as an area, defined by filling an area with circles of radius r_0. The rest of the terms in this component are formed similarly.

Now let $K = A/C$, the number of possible minutiae locations similar to the K defined in Section 14.4.1, and assume K is an integer (i.e., $A \gg C$). Dividing the numerator and denominator of each term in the above expression by C, and substituting $K = A/C$, gives

$$Prob\,(K, R, Q, \rho) = \binom{K}{\rho}\binom{K-R}{Q-\rho} / \binom{K}{Q}, \tag{14.16}$$

which is a hyper-geometric distribution [81]. This distribution relates to sampling without replacement using the hyper geometric law, where elements successively removed affect the probability of the next sample.[1] Expression (14.16) contains the terms—

1. the number of ways ρ minutiae can be placed in a grid of K minutiae locations;
2. the number of ways $(Q-\rho)$ the query minutiae set can be placed in $(K-R)$ locations, the query minutiae that are *not* matched;
3. the number of ways the Q query minutiae can be placed in K locations, the query minutiae that *are* matched.

This balances out, if you will, the effects of differences between the false minutia detection and spurious minutiae generation models of the two sensors used for acquiring $\mathcal{Q}$, the query print, and for acquiring $\mathcal{R}$, the reference print.

Once ρ minutiae positions are matched, the probability that $t \leq \rho$ minutiae among these also have matching directions (as defined in Figure 14.7 and Expression (14.15)) is given by

$$\binom{\rho}{t}\left(\frac{1}{L}\right)^{t}\left(\frac{L-1}{L}\right)^{\rho-t},$$

where L is such that $Prob\,\left(\min\left(\left|\theta_i - \theta_j'\right|, 360 - \left|\theta_i - \theta_j'\right|\right) \leq \theta_0\right) = 1/L$. Here L is related to the number of minutiae directions w in the model of Section 14.4.1. It is a new model variable L, determined by a matching tolerance angle θ_0 in addition to already using a tolerance distance d_0. The probability of matching m minutiae in *both* location as well as direction, using this tolerance angle θ_0, is given by

$$Prob\,(K, R, Q, m) =$$

$$\sum_{\rho=m}^{\min(R,\,Q)} \left(\frac{\binom{R}{\rho}\binom{K-R}{Q-\rho}}{\binom{K}{Q}} \times \binom{\rho}{m}\left(\frac{1}{L}\right)^{m}\left(\frac{L-1}{L}\right)^{\rho-m} \right), \tag{14.17}$$

which is a weighed sum of hyper-geometric distributions and is *a theoretical estimate of the FAR.*

Just as from the probability (14.11) of false association derived by the simple model, there are some conclusions:

1. The probability of a chance match is lower if L is larger. The more features per minutia, the better.

[1] These type of distributions are used for modeling things like software fault tolerance and pre-operational software reliability, e.g., [188].

2. This probability is lower if K, a measure of resolution, is increased.
3. Because (14.17) is a sum from $\rho = m, ..., \min(R, Q)$, the probability of mismatching two prints increases with R and Q.

Because of the balancing effects of the hyper-geometric distribution—

4. A small number Q and a large number R will give lots of False Match possibilities, but (keeping A constant) that can only happen if print $\mathcal{R}$ has a higher spurious minutiae probability than $\mathcal{Q}$, which is probably *not that* true if similar sensors are used for reference and query print acquisition. So R and Q have to increase in tandem. If Q and R increase because of noise, they increase at the same rate because for both prints the probability of detecting spurious minutiae goes up faster than the probability of missing a minutia. This means that the probability of false association goes up fast when the noise in feature detection increases.

   What is clear, is that we have some theoretical upper bound on the probability of False Match: the only way to decrease the probability of (14.17) is to increase Q and R while not increasing minutiae detection error rates. A possible way to achieve this is by matching triangles of minutiae as described by Germain et al. [75,76].

5. This model allows one to work with different numbers, say, $Q < R$, which is typically the case when $\mathcal{Q}$ is a latent print (see Section 14.6).

Note that adding extra features to the minutiae only makes sense when Q and R can be increased or when the probability of spurious minutiae and/or missing minutiae can be decreased. Increasing the resolution K only makes sense under the same conditions.

14.4.4 Model comparison

Let us compare the models of Section 14.4.1 and of this section. Typical fingerprint numbers, say, scanned at 500 dpi and of size 0.75" × 0.6" (metrically: 200 pixels/cm and 1.91 cm × 1.52 cm) are $K = 450$ and $N = Q = R = 40$.

If $m = 25$ minutiae are required to match to declare that two prints are the same, we get—

1. For the first model, using (14.10),

$$Prob\,(K, N, m) = Prob\,(450, 40, 25) = 2.24 \times 10^{-26} \approx 2^{-85};$$

   and hence there are roughly 85 bits of information content in this representation. This is equivalent to a nonsense password like "m4yus78xpmks3bc9" made up of 16 characters if one character has 5.17 bits of information (only taking the lowercase letters and the digits).

2. On the other hand, for the second model, using (14.17),

$$Prob\,(K, R, Q, m) = Prob\,(450, 40, 40, 25) = 2.5 \times 10^{-19} \approx 2^{-62}$$

   and there are approximately 62 bits of information in the second representation, which is equivalent to a random password of 12 characters.

Hence, the more complex model gives an estimate of the probability of a False Match that is quite a bit higher; consequently, the "bit strength" (or key strength) is lower. This is not really a reason to be alarmed since the probability 2.5×10^{-19} is still consistent with the fact that never have two fingers with the same minutiae structure been found: a probability of 10^{-19} represents a probability of $1{:}10^{19}$. What *is important* to note is that apparently when features are quantized, and hence noise is introduced, it is important to model the accidental matching of spurious minutiae and *not* just the matching of correct minutiae. These mismatch probabilities can be used for cohort modeling as in Chapter 9.5.4 and for probabilistic enrollment as described in Chapter 9.5.2 by replacing $Prob\,(O|D)$, the likelihood of the data O, given the world model D.

14.4.5 Imposing structure

Minutiae patterns are generated by the underlying fingerprints that are smoothly flowing oriented textures, and are not random dot patterns. Put differently, because minutiae positions are determined by the ridges, the minutiae points are not spatially randomly distributed (see, e.g., Figure 14.1). This typically implies that the probability of finding sufficiently similar prints from two different fingers is higher than that of finding sufficiently similar sets of random (directed) uniform point patterns. By imposing fingerprint structural (e.g., ridge/valley position, ridge orientation) constraints on the random point configuration space it is possible to derive a more precise estimate of the probability of false association by more carefully enumerating possible fingerprint matches.

How do we impose this structure? The minutiae can lie only on ridges, i.e., along a curve of length $A/(2w)$, where w is the ridge width and $2w$ is the ridge plus valley width. Therefore, the value

$$K = \frac{A}{C} = \frac{A}{\pi\, r_0^2}, \tag{14.18}$$

should be replaced by

$$\kappa = \frac{A}{2w \times 2r_0}.$$

in Equation (14.17). This better expresses the fact that the minutiae cannot all lie in one small region of the print. The distance r_0 in (14.18) is the linear separation in minutiae location along a ridge, as shown in Figure 14.8. We purposely use a new symbol κ, since this number is not just related to matching tolerance but also to an additional fingerprint parameter w, the width of a ridge.

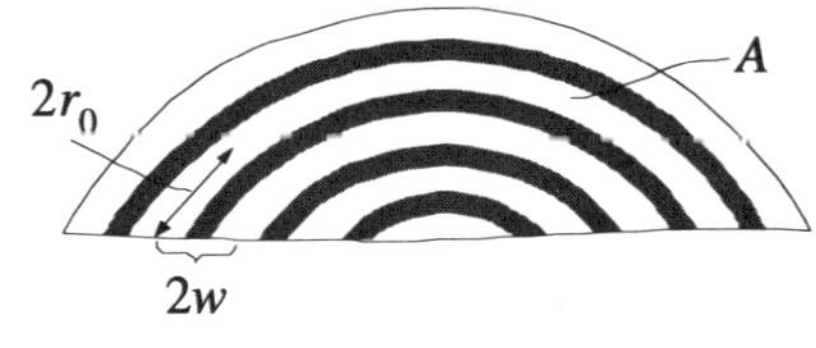

Figure 14.8: Dividing overlapping fingerprint region A by $(2w \times 2r_0)$ roughly gives the number κ of possible minutiae locations.

So we need additional fingerprint parameters such as an estimate of w. These are determined for fingerprint images complying to the FBI standard of 500 dpi:

1. The value of w is taken as reported by Stoney [211] and set to 9.1 pixels (0.018 ") for 500 dpi images.

2. The tolerances r_0 and θ_0 that account for 97.5 percent variation in minutiae location and angle in different impressions of the same finger are determined from a database of 900 fingerprint images where the ground truth of minutiae correspondences are marked by an expert and found to be 15 pixels (0.03 ") and 22.5 degrees.

3. It is further determined in [158] that

$$Prob\ (\min(|\theta - \theta'|, 360 - |\theta - \theta'|) \leq 22.5°) = 0.267$$

for matching different fingers using an automatic fingerprint matching system [95], i.e., $L = 1/0.267 = 3.75$. (Hence, the above assumption $w = 4$ is not an unreasonable one.)

14.4.6 Minutiae distributions

Figure 14.9 examines the assumptions of minutiae uniformity. Figure 14.9a shows the hand-marked minutiae of 89 clean-rolled fingerprints from the NIST 27 database [74] by plotting all 9656 minutiae and smoothing the resulting scatter plot. This plot looks relatively uniform lending credence to our assumption that each of the K possible sites (*within* the standard fingerprint area A) is equally likely. Figure 14.9b shows the scatter plot obtained if all 89 prints are first aligned based on the location of the "core." This plot is less uniform, with many minutiae found near the core (the rightmost prong) and in two branches to the left (roughly the locations of the "deltas"). In general, there are more minutiae near the center of the fingerprint.

It is also assumed in Section 14.4.1 that all w directions are equally likely. In the NIST 27 database each of the hand-marked minutiae also has an angle associated with it. To analyze positional dependence of the minutia angles, each minutia's unit vector is resolved into an x and y component and determined using core alignment. The components are then smoothed, in essence taking the vector sum of all the minutiae in a region. These vector sums are resolved into a magnitude (Figure 14.9c) and direction (Figure 14.9d). In Figure 14.9c areas with high magnitude (dark) show that there is a strong resultant and hence little scatter in the direction of the minutiae. In such areas, Figure 14.9d shows the common direction of minutiae. Note that in this figure strong white values, like 350° degrees, wrap around (e.g., near the middle) to become strong black values, like 10 ^{o}. Interpreting (c) and (d) together we see there are three areas of strong directionality: upward on the left, horizontal in the middle, and downward on the right. So, in general, there is a swirling pattern near the middle of a print.

Thus, the assumptions about K and d in Sections 14.4.1 and 14.4.3 are not exactly right. Locations near the center of a print are more likely to contain minutiae. Also the position of a minutiae relative to the center of the print predisposes it to more likely be one direction than another. What is for sure about these estimates of false association is that in reality the probabilities are somewhat higher and we are erring on the safe side.

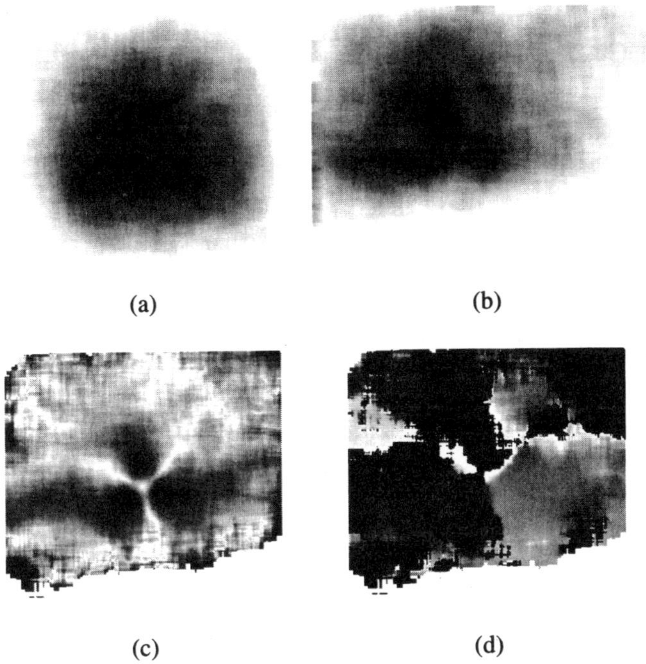

Figure 14.9: Minutia distribution (top) for a portion of the NIST 27 database. Degree of directedness and dominant angles (bottom).

14.5 Standards for legal evidence

Fingerprint identification is routinely used in forensic laboratories and identification units around the world [120] and has been accepted in courts of law for nearly a century [45]. Similarly, handwriting idiosyncrasies (e.g., from ransom notes) and signature forgery are often used in criminal investigation and expert judgements are passed with uncanny certainty. Forensic scientists have evolved elaborate rules for making identity comparisons based on biometric information and their testimony is attributed with the same significance as *prima facie* evidence. Consequently, the testimony of forensic experts has been routinely admitted in courts without much scrutiny and challenge.

However, in the 1993 case of Daubert versus Merrell Dow Pharmaceuticals, Inc., the Supreme Court ruled that the reliability of expert scientific testimony must be objectively established. Additionally, the court stated that when assessing reliability, the following five factors should be considered:

1. Whether the particular technique or methodology in question has been subject to a statistical hypothesis testing.

2. Whether its error rate has been established.

3. Whether the standards controlling the technique's operations exist and have been maintained.

4. Whether it has been peer reviewed, and published.

5. Whether it has a general widespread acceptance.

This is an important decision because it questions the assessments of the forensic experts who have undergone thorough rigorous training and have years of experience.

In sum, expert testimony on person identity based on forensic evidence such as fingerprints and writings has been challenged under *Daubert* [105], and consequently efforts have been made toward establishing "some number" that expresses the individuality of biometric identifiers. To explain these efforts, we have followed the papers by Ratha et al. [181] and Pankanti et al. [158]. These are about fingerprint individuality, but similar analyses for other biometrics may be done. In fact, in Section 14.3 we develop a model for iriscodes.

14.6 Expert fingerprint testimony

Previously we have modeled the probabilities of a chance match for fingerprint pairs that are of reasonable size and have a good number of matching minutiae. This might be the case for well-designed and well-operated fingerprint verification and identification systems, where an input (query) print is matched to enrolled (reference) prints. But what about matching a latent query print $\mathcal{Q}$, found at some crime scene, to some reference print $\mathcal{R}$?

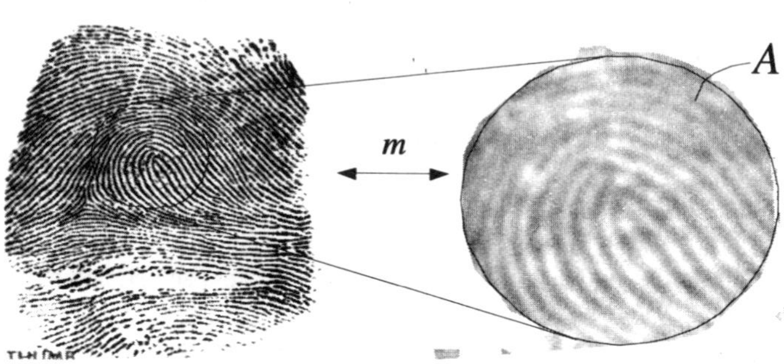

Figure 14.10: Reference print $\mathcal{R}$ on the left, query latent print $\mathcal{Q}$ on the right; in area of overlap A, there are R and Q minutiae, respectively.

The typical situation of matching a latent print with a reference print is shown in Figure 14.10. On the right we have the latent print, which is only a small portion (region A, the overlap) of the reference print. We have three phenomena that cooperate in increasing the probability of false association:

1. The number Q of query minutiae in the latent print $\mathcal{Q}$ is smaller than the number R of reference minutiae in reference print $\mathcal{R}$.

Country	Population	Minutiae
Italy	57.5M	16
Cyprus	0.8M	
Gibraltar	0.03M	
Malta	0.4M	14
Germany	82.0M	12
Turkey	66.0M	
France	59.0M	
Ukraine	49.0M	
Poland	39.0M	
Romania	22.5M	
Greece	11.0M	
Czech Republic	10.5M	
Belgium	10.0M	
Portugal	10.0M	
Sweden	9.0M	
Austria	8.0M	
Finland	5.0M	
Eire (Ireland)	3.9M	
Slovenia	2.0M	
Spain	40.5M	10
Netherlands	16.0M	
Hungary	10.2M	
Denmark	5.3M	
Bulgaria	8.0M	8

Table 14.1: Different numbers of minutiae are required for positive identification in various countries (from [36] and BBC Web site).

2. The print $\mathcal{Q}$ is of lesser quality and the minutiae false detection/false miss rates are higher.

3. The Q minutiae of print $\mathcal{Q}$ have larger localization errors.

Let us determine an upper bound on the probabilities or equivalently a lower bound on the chance, expressed by saying "more than a chance of *one in a* M." In theory the probability of a false conviction is bounded from above by the estimate of false association and, expressed in M, is bounded from below (assuming that a latent fingerprint is the *only* evidence). In these cases, the complex model is more appropriate because R and Q are different.

A match consisting of 12 minutiae points (*the 12-point rule*) is considered sufficient evidence in many courts of law (see Table 14.1). At the minimum $m = 12$ minutiae need to be matched between the prints in Figure 14.10 "beyond a reasonable doubt." But can we say something more specific about the probability of a false association using the 12-point rule?

Q \ m	8	9	10	11	**12**
12	2.26×10^{-7}	4.25×10^{-9}	4.13×10^{-11}	1.65×10^{-13}	$\mathbf{1.55\times10^{-16}}$
13	5.64×10^{-7}	1.34×10^{-8}	1.75×10^{-10}	1.06×10^{-12}	2.01×10^{-15}
14	1.26×10^{-6}	3.63×10^{-8}	5.99×10^{-10}	4.91×10^{-12}	1.41×10^{-14}
15	2.59×10^{-6}	8.77×10^{-8}	1.76×10^{-9}	1.82×10^{-11}	7.05×10^{-14}
16	4.96×10^{-6}	1.94×10^{-7}	4.58×10^{-9}	5.75×10^{-11}	2.82×10^{-13}

Table 14.2: The effects of the fingerprint expert misjudgments (missing, spurious minutiae) in using the 12-point rule.

The value of A was computed for 500 dpi fingerprint images from the minutiae density of 0.246 minutiae/mm^2 ($\approx$ 154 minutiae/inch^2) estimated by Kingston (cf. [211]) from 100 fingerprints; thus $R = 35$. Since latent prints are typically of very poor quality, it is possible that there could be an error in judgment of existence of minutiae in the latent or their possible match to the minutiae in the reference print. The effect of such misjudgments on the chance of false associations is rather dramatic.

Table 14.2 shows the effects of the fingerprint expert misjudgments when using this 12-point rule and the false association probability of Equation (14.17; the number of reference minutiae $R = 12$ for all entries. Let us examine a few of these entries:

- ($Q = 12$, $m = 12$): All R reference minutiae are matched and the chance of a false association is 1.55×10^{-16}. This means that the 12-point rule is wrong once in 0.65×10^{16}.

  Assuming that an expert can correctly glean all the minutiae in the query latent print, a 12-point match with the full-print template is an overwhelming amount of evidence, *provided* that there is no contradictory minutiae evidence in the overlapping area.

- ($Q = 16$, $m = 12$): When there are four more query minutiae, the probability of false association is 2.82×10^{-13}, i.e., 1,000 times more likely.

  The effect of ignoring the four extra query minutiae by explaining them as chance minutiae or false minutiae decreases the probability roughly by a factor 1,000.

- ($Q = 14$, $m = 8$): The probability that by chance $m = 8$ of the $R = 12$ reference minutiae match 8 of the $Q = 14$ query minutiae is 1.26×10^{-6}, which is around 1:1 million.

In general, erroneously pairing minutiae has significantly more impact than missing genuine minutiae in the query latent print. Note that, if applied well, the 12-point rule makes a lot of sense. It is around $m = 12$, that the probability of False Match is around 10^{-16}, around a chance of 1:10^{17}. This is, a magnitude that where for most people find hard to imagine, or understand, how small these probabilities really are.

14.7 Remarks

Realistic error estimates of automated matching technology are those found in properly designed and executed technology evaluations (see Chapter 7). All speculation, e.g., error rates determined using modeling methods such as the above, about intrinsic error rates should be carefully scrutinized. These studies can be overly optimistic about biometric individuality, either due to unrealistic assumptions or due to insufficient realistic data to estimate model parameters.

In other words, the probabilities of False Accepts of biometric matchers and *not* the biometric itself are best estimated using *bona fide* technology evaluations. Other estimates of biometric error rates are based on assumptions about biometric representations and distance measures between these representations. As we will see in Chapter 15 the path to proper design of such tests has plenty of pitfalls and stumbling blocks.

However, the theoretical models *can* establish upper bounds for the performance of a particular biometric. For instance, no matter how good your image acquisition and feature extraction procedures are, you will never get a FAR less than about 10^{-16} from a single finger or less than about 10^{-37} from an iris. While this holds for the representations we have analyzed and a typical operating point in terms of thresholds, it is always possible that some new method or representation might be developed that gives superior results.

15

System Errors Revisited

Earlier in this book (Chapter 5) we described the evaluation of biometric systems in terms of their error rates. But two important questions to consider when comparing these measures, principally the False Accept and False Reject Rates, is how accurately we know the error rates, and how much can they be trusted for making decisions.

In particular, in the report "Best practices in testing and reporting performance of biometric devices" [132], the authors note that it is important to relate the precision of the error estimates to the amount of data that is used. That is, given the quality and quantity of the data, what are the confidence intervals of our error estimates?

To estimate the error rates, it is useful first to examine what data one has to work with. Biometric data are collected from people (subjects) by measuring and digitizing the real-world biometrics, denoted by $\mathcal{B}$.

$\mathcal{B}_1$	$\mathcal{B}_2$	$\dots$	$\mathcal{B}_\mathcal{D}$
B_{11}	B_{21}	$\dots$	$B_{\mathcal{D}1}$
B_{12}	B_{22}	$\dots$	$B_{\mathcal{D}2}$
$\vdots$	$\vdots$	$\ddots$	$\vdots$
B_{1d}	B_{2d}	$\dots$	$B_{\mathcal{D}d}$

Table 15.1: A collection $\mathbf{B}_\mathcal{D}$ of biometric samples B_{ij} obtained from $\mathcal{D}$ biometrics, d samples per biometric.

Suppose we have a collection $\mathbf{B}_\mathcal{D}$ of biometric samples (Table 15.1) acquired from $\mathcal{D}$ biometrics (meaning, these are real-world biometrics, $\mathcal{B}_1, ..., \mathcal{B}_\mathcal{D}$) and d samples are acquired per biometric. In practice the number of samples per biometric can vary, but our notation (and the practicalities of evaluation) become more manageable if we make this simplification. The number $\mathcal{D}$ of biometrics may be larger than the number $\mathcal{P}$ of volunteers from whom the samples are collected, since people may have more than one of the particular biometric (e.g., finger). In any case, the database contains $d \times \mathcal{D}$ biometric samples (see Table 15.1) and given a *biometric match engine*, one can compute test score sets:

- a set of *genuine* (match) scores—

$$\mathbf{X} = \{X_1, X_2, ..., X_M\} = \{s(B_{kp}, B_{kq}) \ \forall \ k;\ p \neq q\};$$

- a set of *imposter* (mismatch) scores—

$$\mathbf{Y} = \{Y_1, Y_2, ..., Y_N\} = \{s(B_{ip}, B_{jq}) \ \forall \ i, j, p, q;\ i \neq j\}.$$

Matching pairs of samples taken from the same biometric in $\mathbf{B}_\mathcal{D}$ gives the $M = d\,(d-1)\,\mathcal{D}$ sample match scores (genuine scores) in set $\mathbf{X}$; matching samples in $\mathbf{B}_\mathcal{D}$ from different biometrics gives the $N = d^2\,\mathcal{D}\,(\mathcal{D}-1)$ mismatch (imposter) scores in set $\mathbf{Y}$. We do not here assume that the matcher is symmetric, and we include both $s(B_{ip}, B_{jq})$ and $s(B_{pq}, B_{ip})$ in $\mathbf{X}$ and in $\mathbf{Y}$.[1]

15.1 Estimating the match score mean

A matcher is completely specified by the match distribution $F(x)$ and the mismatch distribution $G(x)$, so we are interested in estimating characteristics of these distributions to determine the quality of the biometric matcher. These characteristics can be estimated from the match scores $\mathbf{X}$ and mismatch scores $\mathbf{Y}$. How *accurate* these estimates are will depend on the size of the collection $\mathbf{B}_\mathcal{D}$, i.e., the number of samples $d \times \mathcal{D}$, but also on the quality of the data. If one wants to be able say something about the accuracy or the *significance* of the error estimates, this is expressed in terms of *statistical confidence intervals*.

In this chapter we use fingerprint databases of samples (as in Table 15.1) to illustrate the subtleties of estimating False Reject and False Accept error rates from sets of match and non-match scores. As an example, we first concentrate on estimating a particular statistic $\theta = E(x)$ of $F(x)$, the expectation of x. Later we will extend this to other statistics of F, for example, the error rates as a function of threshold T.

Given $\mathbf{X}$, the set of match scores, an (unbiased) estimate $\hat{\theta}$ of the statistic $\theta = E(x)$ is the sample mean

$$\overline{X} = \frac{1}{M} \sum_{i=1}^{M} X_i. \tag{15.1}$$

The mean can only be estimated statistically because we only have *one finite* set of M match scores $\mathbf{X} = \{X_1, X_2, ..., X_M\}$. However, an estimate of the mean by itself might be inaccurate — we need to know how good this estimate of the mean is. We need to know more about the error in our estimate $\hat{\theta} - \theta$, or in the case of estimating the mean, more about $\overline{X} - E(x)$.

To be more precise, we would like to construct a confidence interval for the true $E(x)$ based on the estimate $\overline{X}$. It cannot be guaranteed that $E(x)$ lies with 100 % certainty within this confidence interval; it can only be guaranteed that $E(x)$ lies within the confidence interval with a chance of $(1 - \alpha)\,100\%$. A confidence for $E(x)$ interval takes on the form

$$E(x) \in [\overline{X} - \Delta X_1(\alpha, \mathbf{X}), \overline{X} - \Delta X_2(\alpha, \mathbf{X})], \tag{15.2}$$

where ΔX_1 and ΔX_2 are functions of α and the data.

[1] We will later see that it does not matter whether the matcher is symmetric.

15.1.1 Confidence intervals

First, let us determine the confidence interval of (15.2) assuming we know the distribution $F(x)$. Then, of course, we know the distribution of (15.1), but in addition we know the distribution of $\overline{X} - E(x)$:

$$Dist_{\overline{X}-E(x),F}(x) = Prob_F(\overline{X} - E(x) \leq x), \tag{15.3}$$

which is the distribution of the true error, with $F(x) = Prob\,(X \leq x)$ the match score distribution. What then is a confidence interval of estimate $\overline{X}$?

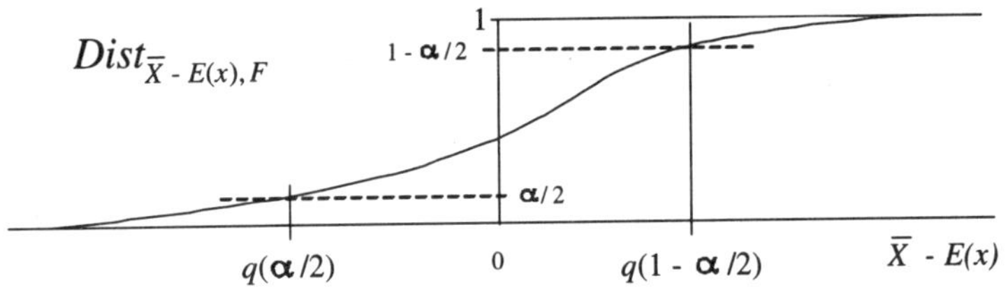

Figure 15.1: Knowledge of the distribution $Dist_{\overline{X}-E(x),F}(x)$ allows the estimation of a region of confidence.

Since we know the distribution of $\overline{X}-E(x)$, we can compute a $(1-\alpha)$ 100% confidence interval for $\overline{X}$

$$\overline{X} - E(x) \in [q(\alpha/2), q(1 - \alpha/2)], \tag{15.4}$$

as in Figure 15.1. Here $q(\alpha/2)$ and $q(1 - \alpha/2)$ are the $\alpha/2$ and $1 - \alpha/2$ quantiles of distribution of (15.3), respectively (see Figure 15.1). By discounting a portion $\alpha/2$ on the lower end and a portion $\alpha/2$ on the higher end, we obtain a region of confidence that contains $E(x)$ with probability $(1 - \alpha)$. We need a confidence interval, determined by $\overline{X}$, that contains the true mean $E(x)$ with probability $(1 - \alpha)$, and (15.4) gives

$$E(x) \in [\overline{X} - q(1 - \alpha/2), \overline{X} - q(\alpha/2)]. \tag{15.5}$$

The size of this confidence region gives an indication of the precision of the estimate of $E(x)$, i.e., how much trust or confidence we can have in $\overline{X}$.

However, to determine (15.5) we need to make some parametric assumption or have some other knowledge about the distribution $Dist_{\overline{X}-E(x),F}(x)$ of (15.3).

15.1.2 Parametric method

The parametric method for confidence interval estimation assumes that the distribution of (15.1) is Normal, which is not a bad assumption because it simply invokes the central limit theorem that the sum of random variables X_i with any distribution approaches a Normal distribution. This, however, means that it is assumed that the match scores, X_i, are independently, identically distributed (i.i.d.).

The parametric assumption is simply that the distribution (15.3) is Normal

$$Dist_{\overline{X}-E(x),F(x)} \sim \mathcal{N}(0,\sigma_{\overline{X}}^2/M), \tag{15.6}$$

i.e., the estimate $\overline{X}$ is Normally distributed around the true sample mean $E(x)$. An unbiased estimator of the variance of this distribution is

$$\sigma_{\overline{X}}^2 = \frac{1}{M-1}\sum_{i=1}^{M}(X_i - \overline{X})^2. \tag{15.7}$$

Estimates of the quantiles of (15.5) are then given by $z(x)$, the inverse of integrals of the Normal curve (see Figure 15.2 and, e.g., [148]) multiplied by the estimate of the standard deviation $\sigma_{\overline{X}}$

$$\hat{q}(\alpha/2) = \sigma_{\overline{X}} \times z(\alpha/2), \tag{15.8}$$
$$\hat{q}(1-\alpha/2) = \sigma_{\overline{X}} \times z(1-\alpha/2). \tag{15.9}$$

Here $z(\alpha/2)$ and $z(1-\alpha/2)$ are determined by the area x under the standard Normal curve as in Figure 15.2.

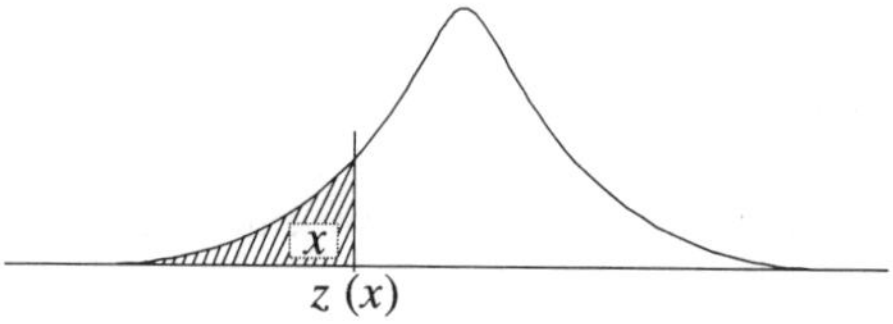

Figure 15.2: Cumulative probabilities of the standard Normal curve.

The estimated parametric confidence interval for $E(x)$, i.e., an estimate for (15.5) is

$$\begin{aligned} E(x) &\in [\overline{X} - \hat{q}(1-\alpha/2), \overline{X} - \hat{q}(\alpha/2)] \\ &= [\overline{X} - \sigma_{\overline{X}}\, z(1-\alpha/2)\, , \overline{X} - \sigma_{\overline{X}}\, z(\alpha/2)], \end{aligned}$$

with the estimates for the quantiles $\hat{q}(\alpha/2)$ and $\hat{q}(1-\alpha/2)$ as in (15.8) and (15.9), respectively.

15.1.3 Nonparametric methods

The Bootstrap, introduced by Efron [59], is an example of a nonparametric method to estimate confidence intervals. Before delving into the bootstrap principle, however, we first address nonparametric methods in general.

Figure 15.3 proposes a way to estimate the distribution of $\overline{X} - E(x)$. The biometric match engine has an underlying distribution $F(x) = \int_0^x p_m(s)\,ds$ from which *first* M match scores $\mathbf{X} = \{X_1, X_2, ..., X_M\}$ are drawn which have mean $\overline{X}$. This is indicated by the horizontal arrow in Figure 15.3.

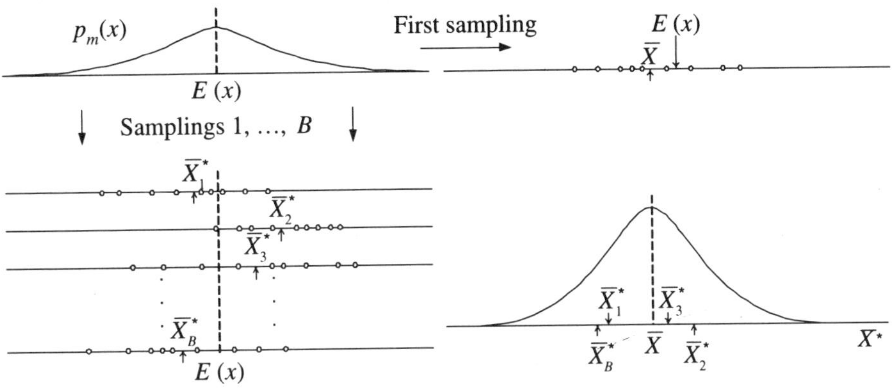

Figure 15.3: The estimate $\overline{X}$ is only *one* estimate of the mean computed from the samples $\mathbf{X}$. True underlying match scores density distribution $p_m(x)$ (top left); a first sampling and sample mean $\overline{X}$ (top right); multiple samplings $\mathbf{X}_i^\star$ of $p_m(x)$ and their respective sample means, $\overline{X}_i^\star$ (bottom left); and $\overline{X}_i^\star$ and their density distribution (bottom right).

Now suppose an additional set of M match scores $\mathbf{X}^\star = \{X_1^\star, X_2^\star, ..., X_M^\star\}$ with distribution $F(x)$ can be observed, having sample mean $\overline{X}^\star$. The question now is how close $\overline{X}^\star$ will be to $\overline{X}$ on average.

Suppose, in addition, that indeed another B sets of match scores $\mathbf{X}_1^\star, ..., \mathbf{X}_B^\star$ with distribution $F(x)$ can be observed with corresponding means $\overline{X}_1^\star, ..., \overline{X}_B^\star$. This is indicated by the downward-pointing vertical arrows in Figure 15.3; and as shown, the average of these means will approach $E(x)$ (and not $\overline{X}$) for large B. The sample mean $\overline{X}$ is an estimate of $E(x)$. What we desire to determine is the distribution of $\overline{X}^\star - \overline{X}$:

$$Dist_{\overline{X}^\star - \overline{X},\, F}(x) \;=\; Prob_F(\overline{X}^\star - \overline{X} \leq x). \tag{15.10}$$

Suppose for the moment that we know $F(x)$ and, hence, $E(x)$. We can then generate as many sets of match score sets $\mathbf{X}_k^\star$, $k = 1, ..., B$ as is desired. The corresponding estimates of the mean $\overline{X}_k^\star$, $k = 1, ..., B$, or rather of $\overline{X}_k^\star - \overline{X}$, are distributed as

$$Dist_{\overline{X}^\star - \overline{X}, F}(x) \;=\; Prob_F(\overline{X}^\star - \overline{X} \leq x),$$

with $\overline{X}$ the sample mean of $\mathbf{X} = \{X_1, ..., X_M\}$. That is, we can randomly sample $Dist_{\overline{X}^\star - \overline{X},\, F}(x)$ of (15.10) over and over again to obtain the B samples. This is an example of a so-called "Monte Carlo" technique, which allows us to estimate distribution (15.10) by generating samples that have this distribution.

Essentially, if we can realize sequences of sets of match scores, we can generate a set of estimates of the mean $\overline{X}_k^\star$, $k = 1, ..., B$. Equivalently, we generate a sequence of $\overline{X}_k^\star - \overline{X}, k = 1, ..., B$. The empirical distribution of the $\overline{X}^\star - \overline{X}$ can then simply be computed as a sorting and counting exercise as

$$
\begin{aligned}
Dist_{\overline{X}^\star - \overline{X}, F}(x) &= \frac{1}{B}\left(\#\,(\overline{X}^\star_k - \overline{X}) \le x\right) \\
&= \frac{1}{B}\sum_{k=1}^{B} \mathbf{1}\left((\overline{X}^\star_k - \overline{X}) \le x\right),
\end{aligned}
\tag{15.11}
$$

as shown in Figure 15.4. We then have an estimate of the distribution of the error $\overline{X}^\star - \overline{X}$ and we can compute statistics, like confidence intervals, of this distribution. Moreover, we can apply the same techniques to other statistics θ, like the error rates, of the match score distribution F and the non-match score distribution G.

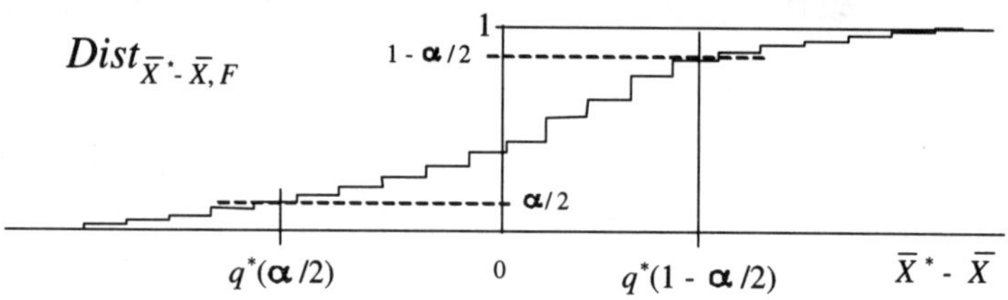

Figure 15.4: The distribution $Dist_{\overline{X}^\star - \overline{X}, F}(x)$ also allows for the estimation of a region of confidence.

15.1.4 Estimating the quantiles

First, without worrying how the estimates $\overline{X}^\star_k, k = 1, ..., B$ are obtained, the confidence interval can be determined by estimating quantiles $q^\star(\alpha/2)$ and $q^\star(1-\alpha/2)$ of the empirical distribution (15.11) of Figure 15.4.

The confidence interval can be constructed by numbering the estimates $\overline{X}^\star_k, k = 1, ..., B$ in ascending order as $\overline{X}^\star_{(1)} \le \overline{X}^\star_{(2)} \le ... \le \overline{X}^\star_{(B)}$. The empirical distribution of (15.11) can be rewritten as

$$
Dist_{\overline{X}^\star - \overline{X}, F}(x) = \frac{1}{B}\left(\#\,(\overline{X}^\star_{(k)} - \overline{X}) \le x\right),
$$

and we have

$$
Dist_{\overline{X}^\star - \overline{X}, F}(\overline{X}^\star_{(k)} - \overline{X}) = \frac{1}{B}\left(\#\,\overline{X}^\star_{(i)} \le \overline{X}^\star_{(k)}\right) = k/B. \tag{15.12}
$$

By construction, exactly k out of the B values of $\overline{X}^\star_{(i)}$ are less than or equal to $\overline{X}^\star_{(k)}$. The two approximate quantiles are

$$
\hat{q}^\star(\alpha/2) \approx \overline{X}^\star_{(k_1)} \text{ and } \hat{q}^\star(1-\alpha/2) \approx \overline{X}^\star_{(k_2)}, \tag{15.13}
$$

where $k_1 = \lfloor(\alpha/2)\,B\rfloor$ and $k_2 = \lfloor(1-\alpha/2)\,B\rfloor$.

This process of estimating the quantiles is illustrated as follows. The estimates $\overline{X}^\star_k$ are, for instance, distributed as in Figure 15.5. A $(1-\alpha)$ 100% confidence interval is determined by—

1. selecting the bottom $(1-\alpha/2)$ 100% $= \lfloor (1-\alpha/2)\, B \rfloor$ of the estimates $\overline{X}^\star_k$; and
2. subtracting from this set the bottom $(\alpha/2)$ 100% $= \lfloor (\alpha/2)\, B \rfloor$ of the estimates $\overline{X}^\star_k$.

This is shown in Figure 15.5: the minimum estimate $\overline{X}^\star_{(k_1)}$ and maximum estimate $\overline{X}^\star_{(k_2)}$ in this leftover set of estimates $\widehat{\mathbf{X}}^\star(\alpha)$ determine the confidence interval

$$[\overline{X}^\star_{(k_1)}, \overline{X}^\star_{(k_2)}].$$

Essentially, the task of confidence estimation is reduced to an exercise in sorting and counting.

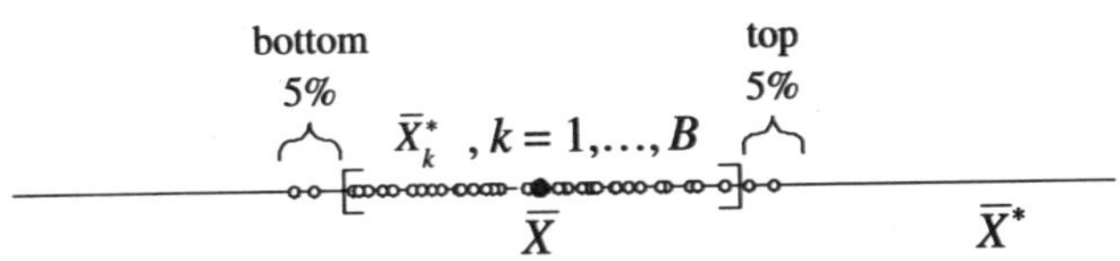

Figure 15.5: The estimates $\overline{X}^\star_k, k = 1, ..., B$ are distributed around $\overline{\mathbf{X}}$.

15.1.5 The Bootstrap

The bootstrap [59] is a recipe for obtaining the sequence of sample mean estimates

$$\overline{X}^\star_1, \overline{X}^\star_2, ..., \overline{X}^\star_k, ..., \overline{X}^\star_B,$$

which, until now, we have taken for granted. The argument proceeds with the observation that the *only* data we have is the set of match scores $\mathbf{X} = \{X_1, ..., X_m\}$ and, therefore, proposes the technique of *resampling with replacement* from the original set $\mathbf{X}$ to obtain the *bootstrap sets*

$$\mathbf{X}^\star_k, \quad k = 1, ..., B.$$

A set of bootstrap sets, for example, is

$$\begin{aligned} \mathbf{X}^\star_1 &= \{X^\star_{11} = X_{10}, X^\star_{12} = X_{10}, ..., X^\star_{1M} = X_{M-1}\} \\ \mathbf{X}^\star_2 &= \{X^\star_{21} = X_{M-2}, X^\star_{22} = X_5, ..., X^\star_{2M} = X_{M-1}\} \\ &\vdots \\ \mathbf{X}^\star_B &= \{X^\star_{B1} = X_{17}, X^\star_{B2} = X_1, ..., X^\star_{BM} = X_{M-9}\}. \end{aligned} \tag{15.14}$$

For each set $\mathbf{X}^\star_k$ a corresponding bootstrap estimate $\overline{X}^\star_k, k = 1, ..., B$ is computed, using the sample mean of (15.1).

More formally, to obtain bootstrap estimates $\hat{\theta}^\star$ of any statistic θ of the distribution $F(x)$, the bootstrap prescribes sampling, with replacement, the set $\mathbf{X}$ many (B) times:

1. *Sampling.* Create a bootstrap sample $\mathbf{X}^\star = \{X_1^\star, ..., X_M^\star\}$ by sampling $\mathbf{X}$ with replacement.

   In bootstrap sample $\mathbf{X}^\star$ any match score X_i may be represented multiple times or may not be represented at all, e.g.,

$$X_1^\star = X_i, X_2^\star = X_i, X_k^\star \neq X_j \ \forall\, k, j \neq i.$$

2. *Estimation.* Calculate sample ***bootstrap estimate*** $\hat{\theta}^\star$ from the bootstrap sample $\mathbf{X}^\star$.

3. *Repetition.* Repeat steps 1 and 2 B times (B large), resulting in a set $\hat{\Theta}^\star$ of B bootstrap estimates $\hat{\theta}_1^\star$, $\hat{\theta}_2^\star$ ..., $\hat{\theta}_B^\star$.

If B is large enough, the empirical distribution $\hat{F}_B(x)$ of the B bootstrap sets of (15.14) obviously converges to the empirical distribution of the original samples $X_1, ..., X_M$:

$$\hat{F}(x) = \frac{1}{M}\,(\# X_i \leq x) = \frac{1}{M}\sum_{i=1}^{M} \mathbf{1}(X_i \leq x).$$

The distribution of the match scores that are used to compute the bootstrap estimates is the same as the empirical distribution of the original samples $\{X_1, ..., X_M\}$. The bootstrap, of course, is only valid if the match scores are i.i.d. and that assumption is surely violated. We use a set of match scores to see what happens.

The match scores are summarized in Table 15.2: a set $\mathbf{X}_1$ of $M_1 = 1{,}120$ scores and a set $\mathbf{X}_2$ of $M_2 = 5{,}600$ scores. If it does not matter that the i.i.d. assumption is violated, this description of the data is enough. Later on we will see that it is important that the data be better described.

$\mathbf{X}_1$	$\mathbf{X}_2$
$M = 1{,}120$	$M = 5{,}600$

Table 15.2: Two sets of match scores $\mathbf{X}_1$, $\mathbf{X}_2$ and their size M.

Figure 15.6 shows the 90% bootstrap confidence intervals for the mean estimate $\overline{X}_1$ of set $\mathbf{X}_1$ and $\overline{X}_2$ of set $\mathbf{X}_2$. One can make two observations:

1. The confidence intervals for $\overline{X}_1$ and $\overline{X}_2$ are about the same. Using more data, $M = 5{,}600$ *versus* $M = 1{,}120$, does not narrow the confidence interval as might have been expected.

2. The confidence interval for $\overline{X}_2$ is not contained in the confidence interval for $\overline{X}_1$, while $M_2 > M_1$. We would expect this since in fact $\mathbf{X}_1 \subset \mathbf{X}_2$. This is not the case in Figure 15.6 because, for various reasons, the match scores are dependent.

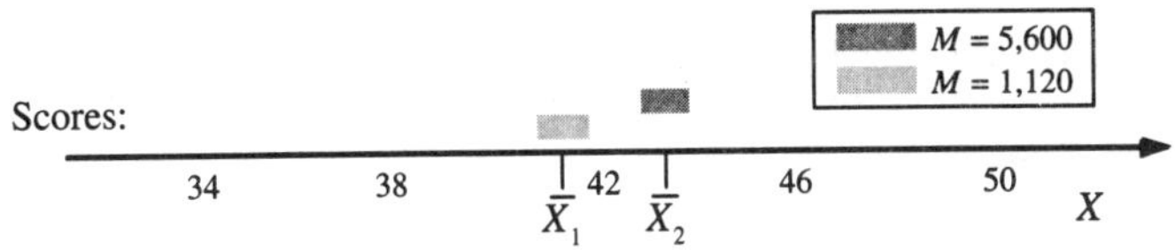

Figure 15.6: The 90% bootstrap confidence intervals for the means $\overline{X}_1$ and $\overline{X}_2$.

15.2 The data are not independent

There is statistical dependence among match scores $X_1, X_2, ...$ (and mismatch scores, as we will see later) because of the way test databases of biometric samples are collected. Subsequent finger impressions are obtained by successive dabbing of the finger on an input device. That is, given a first impression I of a finger plus an additional two impressions I_t and $I_{t+\Delta}$ of the same finger, the match scores $X_i = s(I, I_t)$ and $X_{i+1} = s(I, I_{t+\Delta})$ are certainly correlated. There are further sources of dependence among the scores that are due to subtleties of the collection process of test data. In general, fingerprint image formation is a complex process and a function of many random variables (finger pressure, finger moisture, etc.); for a given individual, these random variables are dependent from one impression to the next.

The problem is the way the bootstrap sets $\mathbf{X}^\star$ of (15.14) are obtained from the original set of match scores $\mathbf{X}$. This sampling with replacement technique *does not* replicate the dependence between subsequent match scores X_i and X_{i+1}. There is less interdependence among match scores in bootstrap set $\mathbf{X}^\star$ than there is in set $\mathbf{X}$. Therefore, the bootstrap estimates $\overline{X}^\star_1, ..., \overline{X}^\star_B$ have lower variance than independent estimates $\overline{X}$ and $\overline{X}'$ would have. In fact, the bootstrap sets, and therefore the bootstrap estimates $\overline{X}^\star$, are sufficiently independent that there is little difference between the parametric method of Section 15.1.2 and the regular bootstrap method. Parametric confidence intervals and bootstrap confidence intervals are studied in [22,23].

15.2.1 How are the data obtained?

This section uses the sets of fingerprint match scores already introduced in Table 15.2 to illustrate the ideas behind the "*subsets bootstrap*." Biometric match scores need to be carefully described according to which match scores are associated with which biometric of which subject. This becomes more important when a subject has more than one instance of the biometric (i.e., fingers).

	M	$\mathcal{D}$	$\mathcal{P}$
$\mathbf{X}_1$	1,120	20	10
$\mathbf{X}_2$	5,600	100	50

Table 15.3: Two sets of match scores used in this chapter, with $\mathbf{X}_1 \subset \mathbf{X}_2$.

Table 15.3 gives the number of volunteers $\mathcal{P}$ and the number of fingers $\mathcal{D}$, where the number of fingers per subject is $c = 2$ (left and right index finger) so that $\mathcal{D} = c\mathcal{P}$. Per finger $d = 8$ impressions have been acquired, which gives $d\,(d-1) = 8 \times 7 = 56$ match scores per finger in the sets $\mathbf{X}_1$ and $\mathbf{X}_2$:

1. A first set $\mathbf{X}_1$ of $M_1 = 1{,}120$ match scores is computed by using samples of $\mathcal{P} = 10$ subjects, i.e., $\mathcal{D} = c\mathcal{P} = 20$ fingers.

   We have $\mathcal{D} = 20$ sets of 56 match scores, or, alternatively, $\mathcal{P} = 10$ sets of 112 match scores.

2. Set $\mathbf{X}_2$ of M_2 =5,600 match scores is computed using $\mathcal{P} = 50$ subjects, i.e. $\mathcal{D} = 100$ fingers; hence, we have $\mathcal{D} = 100$ sets of 56 match scores and $\mathcal{P} = 50$ sets of 112 match scores.

As mentioned before, $\mathbf{X}_1 \subset \mathbf{X}_2$.

If the matcher is symmetric, i.e., $s(B_1, B_2) = s(B_2, B_1)$, only $d\,(d-1)/2$ distinct match scores can be computed from d impressions of a finger. It turns out that for robust confidence interval estimation as the subsets bootstrap described in this chapter, all match scores can be included in $\mathbf{X}$ irrespective of whether the matcher is symmetric [23].

15.2.2 The subsets bootstrap

In [22,23] the "*subsets bootstrap*" is introduced to compute confidence intervals of biometric error estimates, and incorporates the dependency of the match scores $\{X_i, i = 1, ..., M\}$ in the process of generating B estimates $\overline{X}^\star$. The method is motivated by the "moving blocks" bootstrap [123].

Figure 15.7 shows this process. As in Section 15.1.3, we first assume that we know $F(x)$ and, hence, $E(x)$. The distribution F is a peculiar one and the set $\mathbf{X}$ has been grouped in subsets $\mathcal{X}_i, i = 1, ..., \mathcal{M}$ that are independent of one another. As many sets of resampled match scores $\mathbf{X}_k^\star, k = 1, ..., B$ as needed can be generated and the corresponding estimates of the mean $\overline{X}_k^\star, k = 1, ..., B$ have the distribution of (15.3)

$$Dist_{\overline{X}^\star - \overline{X}, F}(x) = Prob_F(\overline{X}^\star - \overline{X} \le x).$$

The estimate $\overline{X} = \theta = E(F(x))$ is the sample mean (15.1) of $\mathbf{X} = \{X_1, ..., X_M\}$. We can again Monte Carlo sample this distribution over and over again to obtain the B samples as in Figure 15.7.

As discussed in Section 15.2.1, the set $\mathbf{X}$ of M match scores is obtained from $\mathcal{P}$ subject or from $\mathcal{D}$ fingers. Then the set of match scores $\mathbf{X} = \{X_1, ..., X_M\}$ can be grouped:

1. by *subject*: $\mathbf{X} = \{\mathcal{X}_1, \mathcal{X}_2, ..., \mathcal{X}_\mathcal{P}\}$, where each set contains $c\,d\,(d-1)$ match scores;

2. by *finger* (biometric): $\mathbf{X} = \{\mathcal{X}_1, \mathcal{X}_2, ..., \mathcal{X}_\mathcal{D}\}$, the sets contain $d\,(d-1)$ match scores each.

Instead of individual match scores $X_i, i = 1, ..., M$, we have sets of match scores:

$$\mathcal{X}_i, i = 1, ..., \mathcal{M}; \;\; \mathcal{M} = \mathcal{D} \text{ or } \mathcal{M} = \mathcal{P}.$$

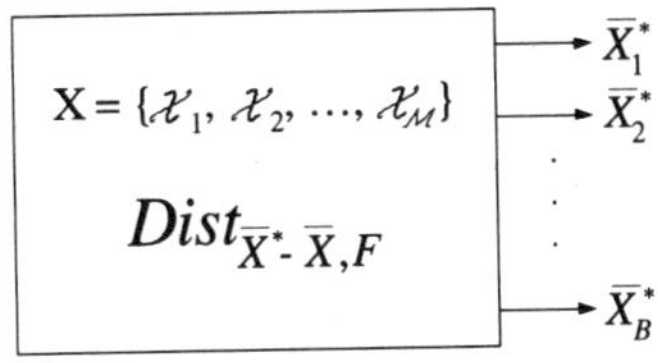

Figure 15.7: Generating B bootstrap estimates $\overline{X}_1^\star, ..., \overline{X}_B^\star$ distributed as F, with expectation $E(F(x)) = \overline{X}$.

The subsets bootstrap proceeds by resampling $\mathbf{X} = \{\mathcal{X}_1, \mathcal{X}_2, ..., \mathcal{X}_M\}$ to obtain bootstrap estimates $\overline{X}^\star$:

1. *Sampling.* Create a bootstrap sample $\mathbf{X}^\star = \{\mathcal{X}_1^\star, ..., \mathcal{X}_M^\star\}$ by sampling whole subsets $\mathcal{X}_i$ from $\mathbf{X}$ with replacement. Here set $\mathbf{X}^\star$ may contain any subset $\mathcal{X}_i$ multiple times or no times.

2. *Bootstrap estimate.* Calculate the bootstrap mean estimate $\overline{X}^\star$ from bootstrap set $\mathbf{X}^\star$:
$$\overline{X}^\star = \frac{1}{M} \sum_{X_i \in \mathbf{X}^\star} X_i.$$

3. *Repetition.* Repeat steps 1 and 2 B times (B large), to obtain estimates $\overline{X}_1^\star$, $\overline{X}_2^\star$ ... $\overline{X}_B^\star$ as in the regular bootstrap.

The B bootstrap estimates in set $\overline{\mathbf{X}}^\star = \{\overline{X}_i^\star, i = 1, ..., B\}$ are now distributed as shown in Figure 15.8 with the variance of the mean estimates higher than for the traditional bootstrap of Figure 15.5. The confidence intervals are obtained in the same fashion as above. One can compute a $(1 - \alpha)\, 100\%$ confidence interval by subtracting the lower $(\alpha/2)\, 100\%$ and the higher $(\alpha/2)\, 100\%$ of the estimates and from the total set of bootstrap estimates $\overline{\mathbf{X}}^\star$, as illustrated in Figure 15.8.

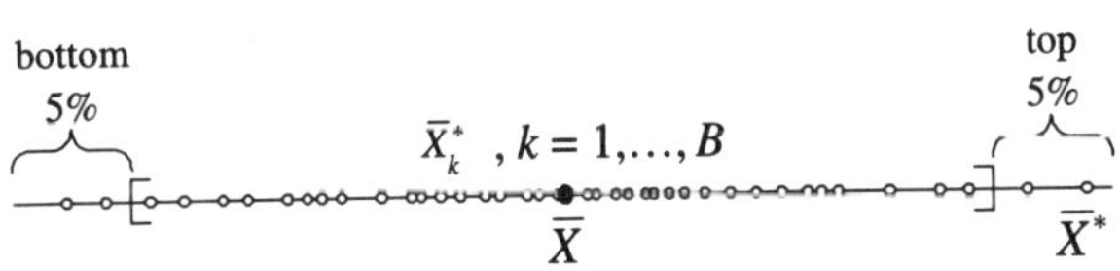

Figure 15.8: The bootstrap estimates $\overline{X}^\star$ are again distributed around $\overline{X}$ but now with larger variance.

15.2.3 Estimating the match score mean (continued)

Figure 15.9 shows the confidence intervals for the mean estimates $\overline{X}_1$ and $\overline{X}_2$ of the sets $\mathbf{X}_1$ and $\mathbf{X}_2$ described in Table 15.3.

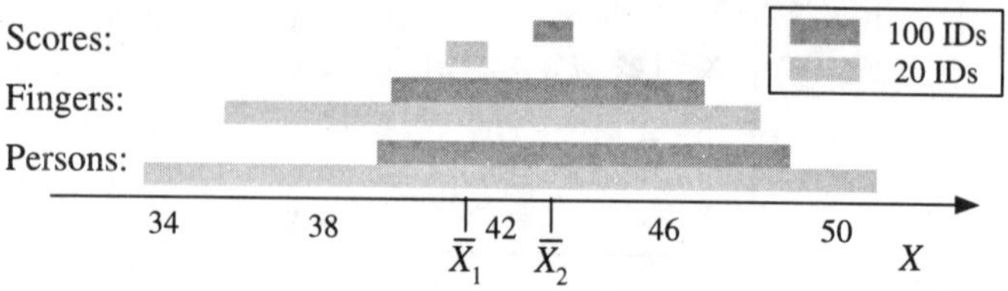

Figure 15.9: The subsets bootstrap confidence intervals for mean estimates $\overline{X}_1$ and $\overline{X}_2$ using the traditional bootstrap and the subsets bootstrap.

The top two are the confidence intervals computed with the traditional bootstrap as in Figure 15.6. The confidence intervals in the middle are computed by sampling with replacement from the subsets $\mathcal{X}_1, ..., \mathcal{X}_\mathcal{D}$. The lower two confidence intervals are computed by sampling with replacement of $\mathcal{X}_1, ..., \mathcal{X}_\mathcal{P}$ There is quite a bit of difference between the finger and the person subsets bootstrap, which indicates that there is dependence between match scores of different fingers from the same individual.

15.3 Confidence intervals on the FRR and FAR

Up till now, we have talked about the estimation of the mean $E(x)$ of the match score distribution $F(x)$. This is only one particular example of a *statistic*, denoted as $\theta(F)$, of the match score distribution. The False Reject Rate $\text{FRR}(t_o)$ at some operating point t_o is a statistic too. Similarly, the False Accept Rate $\text{FAR}(t_o)$ is a statistic of the mismatch score distribution $G(x)$. It gets confusing, though: we want to estimate the error rates, which are statistics of F and G, using biometric test data $\mathbf{X}$ and $\mathbf{Y}$. In addition, we want to estimate the error in our estimate (*of the error*).

15.3.1 Distributions F and a G specify the engine

The accuracy of a biometric match engine is in theory completely specified by its $F(x)$, the genuine score distribution, and its $G(y)$, the imposter score distribution. Often a given biometric match engine is tuned to a specific biometric group or *user population*, $\text{B}_\mathcal{U}$, and ideally also trained somehow to some "imposter" population $\text{B}_\mathcal{I}$ (see Chapter 9). However, this simply means that the match engine is specified by some $F_\mathcal{U}(x)$ and some $G_\mathcal{I}(y)$, which could be viewed as conditionals $F(x|\mathcal{U})$ and some $G(y|\mathcal{I})$, if you will.

A biometric matcher can also be specified in terms of a False Reject Rate $\text{FRR}(T)$ and False Accept Rate $\text{FAR}(T)$ as noted in Chapter 5. These error rates and probability distributions are related to $F(x)$ and $G(y)$ as

$$\text{FRR}(x) \quad = \quad Prob\,(s_m \leq x|H_o) = F(x), \tag{15.15}$$

where s_m is a random variable, denoting the match score; and,

$$\text{FAR}(y) \quad = \quad Prob\,(s_n > y|H_a) \tag{15.16}$$

$$= \quad 1 - Prob\,(s \le y|H_a) = 1 - G(y),$$

where s_n is a random variable called the mismatch (non-match) score.

The biometric matcher will be operating at some threshold $T = t_o$ and ultimately the only really important errors are FRR(t_o) and FAR(t_o). Therefore, first consider the error estimates at operating point $T = t_o$:

$$\text{FRR}(t_o) \quad = \quad \frac{1}{M}\sum_{i=1}^{M} \mathbf{1}\,(X_i \le t_o) = \frac{1}{M}(\#\, X_i \le t_o), \qquad (15.17)$$

$$\text{FAR}(t_o) \quad = \quad \frac{1}{N}\sum_{j=1}^{N} \mathbf{1}\,(Y_j > t_o) = \frac{1}{N}(\#\, Y_j > t_o). \qquad (15.18)$$

FRR(t_o) and FAR(t_o) are empirical estimates of the true FRR and FAR at some $T = t_o$. The FAR is not a probability distribution (being 1 at 0 and 0 at ∞) but since the notation is more common we continue with this term rather than the more strictly correct use of the error distribution $G(y)$. Further, to be consistent we should denote FRR and FRR of (15.17) and (15.18) as $\widehat{\text{FRR}}$ and $\widehat{\text{FRR}}$, respectively, because these are estimates.

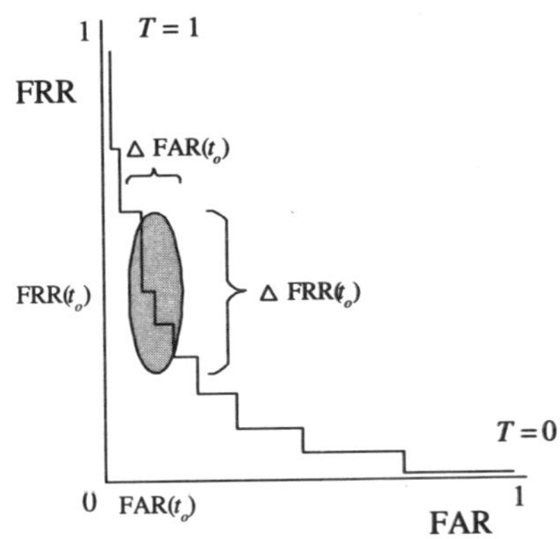

Figure 15.10: An estimate of the ROC(T) and confidence intervals around the operating point t_o.

Figure 15.10 gives an illustration of what is needed in terms of the ROC as defined in Chapter 5. We need estimates of FRR(t_o) and FAR(t_o) at operating point t_o. We also need confidence intervals, denoted as $\Delta\,\text{FRR}(t_o;\alpha)$ and $\Delta\,\text{FAR}(t_o;\alpha)$, on these error estimates, where α is the confidence, with the true False Accept Rate falling in the interval $\Delta\,\text{FAR}(t_o;\alpha)$ with probability $(1-\alpha)$, likewise for the FRR. Hence, what we end up with is an area of confidence as indicated in Figure 15.10. This area is an elliptical region bounded by the confidence intervals of the error estimates:

$$\Delta\text{FRR}(t_o;\alpha) = [\text{FRR}^{\star}_{(k_1)}(t_o;\alpha), \text{FRR}^{\star}_{(k_2)}(t_o;\alpha)] \qquad (15.19)$$

and

$$\Delta\text{FAR}(t_o;\alpha) = [\text{FAR}^\star_{(k_1)}(t_o;\alpha), \text{FAR}^\star_{(k_2)}(t_o;\alpha)]. \tag{15.20}$$

Here the confidence intervals are estimated bootstrap intervals determined by a variation on the subsets bootstrap discussed in Section 15.2.2. The following method of obtaining the estimates of these confidence intervals of the error estimates is basically a rehash of constructing the bootstrap confidence interval for the mean estimate $\overline{X}$ of the match distribution $F(x)$.

15.3.2 Error in the FRR estimate

Remember, we have M match scores $\mathbf{X} = \{X_1, ..., X_M\}$ computed using mated fingerprint pairs of biometric data $\mathbf{B}_\mathcal{D}$ as shown in Table 15.1. The samples are acquired from $\mathcal{M}$ entities, where $\mathcal{M} = \mathcal{P}$ subjects or $\mathcal{M} = \mathcal{D}$ real-world biometrics $\mathcal{B}$.

Dividing up the match scores

As seen in Section 15.2.1, the set $\mathbf{X}$ can be divided into $\mathcal{M}$ independent subsets $\mathbf{X} = \{\mathcal{X}_1, ..., \mathcal{X}_\mathcal{M}\}$ with $\mathcal{M} = \mathcal{D}$, the number of fingers, or $\mathcal{M} = \mathcal{P}$, the number of subjects. The estimate of the False Reject Rate is already given in (15.17), and stays the same irrespective of whether the match scores are divided up in sets or not. Repeating it here,

$$\text{FRR}(t_o) \approx \frac{1}{M}\sum_{i=1}^{M} \mathbf{1}\,(X_i \leq t_o) = \frac{1}{M}(\#\, X_i \leq t_o).$$

A bootstrap estimate of a $(1-\alpha)100\,\%$ confidence interval for the estimate $\text{FRR}(\text{t}_\text{o})$ is obtained in the same fashion as a confidence interval for the estimate of the mean $\overline{X}$.

Being precise, the subsets bootstrap recipe for obtaining a confidence estimate on the False Reject Rate is as follows:

1. Divide the set of match scores $\mathbf{X}$ into $\mathcal{M}$ independent subsets $\mathcal{X}_1, ..., \mathcal{X}_\mathcal{M}$; here the sets are "independent" because they are from more-or-less independent biometric entities.

2. Many (B) times do—

   Generate a bootstrap set $\mathbf{X}^\star$ by sampling $\mathcal{M}$ subsets from $\mathbf{X} = \{\mathcal{X}_1, ..., \mathcal{X}_\mathcal{M}\}$ with replacement (each set still contains M samples).

   Compute the corresponding bootstrap estimate $\text{FRR}^\star(t_o)$ using set $\mathbf{X}^\star$.

This gives a set $\{\text{FRR}^\star_k(t_o), k = 1, ..., B\}$ of B bootstrap estimates. Rank the estimates $\text{FRR}^\star(t_o)$ as

$$\text{FRR}^\star_{(1)}(t_o) \leq \text{FRR}^\star_{(2)}(t_o) \leq ... \leq \text{FRR}^\star_{(B)}(t_o).$$

As in (15.12), the distribution of estimates $\text{FRR}^\star(t_o)$ of the subset bootstrap sets $\mathcal{X}_k, k = 1, ..., B$ is by construction

$$\begin{aligned} Dist_{\text{FRR}^\star(t_o),\hat{F}}(\text{FRR}^\star_{(k)}(t_o)) &= \frac{1}{B}\left(\#\,\text{FRR}^\star_{(i)}(t_o) \leq \text{FRR}^\star_{(k)}(t_o)\right) \\ &= k/B. \end{aligned} \tag{15.21}$$

Exactly k out of the B values of $\text{FRR}^{\star}_{(i)}(t_o)$ are less than or equal to $\text{FRR}^{\star}_{(k)}(t_o)$. The two approximate quantiles are

$$\hat{q}(\alpha/2) \approx \text{FRR}^{\star}_{(k_1)} \text{ and } \hat{q}(1-\alpha/2) \approx \text{FRR}^{\star}_{(k_2)}, \tag{15.22}$$

with $k_1 = \lfloor(\alpha/2)\,B\rfloor$, $k_2 = \lfloor(1-\alpha/2)\,B\rfloor$. Precisely as for the estimate of the mean in Sections 15.1 and 15.2.

As shown in Figure 15.11, at the risk of being repetitive, the confidence interval of error rate estimate $\text{FRR}(t_o)$ is obtained by eliminating the bottom $(\alpha/2)\,100\%$ and the top $(\alpha/2)\,100\%$ of the bootstrap estimates $\text{FRR}^{\star}_{(k)}(t_o)$.

Figure 15.12 illustrates the results of FRR confidence interval estimation using the finger and person bootstrap methods. 5,600 match scores involving 50 IDs (100 fingers) were used for this experiment. It is readily observed that the confidence intervals estimated by the finger subset are contained in the corresponding confidence intervals estimated using the person subset bootstrap. This result demonstrates that the dependence among the match scores is better isolated by the person subset than by the finger subset bootstrap method.

15.3.3 Error in the FAR estimate

For estimating the mismatch, or False Accept, error FAR, a set of mismatch scores $Y_j, j = 1, ..., N$ is computed using the biometric sample data $\mathbf{B}_{\mathcal{D}}$ of Table 15.1. In total, there are $\mathcal{D}\,d$ samples, from among which a set $\mathbf{Y}$ of N non-match scores can be computed. (We will get back to this set later.) With set $\mathbf{Y} = \{Y_1, ..., Y_N\}$ mismatch scores, we repeat from (15.18) the False Accept Rate estimate

$$\text{FAR}(t_o) = \frac{1}{N}\sum_{i=1}^{N} \mathbf{1}\,(Y_i > t_o) = \frac{1}{N}(\#Y_i > t_o),$$

where we are interested in determining the $(1-\alpha)\,100\%$ confidence interval introduced in Expression (15.20)

$$\Delta\text{FAR}(t_o; \alpha) \approx [\text{FAR}^{\star}_{(k_1)}(t_o; \alpha), \text{FAR}^{\star}_{(k_2)}(t_o; \alpha)],$$

based on (subsets bootstrap) estimates $\text{FAR}^{\star}_k, k = 1, ..., B$ of $\text{FAR}(t_o)$.

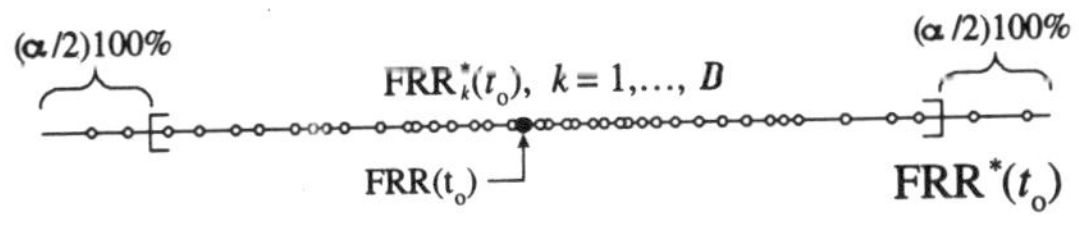

Figure 15.11: The spread of the bootstrap samples $\text{FRR}^{\star}_k(t_o)$ around $\text{FRR}(t_o)$ gives the confidence interval $[\text{FRR}^{\star}_{(k_1)}(t_o; \alpha), \text{FRR}^{\star}_{(k_2)}(t_o; \alpha)]$.

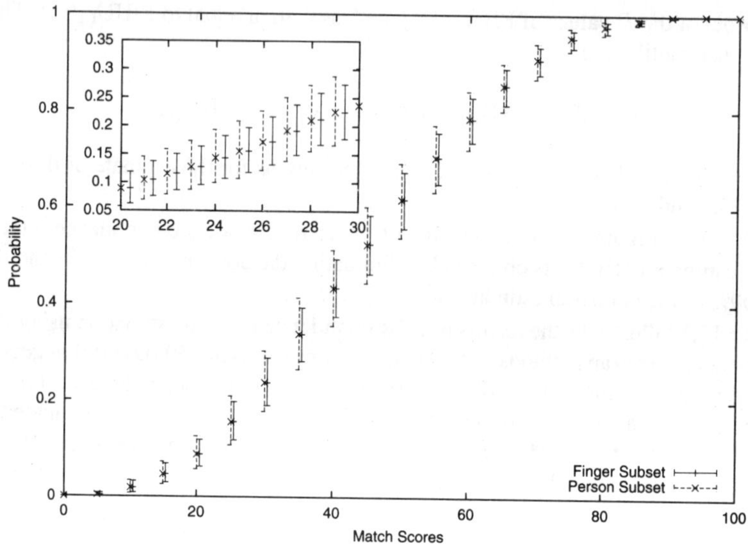

Figure 15.12: The FRR confidence intervals estimated using finger and person subset bootstrap using 5,600 match scores. The inset shows an enlargement of one portion of the overall plot.

The mismatch scores are not independent

For estimating the False Reject Rate $\text{FRR}(t_o)$, the match scores $\mathbf{X}$ would be independent if only one mated pair per finger were used to determine the estimates, requiring d times more fingers for the same number of scores. Since we also use multiple impressions of each finger for estimating False Accept error rates $\text{FAR}(t_o)$, we again suspect some dependence among the data $\mathbf{Y} = \{Y_1, ..., Y_N\}$. Further, let I be a fingerprint from a first subject and I_a and I_b be impressions of two *different* fingers from a second subject. The mismatch scores $s(I, I_a)$ and $s(I, I_b)$ will also not be statistically independent. To obtain independent (in the sense of never using the same finger twice) mismatch scores, many different non-mated fingerprint samples would need to be collected, which is an impossible burden on the collection process—a database of $\mathcal{D}$ people would yield only $\mathcal{D}/2$ such "independent" scores instead of the $d^2\,\mathcal{D}\,(\mathcal{D}-1)$ dependent mismatch scores we could obtain.

Incidentally, there is dependency between the sets $\mathbf{X}$ and $\mathbf{Y}$ too, but we will worry about that in Chapter 16 when we look at the EER estimate.

The subsets bootstrap FAR estimate

We partition the mismatch scores $\mathbf{Y}$ into subsets $\mathcal{Y}_1, ..., \mathcal{Y}_\mathcal{N}$ such that these sets are as independent as possible. The subsets bootstrap for obtaining a confidence interval estimate on the False Accept Rate $\text{FAR}(\text{t}_\text{o})$ is then the same story again.

In the next section we discuss how to construct the $\mathcal{N}$ independent subsets, which is more cumbersome than for the FRR and only approximately possible. The subsets bootstrap interval is determined the same way as for the FRR and we only present it here for completeness:

1. Divide the set of match scores $\mathbf{Y}$ into $\mathcal{N}$ "independent subsets" $\mathcal{Y}_1, ..., \mathcal{Y}_{\mathcal{N}}$.

2. B times generate a bootstrap set $\mathbf{Y}^\star$ by sampling with replacement $\mathcal{N}$ subsets from $\mathbf{Y} = \{\mathcal{Y}_1, ..., \mathcal{Y}_{\mathcal{N}}\}$ (each $\mathbf{Y}^\star$ contains N samples). Compute the corresponding bootstrap estimate $\text{FAR}^\star(t_o)$ using set $\mathbf{Y}^\star$.

This gives a set $\{\text{FAR}^\star_k(t_o), k = 1, ..., B\}$ of B bootstrap estimates that can be ordered as

$$\text{FAR}^\star_{(1)}(t_o) \leq \text{FAR}^\star_{(2)}(t_o) \leq ... \leq \text{FAR}^\star_{(B)}(t_o).$$

By construction, the distribution of False Accept Rate estimates is

$$Dist_{\text{FAR}^\star(t_o),\hat{G}}(\text{FAR}^\star_{(k)}(t_o)) = \frac{1}{B}\left(\#\,\text{FAR}^\star_{(j)}(t_o) \leq \text{FAR}^\star_{(k)}(t_o)\right) = \frac{k}{B},$$

where $\hat{G}$ is the distribution of the samples in the bootstrap sets $\mathbf{Y}^\star$. Exactly k out of the B values of $\text{FAR}^\star_{(j)}(t_o)$ are less than or equal to $\text{FAR}^\star_{(k)}(t_o)$. The two approximate quantiles are $\hat{q}(\alpha/2) \approx \text{FAR}^\star_{(k_1)}$ and $\hat{q}(1-\alpha/2) \approx \text{FAR}^\star_{(k_2)}$, with $k_1 = \lfloor(\alpha/2)\,B\rfloor$, $k_2 = \lfloor(1-\alpha/2)\,B\rfloor$. Just like the estimate of the mean in Sections 15.1 and 15.2 and the estimate of FRR of Section 15.3.2.

The mechanics of dividing the mismatch scores

We have $\mathcal{D}$ fingers and d samples per finger. Matching the samples of one finger against those of a non-mated finger give d^2 mismatch scores; the number of pairs of fingers is $\mathcal{D}\,(\mathcal{D}-1)$; hence the total number N of mismatch scores is

$$N = d^2\,\mathcal{D}\,(\mathcal{D}-1). \tag{15.23}$$

Further, there are $\mathcal{P}$ subjects and c fingers per subject, hence $\mathcal{D} = c\,\mathcal{P}$. Substituting this in (15.23) and a little manipulation gives

$$N = d^2\,c\mathcal{P}\,(c\mathcal{P}-1) = \mathcal{P}\left[(dc)^2\,(\mathcal{P}-1) + d^2\,c\,(c-1)\right]. \tag{15.24}$$

From (15.23) and (15.24) it is seen that the mismatch scores $\mathbf{Y}$ can be divided as shown in Table 15.4:

# subsets	Subsets	Set size
$\mathcal{D}$	$\mathcal{Y}_1 \cdots \mathcal{Y}_\mathcal{D}$	$d^2\,(\mathcal{D}-1)$
$\mathcal{P}$	$\mathcal{Y}_1 \cdots \mathcal{Y}_\mathcal{P}$	$(dc)^2\,(\mathcal{P}-1) + d^2\,c\,(c-1)$.

Table 15.4: Alternative ways of dividing the samples for the subset bootstrap to estimate FAR confidence intervals. In both cases $||\cup_j \mathcal{Y}_j|| = N$.

To construct these sets, we have to establish some more notation. Suppose we are dealing with $\mathcal{N}$ biometric "entities," (either people or fingers) $\mathcal{B}_1, \mathcal{B}_2, ..., \mathcal{B}_\mathcal{N}$. Each entity

$j = 1, ..., \mathcal{N}$ is associated with a set of samples $\mathcal{I}_j$. Denote an all-against-all match of the sets of samples $\mathcal{I}_j$ and $\mathcal{I}_k$ as

$$S(\mathcal{I}_j, \mathcal{I}_k) = \{s(B_m, B_n);\ B_m \in \mathcal{I}_j,\ B_n \in \mathcal{I}_k,\ j \neq k\}, \tag{15.25}$$

where B_m and B_n are biometric samples. The set of mismatch scores between set $\mathcal{I}_j$ and all other sets $\mathcal{I}_k$ is then

$$\mathcal{Y}_j = \bigcup_{k \neq j} S(\mathcal{I}_j, \mathcal{I}_k), \tag{15.26}$$

and we have subsets $\{\mathcal{Y}_1, \mathcal{Y}_2, ..., \mathcal{Y}_\mathcal{N}\}$.

- A first way to obtain these subsets is by choosing the $\mathcal{I}_j$ to be the sets of d samples of finger $\mathcal{B}_j$, for $j = 1, ..., \mathcal{D}$. Thus $\mathcal{N} = \mathcal{D}$, the number of fingers. The set $S(\mathcal{I}_j, \mathcal{I}_k)$ has d^2 mismatch scores and hence, as shown in Table 15.4, $\mathcal{Y}_j$ consists of the d^2 $(\mathcal{D} - 1)$ mismatch scores $s(B_{j\star}, B_{k\star}), k \neq j$ in the notation of Table 15.1.

- A second way is to divide the mismatch scores $\mathbf{Y}$ into the $\mathcal{N} = \mathcal{P}$ sets $\mathcal{I}_j$ of the $c \times d$ samples associated with subject j, for $j = 1, ..., \mathcal{P}$. Our subsets $\mathcal{Y}_j$ are still as defined by (15.26) but (15.25) is redefined as

$$S(\mathcal{I}_j, \mathcal{I}_k) = \{s(B_m, B_n);\ B_m \in \mathcal{I}_j,\ B_n \in \mathcal{I}_k\}, \tag{15.27}$$

  where B_m and B_n are biometric samples *not* from the same biometric $\mathcal{B}$.

  Now $\mathcal{Y}_j$ consists of the $[(dc)^2\ (\mathcal{P} - 1)]$ "inter-subject" mismatch scores together with $[d^2 c(c-1)]$ "intra-subject" mismatch scores —mismatches between different fingers of the same subject, as shown in the second line of Table 15.4.

Note, however, that Expression (15.26) does not result in subsets $\mathcal{Y}_j$ that are completely independent. That is, the samples $\mathcal{I}_j$ also contribute mismatch scores to the sets $\mathcal{Y}_k, k \neq j$. In that sense, resampling with replacement subsets of $\mathbf{Y}$ only isolates some of the dependence between the Y_j. One way to construct truly independent subsets $\mathcal{Y}_j$ is by selecting $\mathcal{Y}_j = S(\mathcal{I}_{2j-1}, \mathcal{I}_{2j})$. However, only a relatively small number $d^2\ \mathcal{D}/2$ of mismatch scores are then being used for the estimation of confidence regions.

Figure 15.13 illustrates the results of FAR confidence interval estimation using finger and person bootstrap methods. A number of 633,600 non-match scores involving 50 people (100 fingers) are used. It is readily observed that the confidence intervals estimated by the finger subset are contained in the corresponding confidence intervals estimated using the person subset bootstrap. This result again demonstrates that the dependence among the non-match scores is better captured by the person subset than by the finger subset bootstrap method.

15.4 How good are the confidence intervals?

Suppose we have a biometric matcher with match score distribution $F(x)$. When estimating a statistic $\theta(F)$, and its confidence interval, a number of things need to be properly specified:

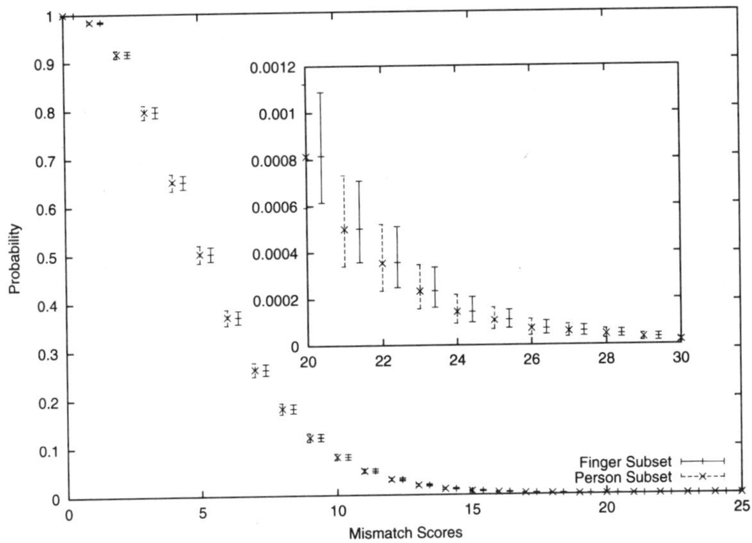

Figure 15.13: The FAR confidence intervals estimated using finger and person subset bootstrap using 633,600 non-match scores. The inset shows an enlargement of a portion of the overall plot.

1. The amount of data to determine the estimate. For an estimate of $\theta(F)$, biometric scores of mated samples are needed.

2. An unbiased statistical estimator of $\theta(F)$. (If the estimate is biased, confidence interval estimation becomes complicated [169].)

3. The desired confidence $(1 - \alpha)$ in the estimate. The true θ will then be in the confidence interval $(1 - \alpha)\,100\%$ of the time.

There exists a *true* confidence interval, and the estimated confidence interval is just *one* estimate. We have been talking about how to estimate this confidence interval; now we turn to the accuracy of the confidence interval estimates.

For simplicity, even though the mean of the match score distribution may not be a very important matcher statistic in itself, we focus on estimating, with confidence $(1 - \alpha)$, the mean $E(x)$ of the this distribution $F(x)$. A set of M samples from $F(x)$ is used to obtain the estimate. This set of match scores $\mathbf{X} = \{X_1, X_2, ..., X_M\}$ is generated by offering M mated biometric sample pairs to the biometric matcher.

15.4.1 The set of samples X

The match score set $\mathbf{X}$ are the only samples we have. The sample mean $\overline{X}$ of this set is our best estimate of $E(x)$. The *true* confidence interval of $\overline{X}$, as specified in (15.4), can be rewritten as

$$\overline{X} - E(x) \in [q(\alpha/2; M),\ q(1-\alpha/2; M)], \tag{15.28}$$

to indicate that the true underlying confidence interval is a function of M. Here $q(\alpha/2; M)$ and $q(1-\alpha/2; M)$ are the true quantiles of the distribution of $\overline{X}$.

This $(1-\alpha)$ 100% confidence interval depends on *both* parameters M and α and both influence the size of the confidence interval

$$\mid Q(\alpha, M) \mid = q(1-\alpha/2; M) - q(\alpha/2; M), \tag{15.29}$$

where $Q(\alpha, M)$ denotes the $(\alpha - 1)$ 100% confidence interval. The parameters of α and M are part of the above basic specifications, but influence (15.29) in different ways:

1. Obviously, the estimate $\overline{X}$ of $E(x)$ becomes better as M becomes larger. When increasing the number M of match scores $\mathbf{X}$, the error in the estimate $\overline{X}$ decreases and $Q(\alpha, M) \to 0$ as $M \to \infty$.

   However, this is only the case if additional match scores X are independent, or at least only weakly dependent on those in the existing set $\mathbf{X}$.

2. The parameter α in (15.29) is a probability $\alpha \in [0, 1]$. In fact, $(1-\alpha)$ is the probability that $\overline{X} - E(x) \in [q(\alpha/2; M), q(1-\alpha/2; M)]$, the true confidence interval.

Repeating it differently, the illustrative example with $\theta(F) = E(x)$ we have been addressing throughout this chapter is:

1. Given the set of match scores $\mathbf{X}$ (and hence, given some M), and some desired confidence measure $(1-\alpha)$,

2. Use set $\mathbf{X}$ to compute the sample mean $\overline{X}$, the estimate of $E(x)$. Further use set $\mathbf{X}$ to estimate the confidence interval, or quantiles.

Suppose we can somehow obtain estimates $\hat{q}(\cdot)$ of the quantiles of the distribution of the error $\overline{X} - E(x)$. Then from (15.5), we get

$$E(x) \in [\overline{X} - \hat{q}(1-\alpha/2; M), \overline{X} - \hat{q}(\alpha/2; M)]. \tag{15.30}$$

If these quantile estimates $\hat{q}(1-\alpha/2; M)$ and $\hat{q}(\alpha/2; M)$ are correct, then with probability $(1-\alpha)$ 100% the true mean $E(x)$ lies in the confidence interval.

Independent of how these quantiles are estimated, the question is whether these estimated confidence intervals are correct. But how can we test the correctness of the interval estimate if the only data we have is $\mathbf{X}$ and we have already used it all to compute the estimates?

15.4.2 More independent samples $\mathbf{X}_1, \mathbf{X}_2, \ldots$

Suppose we have the capability of sampling additional sets of M samples $\mathbf{X}$ from distribution $F(x)$, and label these as $\mathbf{X}_1, \mathbf{X}_2, \ldots, \mathbf{X}_k, \ldots, \mathbf{X}_K$. Simply repeating (15.30) for set $\mathbf{X}_k$ we get

$$E(x) \in [\overline{X}_k - \hat{q}_k(1-\alpha/2; M), \overline{X}_k - \hat{q}_k(\alpha/2; M)], \tag{15.31}$$

which is true with $(1-\alpha)\,100\%$ confidence (if the $\hat{q}_k(\cdot)$ are also correct). This implies

$$\overline{X}_k - E(x) \in [\hat{q}_k(\alpha/2; M), \hat{q}_k(1-\alpha/2; M)], \tag{15.32}$$

with $(1-\alpha)\,100\%$ confidence. Manipulating the confidence interval of (15.30), we also have

$$\overline{X} - E(x) \in [\hat{q}(\alpha/2; M), \hat{q}(1-\alpha/2; M)].$$

with $(1-\alpha)\,100\%$ confidence, a different confidence interval for $E(x)$.

We now have two confidence intervals for $E(x)$ determined by the two different sets of match scores $\mathbf{X}$ and $\mathbf{X_k}$. We would like to test the correctness of the confidence intervals. A possible way to do this is to assume $E(x) = \overline{X}$, and to test if $\overline{X}_k - \overline{X}$ lies in the confidence interval for $E(x)$ determined by $\overline{X}$, the first estimate. This amounts to testing whether

$$\overline{X}_k - \overline{X} \in [\hat{q}(\alpha/2; M), \hat{q}(1-\alpha/2; M)]. \tag{15.33}$$

Checking the validity of (15.33) for one estimate $\overline{X}_k$ does not mean much and we need to generate many estimates $\overline{X}_k, k = 1, ..., K$ and test (15.33) many (K) times.

What percentage of the time will (15.33) be true? How well does the confidence interval $Q(\alpha; M)$ around $\overline{X}$ predict independent estimates $\overline{X}_k$ of $E(x)$?

15.4.3 The match score set X

How do we test (15.33) with the data $\mathbf{X}$ we have? Data is costly to obtain and one would like to use all the data for estimating a particular statistic and its confidence interval. One would rather not give up data for the above independent test data sets $\mathbf{X}_k, k = 1, ..., K$. However, the only way to construct these additional sets is by using match scores from the original data, the match scores with distribution $F(x)$. The match score sets $\mathbf{X}$ that are used for testing the validity of the confidence interval should not be used in the estimation, of course.

Here we are not that concerned with minimizing the error in the estimate of $E(x)$ and it is not that important that M is large. We are just concerned with testing the correctness of the confidence interval estimates, but still there are certainly not enough match scores to construct a sufficient number K of sets $\mathbf{X}_k$. We have to find another way to test

$$X_k - \overline{X} \in [\hat{q}(\alpha/2; M), \hat{q}(1-\alpha/2; M)],\ k = 1, 2, ..., K. \tag{15.34}$$

This can be accomplished by splitting the data $\mathbf{X}$ in half many times. The first half is used to estimate the quantiles $\hat{q}(\alpha/2; M)$, $\hat{q}(1-\alpha/2; M)$ and to compute sample mean $\overline{X}$; the second half of the data is used to compute sample mean estimate $\overline{X}_k$.

Assume that the data set $\mathbf{X}$ of samples from $F(x)$ this time contains $2\,M$ match scores. Then given confidence $(1-\alpha)$ and M—

1. Randomly split $\mathbf{X}$ into two sets, $\mathbf{X}_a$ and $\mathbf{X}_b$, each set containing M match scores.

2. Use set $\mathbf{X}_a$ to compute sample mean $\overline{X}_a$, an estimate of $E(x)$, and construct the confidence interval

$$\overline{X}_a - E(x) \in [\hat{q}_a(\alpha/2; M), \hat{q}_a(1-\alpha/2; M)].$$

3. Use set $\mathbf{X}_b$ to estimate sample mean $\overline{X}_b$.

4. Check whether the estimated difference $\overline{X}_b - \overline{X}_a$ is in the above confidence interval:

$$\hat{Q}_a(\alpha; M) = [\hat{q}_a(\alpha/2; M) : \hat{q}_a(1-\alpha/2; M)].$$

By repeating steps 1 through 4 a large number K times, estimate the probability that $\overline{X}_b$ lies in the confidence interval $\hat{Q}_a(\alpha; M)$ determined by data $\mathbf{X}_a$. In other words, determine what the predictive power of a confidence interval of $E(x)$ is for independently obtained estimates $\overline{X}_b$ of $E(x)$.

15.4.4 Confidence interval validation

The above procedure describes a way to estimate $Prob\,[\overline{X}_b \in \hat{Q}_a(\alpha; M)]$. That is, it estimates the probability that $\overline{X}_b$ falls in the confidence interval for $E(x)$.

We use the data from Table 15.3 that has been used throughout this chapter to illustrate the influence of subset selection for the bootstrap. Two sets $\mathbf{X}_a$ and $\mathbf{X}_b$, as shown in Table 15.5, each of $M = 1{,}120$ match scores are selected. The set $\mathbf{X}_1$ is used directly as $\mathbf{X}_a$; the set $\mathbf{X}_b$ is selected as an independent subset of the $\mathbf{X}_2$ in Table 15.3. The table includes the number of fingers $\mathcal{D}$ and the number of subjects $\mathcal{P}$.

	M	$\mathcal{D}$	$\mathcal{P}$
$\mathbf{X}$	2,240	40	20
$\mathbf{X}_a = \mathbf{X}_1$	1,120	20	10
$\mathbf{X}_b \subset \mathbf{X}_2 \setminus \mathbf{X}_1$	1,120	20	10

Table 15.5: The independent match scores sets $\mathbf{X}_a$ and $\mathbf{X}_b$ are obtained by randomly dividing the set $\mathbf{X}$.

Note that it has not been specified (yet) how the quantile estimates $\hat{q}_a$ and $\hat{q}_b$ are obtained. A test as described in the previous section could be used to verify any confidence interval estimate, no matter how obtained. We return to the bootstrap method for estimating confidence intervals and therefore rewrite our test as

$$\overline{X}_b - \overline{X}_a \in [\hat{q}_a^\star(\alpha/2; M), \hat{q}_a^\star(1-\alpha/2; M)]. \tag{15.35}$$

The estimates $\hat{q}^\star(\alpha/2)$ and $\hat{q}^\star(1-\alpha/2)$ are constructed by bootstrap sampling set $\mathbf{X}_a$. The set $\mathbf{X}_b$ is not sampled with replacement and only used to compute sample mean $\overline{X}_b$.

Remember that three types of bootstrap sampling with replacement have been used: (i) regular bootstrap, (ii) finger subset bootstrap, and (iii) person subset bootstrap. Further, the parametric method for confidence intervals (see Section 15.1.2) has also been introduced. The data described in Table 15.5, i.e., the 2,240 match scores $\mathbf{X}$, are used to test (15.35) by splitting the complete set $\mathbf{X}$ randomly into two sets $\mathbf{X}_a$ and $\mathbf{X}_b$ a large number (K) times.

15.4.5 Sample means versus the true mean

Before conducting our experiments, we should notice that the confidence intervals that we estimate are confidence intervals on the value of the true mean $E(x)$, and must we must ask ourselves what these confidence intervals tell us about the sample means $\overline{X}_b$ that we are estimating. If the true mean $E(x)$ falls in the confidence interval with probability $1-\alpha$, how often will a sample mean $\overline{X}_b$ fall in the same interval?

To answer this question, we examine the distribution of the difference in the two sample means $\overline{X}_a$ and $\overline{X}_b$

$$Dist_{\overline{X}_b - \overline{X}_a, F}(x) \;=\; Prob_F(\overline{X}_b - \overline{X}_a \leq x). \tag{15.36}$$

Assume for the moment that by the central limit theorem, for large sample sizes, both $\overline{X}_a$ and $\overline{X}_b$ are Normally distributed. We arrive at the estimates $\overline{X}_a$ and $\overline{X}_b$ from an equal number M by the same method and therefore assume that the estimates are independently, identically distributed, i.e.,

$$Dist_{\overline{X}_a, F}(x) \;=\; \mathcal{N}(E(x), \sigma^2) \text{ and } Dist_{\overline{X}_b, F}(x) \;=\; \mathcal{N}(E(x), \sigma^2). \tag{15.37}$$

Then

$$Dist_{\overline{X}_b - \overline{X}_a, F}(x) \;=\; \mathcal{N}(E(x), 2\sigma^2). \tag{15.38}$$

With Normal tables $z(x)$ as in Figure 15.2, we can look up the normalized deviation x_α from the mean that gives $(1 - \alpha)\,100\%$ confidence intervals $z(x_\alpha) = 1 - \alpha/2$. The confidence that the sample mean $\overline{X}_b$ falls within this same confidence interval is

$$\alpha_{\overline{X}_b} \;=\; 2\,(1 - z(x_{\alpha/\sqrt{2}})). \tag{15.39}$$

For $\alpha = 0.1$, $x_\alpha \approx 1.635$ and $\alpha_{\overline{X}_b} = 0.25$, that is the 90 percent confidence interval on the true mean will contain the sample mean 75 percent of the time. This figure can be compared with the outcome of the experiments on real fingerprint data. These results, using the regular bootstrap and the subset bootstrap quantiles $\hat{q}^\star_a$, are given in Table 15.6. These are 90 percent confidence intervals, so $\alpha = 0.1$. Clearly, the subset bootstrap confidence intervals show a great improvement because of the isolation of the most obvious sources of dependence in the match scores.

Estimate \ Error	$\overline{X}_b \in \hat{Q}_a(\alpha; M)$
Regular bootstrap	14.49 %
Finger subset	59.15 %
Person subset	73.10 %

Table 15.6: What percentage of times does $\hat{Q}_a(\alpha; M)$ contain $\overline{X}_b$. Here $\alpha = 0.1$ and $M = 1{,}120$. The true confidence intervals should contain $\overline{X}_b$ 75 percent of the time under the Normal assumptions.

As shown in Figure 15.14, the distribution of (15.36) can also be evaluated as a double integral

$$\begin{aligned} Dist_{\overline{X}_b - \overline{X}_a, F}(x) &= Prob_F(\overline{X}_b - \overline{X}_a \leq x) \\ &= \int\int_{\bar{x}_b - \bar{x}_a \leq x} p(\bar{x}_a, \bar{x}_b)\, \mathrm{d}\bar{x}_a\, \mathrm{d}\bar{x}_b. \end{aligned} \tag{15.40}$$

Here $p(\bar{x}_a, \bar{x}_b)$ is the joint probability density function of the two estimates of the mean $\bar{x}_a$ and $\bar{x}_b$, which are just random variables and both are estimates of the mean of $F(x)$. Figure 15.14 shows the situation, the distribution $Dist_{\overline{X}_b - \overline{X}_a, F}(x)$ is the probability mass of $p(\bar{x}_a, \bar{x}_b)$ on the half-plane $\bar{x}_a - \bar{x}_b \leq x$.

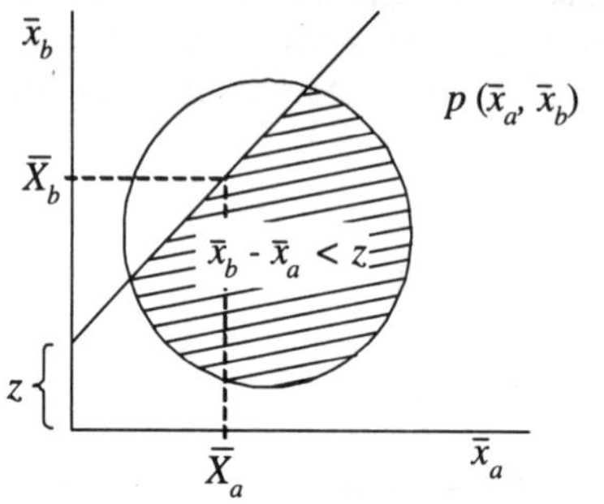

Figure 15.14: The error in the estimate $\overline{X}_b$ depends of $\overline{X}_a$.

If the data $\mathbf{X}_a$ and $\mathbf{X}_b$ are independent, the random variables $\bar{x}_a$ and $\bar{x}_b$ are independent. Since both are estimates of the mean of $F(x)$ obtained from M match scores $\mathbf{X}_a$ and $\mathbf{X}_b$, respectively, we have $p_a(\bar{x}_a) = p_b(\bar{x}_b) = p(\bar{x})$. Therefore, (15.40) becomes

$$Dist_{\overline{X}_b - \overline{X}_a, F}(x) = \int\int_{\bar{x}_b - \bar{x}_a \leq x} p(\bar{x}_a)\, p(\bar{x}_b)\, \mathrm{d}\bar{x}_b\, \mathrm{d}\bar{x}_a, \tag{15.41}$$

which can be developed into a convolution of the probability density function $p(\bar{x})$.

Clearly, the confidence intervals when using subsets instead of match scores for bootstrap sampling with replacement are much more accurate. There is dependency between different match scores X of a biometric match engine, no matter what. For instance, when a set of match scores $\{X_1, X_2, ..., X_M\}$ is obtained with an optical sensor the scores will be consistently higher than if the samples are obtained with a solid-state sensor. This, in itself, may introduce match score dependencies that can never be eliminated with subset sampling. This dependency will also affect the accuracy of the confidence intervals because subsets $\mathcal{X}_i$ and $\mathcal{X}_j$ might be dependent—just because of the sensing device that is used to acquire the biometric samples. Confidence intervals of match and mismatch score should *always* be suspected to be underestimates. But that is just the way it is.

16

Advanced Topics

In this chapter we describe some additional methods for analyzing biometric matchers, and in particular expand on the tools of Chapters 5 and 6, for comparing two or more 1:1 or $1\!:m$ matchers or search engines (Figure 16.1).

Section 16.1 presents matcher comparison prerequisites. Section 16.2 describes what type of test data is needed to compare matchers, then in Section 16.3 we describe the notion of cost functions. After all, minimizing the expected cost of the decisions is a good thing to do if one is able to associate a cost with each error. Section 16.3.1 hints at how one would determine whether the cost associated with matcher a is statistically really different from the cost associated with matcher b. This is a continuation of Chapter 15. Section 16.4 is an exploration of the duality between confidence intervals and hypothesis testing aimed at answering comparative questions about matchers. An interesting aspect is that tests deal with multiple score sets, and the subsets bootstrap sampling principle has to be somewhat modified.

In Section 16.5, we switch gears a little and we further develop a model explaining the relationship between a rank-order statistic (CMC, RPM) and the conventional error rate statistics (FAR/FRR). Section 16.6 explores the rank-order statistics a further by introducing the *finite world model.* The chapter begins, however, with Section 16.1, which summarizes what is needed in general to compare two matchers.

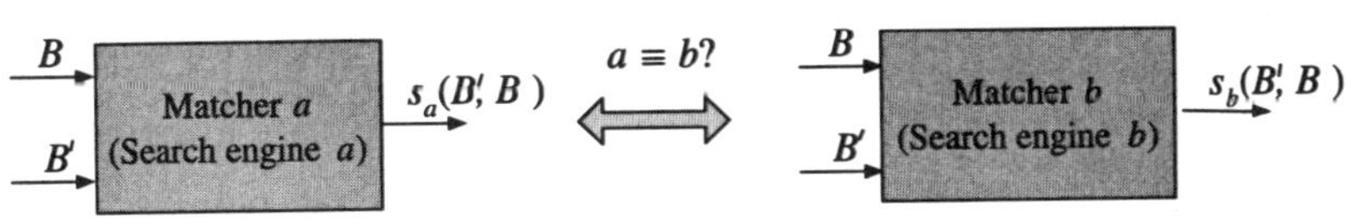

Figure 16.1: Comparing matchers is a matter of comparing many output scores $s_a(B', B)$ and $s_b(B', B)$.

16.1 What is needed to compare matchers?

Four things are needed statistically to compare matchers a and b (and c, d, ...):

Data: Match scores $\mathbf{X}_a = \{X_{a1}, ..., X_{aM}\}$ and mismatch scores $\mathbf{Y}_a = \{Y_{a1}, ..., Y_{aN}\}$, obtained from matcher a are needed. Scores $\mathbf{X}_b$ and $\mathbf{Y}_b$ obtained from matcher b are needed. Hence, a set of match scores and a set of mismatch scores are needed for all matchers a, b, c, and so on that are being compared. This data can be generated using databases of biometric samples acquired from $\mathcal{Q}$. The samples in these databases need to be carefully labeled as to how they are associated to the volunteers $1, ..., \mathcal{Q}$.

Rank sets $\mathbf{K}$ are needed to evaluate the ranking or sorting capabilities of a matcher. In principle, only ranks computed for biometric templates $B'_\ell \in \mathbf{M}$, the enrollment database, should be used under the closed-world assumption. However, it also fine to use input queries B'_ℓ with $\ell > m+1$. The finite world model to express the behavior of a $1{:}\,m$ search engine does not preclude using samples from biometrics that are not in M.

The question is, as noted earlier. How many match and non-match scores are really needed? The numbers are roughly determined by the "*30-error rule*" [171]: if one needs to measure error rates of one in x, roughly $30 \times x$ samples are needed. For example, to measure a FAR in the order of 10^{-5}, some 3,000,000 mismatch scores are needed to generate 30 False Accepts. The 30-error rule also can be used to gauge the number M of match scores that are needed.

Accuracy measures: Given the data, accuracy can be expressed in terms of an ROC curve or by other statistics (for example, the Cumulative Match Curve of Section 16.5). These accuracy measures all express characteristics, and are statistics, of the underlying match score distributions $F_a(x)$ and mismatch score distributions $G_a(y)$ of the matchers that are being compared.

Evaluating a matcher amounts to estimating these statistics using sample sets $\mathbf{X}_a$, $\mathbf{X}_b$, ... and $\mathbf{Y}_a$, $\mathbf{Y}_b$, ... Comparing two matchers a and b is the comparison of two accuracy estimates, the difference between the estimates.

Evaluating a $1{:}\,m$ search engine amounts to estimating CMC statistics using rank sets $\mathbf{K}_a$, $\mathbf{K}_b$, ... Comparing two $1{:}\,m$ search engines a and b involves the comparison of two CMCs or RPMs.

Confidence intervals: Confidence intervals are best determined using nonparametric techniques that make no assumptions about the underlying distributions of the accuracy estimates. The subsets bootstrap, introduced in Chapter 15, is a good choice. However, because of the dependence between scores, bootstrap sets should be generated in tandem, by sampling with replacement from the subject set with labels $i = 1, ..., \mathcal{P}$. Here $\mathcal{P}$ are the number of volunteers used to collect the data.

Hypotheses: With this modified resampling process hypotheses about matchers a and b (and additional matchers) can be formulated.

"The construction of confidence intervals and hypothesis testing are dual problems in statistical theory" [169], that is, hypothesis testing amounts to reasoning about confidence intervals. One-sided and two-sided tests are easily formulated and the corresponding one-sided and two-sided confidence bounds for comparisons of matchers a and b (and other matchers) are computed in a straightforward fashion. The matchers a, b, c... can be compared in a pair-wise fashion.

It should be noted that the regular bootstrap is valid for i.i.d. random variables only. The random variables are the match scores X and mismatch scores Y. The subsets bootstrap of Section 15.2.2 relaxes the independence assumption to a large extent, but the fact remains that the scores are surely *not* identically distributed. Generally, in the statistics community, the bootstrap has been found to be relatively insensitive to the violation of the identically distributed assumption. Techniques such as described in [21] can be used to cluster the scores, or rather the volunteers; bootstrap confidence region analyses on the separate clusters of subjects can then be performed.

16.2 The data needed to compare matchers

We examine what is needed in terms of test data. For both matchers a and b, match scores $\mathbf{X}_a$, $\mathbf{X}_b$ and mismatch scores $\mathbf{Y}_a$ and $\mathbf{Y}_b$ are needed. Usually these score sets can be generated from databases of samples.

Remembering our notation, as shown in Figure 16.1, we have two biometric matchers, a and b, yielding scores $s_a(B', B)$ and $s_b(B', B)$ based upon pairs of samples derived from "real world" biometrics $\mathcal{B}$ and $\mathcal{B}'$. Each of these matchers has the capability to make a decision about the "YES/NO" hypotheses:

$$H_o : \mathcal{B}' \equiv \mathcal{B} \text{ and } H_a : \mathcal{B}' \not\equiv \mathcal{B}.$$

The question we would like to answer is: "On average, which matcher, a or b, more often makes the correct decision?" Let us try to phrase this question more formally. We step back for a moment to determine how to pose this question and then look at what could be measured about matchers a and b to answer it.

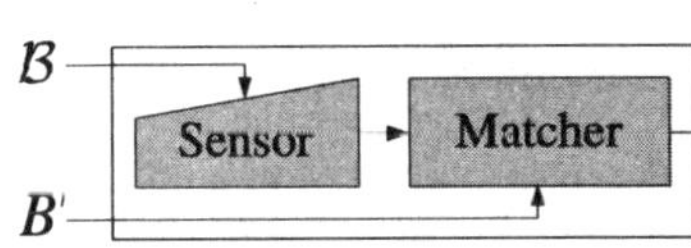

Figure 16.2: A bundled sensor-matcher can only be evaluated while in operation.

Sometimes, matchers come "bundled" with a biometric acquisition device: a fingerprint sensor, a camera for face or iris, a microphone for voice, and so on (see Figure 16.2). In that case, the matcher comparison test is inseparable from the acquisition of the data and indeed other issues such as the selection of the subject populations. Real world biometrics $\mathcal{B}$ and $\mathcal{B}'$, matching and non-matching, are then sampled as part of the comparison test (Table 16.1), and the sampled machine representations (or templates), $B = f(\mathcal{B})$ and $B = f(\mathcal{B}')$ may not be available to the test agency. For these bundled authentication systems,

	$\mathcal{B}$	$\mathcal{B}'$
Sensor a	$B_a = f_a(\mathcal{B})$	$B'_a = f_a(\mathcal{B}')$
Sensor b	$B_b = f_b(\mathcal{B})$	$B'_b = f_b(\mathcal{B}')$

Table 16.1: Sampling the biometrics $\mathcal{B}$ and $\mathcal{B}'$.

the real question that must be answered is then: Which pair is better, (Sensor a, Matcher a) or (Sensor b, Matcher b)?

This can be addressed with a *scenario evaluation* as discussed in Chapter 7. As shown in Table 16.2 on the left, the scores s_a and s_b *and* the *biometric samples* B_a and B_b are functions of the matcher. The biometric samples, therefore, are generated as part of the comparison test. Further discussion of scenario evaluations can be found in [132]. For such a test volunteers must be found to present the biometrics. It is very difficult to acquire truly representative biometric data while at the same time maintaining the rigor needed for a principled test, mainly because the problem of biometric authentication is hard to define in general terms. Biometric systems, today, still are too much "one-of-a-kind" systems developed for specific uses and users. Therefore, direct comparison tests between bundled matcher a and bundled matcher b are *not* very good tests of the core 1:1 matchers.

Scenario evaluation		*Technology evaluation*	
Matcher a:	$s_a(B_a, B'_a)$	Matcher a:	$s_a(B, B')$
Matcher b:	$s_b(B_b, B'_b)$	Matcher b:	$s_b(B, B')$

Table 16.2: The data are collected during the test; alternatively, pre-existing databases can be used.

A technology evaluation (as discussed in Chapter 7), on the other hand, takes the question of comparing matcher a and matcher b more literally and concentrates on the match engines s_a and s_b. The question is then better defined, and therefore the answer is easier to give; so we concentrate on technology evaluations.

In a technology evaluation, both matchers are applied to existing, public or proprietary, databases of many biometric samples B and B', as in Table 15.1; and estimated matcher statistics are compared.

16.3 Cost functions in detail

In Chapter 5 we described the use of cost functions, using cost as a single representative figure of merit for an operating point of a matcher that can be compared to the cost of other operating points of the same or other matchers. Specifying the minimum cost is also a way of choosing the operating point for a particular matcher. Here, we describe the details of how costs are used in the NIST (National Institute for Standards and Technology) speaker identification evaluations, and describe the use of statistical tests using costs for comparison of matchers.

The problem of recognizing the identity of a speaker is a task that has been adopted by the speech community. Speaker recognition algorithms are trained on speech data, which is plentifully available. The 2003 NIST speaker recognition evaluation plan [150], perhaps therefore, also focuses on data. The evaluation procedure includes the specification of the training data—e.g., *limited data* and *extended data* (essentially, small and large training sets). The evaluation of speaker recognition algorithms is then defined in terms of the data that may be used for training or enrollment and in terms of a matcher quality measure based on "cost."

The expected cost, as introduced in Chapter 5, is a refinement of the expected overall error $E(t)$ of (5.13). The NIST speaker recognition evaluation uses cost to express the quality of a speaker recognition system as a *single* number. A numerical measure of cost is defined as

$$Cost = \sum_{\text{allCnd}} \{C_{\text{Error}|\text{Cnd}} \times P_{\text{Error}|\text{Cnd}} \times P_{\text{Cnd}}\}, \tag{16.1}$$

where

$$\begin{aligned} C_{\text{Error}|\text{Cnd}} &= \text{the cost of making an error when condition = Cnd} \\ C_{\text{Error}|\text{Cnd}} &= \text{the probability of an error when condition = Cnd} \\ P_{\text{Cnd}} &= \text{the } a\ priori \text{ probability that condition = Cnd.} \end{aligned}$$

The conditions "Cnd" are what we have been calling hypotheses. Therefore, (16.1) is a technique for weighting the hypotheses which we can write in our standard notation:

Cnd $= H_o$: The speaker (subject *d*) is genuine and the probability of falsely rejecting the subject is FRR. The cost of such an error is C_{FR}.

Cnd $= H_a$: The subject *d* is an imposter and the probability of falsely accepting the subject is FAR, the cost of this error is C_{FA}.

The expected decision cost of (16.1) then becomes the sum of two costs

$$Cost = C_{\text{FR}} \times \text{FRR} \times P_{\mathcal{G}} + C_{\text{FA}} \times \text{FAR} \times (1 - P_{\mathcal{G}}), \tag{16.2}$$

where $P_{\mathcal{G}}$ is the prior probability of a genuine user and $1 - P_{\mathcal{G}}$ the prior probability of an intruder. The best speaker recognition algorithm is then the algorithm with the lowest cost.

The advantage of a cost model is that the model parameters can be applied for different applications by using different parameters. For example, the parameter values in Table 16.3 are used for the primary 2003 evaluation criterion of a speaker recognition system. However, it is not explained why these particular numbers are selected.

C_{FR}	C_{FA}	$P_{\mathcal{G}}$
10	1	0.01

Table 16.3: Speaker recognition cost model parameters for the primary evaluation decision strategy.

It is noted in [150] that a disadvantage of this cost measure is that the actual value $Cost$ of (16.2) lacks intuitive interpretation and therefore is normalized. (Though in other circumstances the costs could be expressed in, say dollar terms, giving the expected cost a real meaning.) Normalization proceeds as follows: suppose, as in Table 16.3,

$$C_{\mathrm{FR}}\, P_{\mathcal{G}} \;<\; C_{\mathrm{FA}}\,(1 - P_{\mathcal{G}}),$$

then by rejecting every speaker, the minimal cost that can be achieved without doing any computations is $Cost = C_{\mathrm{FR}} P_{\mathcal{G}}$. Therefore, the cost is normalized as

$$C_{\mathrm{Norm}} \;=\; \mathrm{FRR} + \frac{C_{\mathrm{FA}}(1 - P_{\mathcal{G}})}{C_{\mathrm{FR}} P_{\mathcal{G}}}\,\mathrm{FAR} \;=\; \mathrm{FRR} \,+\, w_{\mathrm{FA}} \times \mathrm{FAR}, \qquad (16.3)$$

with $w_{\mathrm{FA}} > 1$ and $C_{\mathrm{Norm}} \in [0,\, 1 + w_{\mathrm{FA}}]$. Hence, C_{Norm} of a speaker recognition algorithm, given training and testing data, is on a known scale defined by the model parameters of Table 16.3.

If, on the other hand, the model parameters are specified differently and

$$C_{\mathrm{FR}}\, P_{\mathcal{G}} \;>\; C_{\mathrm{FA}}\,(1 - P_{\mathcal{G}}),$$

the cost is normalized as

$$C_{\mathrm{Norm}} \;=\; w_{\mathrm{FR}} \times \mathrm{FRR} \,+\, \mathrm{FAR}, \qquad (16.4)$$

with $C_{\mathrm{Norm}} \in [0,\, 1 + w_{\mathrm{FR}}]$ $(w_{\mathrm{FR}} > 1)$.

The normalized cost can be expressed as a function of operating threshold t. For example, using (16.3) the weighted errors can be related to each other with a two-dimensional curve

$$\mathrm{CST}_{\mathrm{Norm}}(T) \;=\; (\mathrm{FRR}(T),\; w_{\mathrm{FA}} \times \mathrm{FAR}(T)), \qquad (16.5)$$

which is an "ROC-like" curve with the x-axis scaled to $[0, w_{\mathrm{FA}}]$. Figure 16.3 gives an example. Weighing the errors equally, the minimal achievable cost operating threshold is $T = t_e$. When more weight is given to a False Accept error, the FAR is lower, at the expense of a higher FRR and the minimal achievable cost operating point is $T = t_w$.

The NIST method of selecting the final winner involves receiving a score file and the suggested threshold from each of the participants. NIST computes the FAR and FRR of the system at the threshold supplied and computes the cost for each participant. The system with the computed lowest cost is declared the winner. There are two issues with such a process. First, no confidence intervals are computed. Hence, the costs can be very close and the winner may have a statistically insignificant lower cost than the next participant. Second, the wrong choice of threshold can make a real winner lose the competition. Note that only NIST has the ground truth information on the test data. Hence, from the scores, NIST could compute the cost at various thresholds T for all participating algorithms following (16.3) as

$$C_{\mathrm{Norm}}(T) \;=\; \mathrm{FRR}(T) \,+\, w_{\mathrm{FA}} \times \mathrm{FAR}(T).$$

The lowest cost achieved by the recognition algorithms by varying the threshold T could be used to declare a winner, as this can be in principle less than the error rate observed at the threshold supplied by the participants. That is, each participating algorithm could

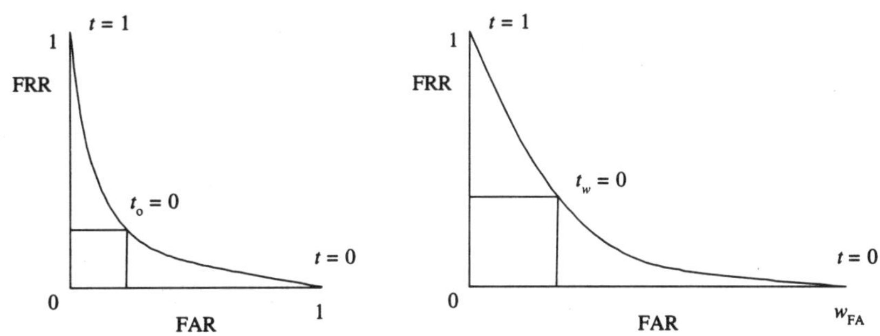

Figure 16.3: Weighing the errors means choosing another operating point.

have another threshold that results in lower cost on the test data, however, they will not be declared the winner. Another method would be to compute the lowest cost achievable by an algorithm and declare the winner based on the "lowest achievable" cost. As the evaluation is phrased, the best threshold for a speaker recognition algorithm is to use the same threshold with lowest cost that is computed from the "training data" for which the ground truth is available. Assuming the test data represents the training data well enough, the evaluation process can be assumed to work well.

We are glossing over some details here that arise if one realizes that the training data here really is meant to be *enrollment data* for database **M** of a $1{:}\,m$ search engine.

16.3.1 Statistical tests using costs

As noted above, when comparing the costs of matcher a and matcher b, ideally confidence intervals on the costs, $Cost_a$ and $Cost_b$ (16.2) for matcher a and b, should be constructed. Or, put better, it should be statistically determined whether

$$(Cost_a - Cost_b) = 0 \text{ or } (Cost_a - Cost_b) \neq 0. \tag{16.6}$$

In other words, is the difference of the estimated costs of two matchers statistically significant? One would like to develop a statistical test to determine if the expected cost, as defined in (16.2), is really different for the two matchers a and b.

Using the NIST normalization procedure, the question of whether the expected costs of matcher a and matcher b are different can be phrased like (16.6). The corresponding hypotheses for the normalized cost C_{Norm} of (16.3) are

$$\begin{array}{lrcl} H_o: & (\text{FRR}_a + w_{\text{FA}} \times \text{FAR}_a) - (\text{FRR}_b + w_{\text{FA}} \times \text{FAR}_b) & = & 0, \\ H_a: & (\text{FRR}_a + w_{\text{FA}} \times \text{FAR}_a) - (\text{FRR}_b + w_{\text{FA}} \times \text{FAR}_b) & \neq & 0. \end{array} \tag{16.7}$$

If costs are not available, we can construct a similar statistical test using the expected overall error (5.13) introduced in Chapter 5.

Remember, the expected overall error for (say) matcher a is given by

$$\mathrm{E}_a(t_a) = \mathrm{FAR}_a(t_a) \times P_{\mathcal{I}} + \mathrm{FRR}_a(t_a) \times P_{\mathcal{G}}, \tag{16.8}$$

and the hypotheses of (16.7) become

$$\begin{array}{llll} H_o: & E_a(t_a) - E_b(t_b) & = & 0, \\ H_a: & E_a(t_a) - E_b(t_b) & \neq & 0. \end{array} \tag{16.9}$$

What is it that we need to do in order to say something statistically meaningful about this? And what is it that we need to have in terms of data?

1. As noted in Section 16.2, we need match scores as well as mismatch scores measured using *both* matcher a and using matcher b. We denote these sets as in Table 16.4. (The sets $\mathbf{X}_a$ and $\mathbf{X}_b$ are now match scores of different matchers and *not* "half" scores sets of the same matcher as in Section 15.4.)

Matcher a		Matcher b	
$\mathbf{X}_a$	$\mathbf{Y}_a$	$\mathbf{X}_b$	$\mathbf{Y}_b$

Table 16.4: Matcher comparison data.

2. We further need to somehow estimate the operating points t_a and t_b for both matchers. We denote these estimates by $\hat{t}_a$ and $\hat{t}_b$.

3. An estimate needs to be constructed for the statistic $E_a(t_a) - E_b(t_b)$, or more precisely for $E_a(\hat{t}_a) - E_b(\hat{t}_b)$. This estimate we denote, maybe a little redundantly, as
$$\hat{E}_a(\hat{t}_a) - \hat{E}_b(\hat{t}_b).$$

Then to decide between the hypotheses (16.9) with $(1-\alpha)\,100\%$ confidence, we need, in addition, a confidence interval on the estimate $\hat{E}_a(\hat{t}_a) - \hat{E}_b(\hat{t}_b)$.

This is an interesting extension of Chapter 15, since we now have four sets of match scores, and again there are dependencies between the scores in each of these sets, but for the estimates developed in that chapter *only one* set of scores, either $\mathbf{X}$ or $\mathbf{Y}$, is needed for each estimate. In Section 16.4, we will develop further statistical tests that use four score sets. First, however, to keep things simpler we look at a statistic of a *single* matcher and its confidence interval, which can be estimated using a set of match scores $\mathbf{X}$ and a set of mismatch scores $\mathbf{Y}$.

16.3.2 Estimating operating thresholds

Even though the Equal Error Rate (EER) may not be a very meaningful matcher statistic, we use it here as an example of estimating an operating point using two sets of match scores, $\mathbf{X}$ and $\mathbf{Y}$. The EER, of course, is reached at the operating point t_e where $\mathrm{FRR}(t_e) =$

$\text{FAR}(t_e)$ and we denote the estimate of t_e as $\hat{t}_e$. This estimate is easily found by determining $\hat{t}_e$ such that

$$\frac{|\{X_i \leq \hat{t}_e\}|}{|\{X_i\}|} = \frac{|\{Y_j > \hat{t}_e\}|}{|\{Y_j\}|}. \tag{16.10}$$

Constructing a confidence interval for this estimate requires us to revisit our subset sampling with replacement.

Until now B bootstrap sets like $\{\mathbf{X}_1^\star, ..., \mathbf{X}_B^\star\}$ have been generated in isolation. The set $\mathbf{X}$ is divided up into $\mathcal{M}$ subsets $\mathcal{X}_i$ and the bootstrap sets $\mathbf{X}^\star$ are obtained by sampling the subsets with replacement $\mathcal{M}$ times. The bootstrap sets $\mathbf{Y}^\star$ are obtained in a similar fashion. Here $\mathcal{M}$ is the number of biometric "entities" (e.g., *either* the number of fingers or the number of volunteers presenting fingers); we continue here with $\mathcal{M} = \mathcal{P}$, the number of volunteers (subjects), since this subdivision gives the more accurate confidence intervals.

We have to rethink the bootstrap sampling a little. The above bootstrap sets are obtained from the samples in Table 15.1 pretty much in an independent fashion. However, there is obvious interdependence between sets $\mathbf{X}$ and $\mathbf{Y}$ because the sets are obtained from the same subjects. There is interdependence between individual match scores X_i and non-match scores Y_j and this dependence has to be somehow replicated into the bootstrap sets $\mathbf{X}^\star$ and $\mathbf{Y}^\star$ when sampling with replacement.

In Chapter 15, it is shown that the sets $\mathbf{X}$ and $\mathbf{Y}$ can be subdivided as

$$\begin{aligned} \mathbf{X} &= \{\mathcal{X}_i,\ i = 1, ..., \mathcal{P}\} \quad \text{and} \\ \mathbf{Y} &= \{\mathcal{Y}_j,\ j = 1, ..., \mathcal{P}\}, \end{aligned} \tag{16.11}$$

respectively, where the subsets $\mathcal{X}_i,\ i = 1, ..., \mathcal{P}$ and subsets $\mathcal{Y}_j,\ j = 1, ..., \mathcal{P}$ are more or less independent. Here $\mathcal{P}$ is the number of volunteers and, therefore, the number of score subsets, since each volunteer is associated with a subset.

Note that there is also much dependence, of course, between the match score set $\mathbf{X}$ and the mismatch scores $\mathbf{Y}$. After all, the score sets ($\mathcal{X}_i$ and $\mathcal{Y}_j$) are associated with the same volunteer for $i = j$. Therefore, we combine the sets of (16.11) into a set of pairs of subsets:

$$(\mathbf{X}, \mathbf{Y}) = \{(\mathcal{X}_i, \mathcal{Y}_i),\ i = 1, ..., \mathcal{P}\}, \tag{16.12}$$

and sample with replacement from $(\mathbf{X}, \mathbf{Y})$ instead. A person subset bootstrap confidence interval for $\hat{t}_e$ is now fairly straightforward to construct. That is, B times do—

1. Generate two bootstrap sets $\mathbf{X}^\star$ and $\mathbf{Y}^\star$ in tandem by simultaneously sampling with replacement $\mathcal{P}$ set pairs from $(\mathbf{X}, \mathbf{Y})$ of (16.12). The two bootstrap sets are

$$(\mathbf{X}^\star, \mathbf{Y}^\star) = \{(\mathcal{X}_1^\star, \mathcal{Y}_1^\star), ..., (\mathcal{X}_\mathcal{P}^\star, \mathcal{Y}_\mathcal{P}^\star)\}.$$

2. Compute the corresponding bootstrap estimate $\hat{t}_e^\star$ using sets $\mathbf{X}^\star$ and $\mathbf{Y}^\star$.

This gives B different bootstrap estimates $\hat{t}_e^\star$ for which we can construct an empirical distribution by ranking the estimates as

$$\hat{t}_{e(1)}^\star \leq \hat{t}_{e(2)}^\star \leq ... \leq \hat{t}_{e(B)}^\star.$$

A bootstrap estimate of a $(1-\alpha)100\,\%$ confidence interval for the estimate $\hat{t}_e^\star$ is obtained in the usual fashion. That is,

$$\hat{t}_e \in [\hat{t}^\star_{e(k_1)}(\alpha;\mathcal{P}), \hat{t}^\star_{e(k_2)}(\alpha;\mathcal{P})], \tag{16.13}$$

where $k_1 = \lfloor(\alpha/2)\,B\rfloor$ and $k_2 = \lfloor(1-\alpha/2)\,B\rfloor$. This time, we write the quantile estimates as $\hat{t}^\star_{e(k_1)}(\alpha;\mathcal{P})$ and $\hat{t}^\star_{e(k_2)}(\alpha;\mathcal{P})$ to remind ourselves that they are functions of both α and $\mathcal{P}$—the number of volunteers used in the comparison.

16.4 Statistical tests at fixed FRR or FAR

A biometric matcher is often tuned to operate at some decision threshold t_o and only the piece of the operating curve around this threshold is of interest. Because of the FRR-FAR trade-off, it is typically specified that the 1:1 matcher should either operate below some fixed FRR or below some fixed FAR. That is, the requirement is given in terms of convenience or security, respectively, and for a given matcher this specification determines the operating threshold t_o.

The value of the particular threshold t_o, however, is not a relevant system parameter when comparing two 1:1 matchers. For fixed FRR (or FAR) matcher a will operate at threshold t_a, $\text{FRR}_a(t_a)$ (or $\text{FAR}_a(t_a)$) while matcher b operates with error rates $\text{FRR}_b(t_b)$ (or $\text{FAR}_b(t_b)$) at threshold t_b. (In the following, when things are clear, we also denote this as $\text{FAR}(t_a)$, $\text{FRR}(t_a)$ or FAR_a, FRR_a.)

Nevertheless, one probably wants to compare 1:1 matcher a and 1:1 matcher b at some specified error rate, a required FRR or FAR:

1. Set the FRR for both matchers a and b to specification, which gives two thresholds t_a and t_b, with in general $t_a \neq t_b$.

   Compare the corresponding False Accept Rates $\text{FAR}(t_a)$ and $\text{FAR}(t_b)$.

2. Set the FAR for both matchers to specification, this gives two new thresholds t_a and t_b (but again with $t_a \neq t_b$).

   The corresponding False Reject Rates $\text{FRR}(t_a)$ and $\text{FRR}(t_b)$ are then compared.

The better 1:1 matcher is then the matcher with the lower FAR or the lower FRR, respectively. With this, accuracy comparison of matcher a and matcher b becomes a matter of hypothesis formulation and testing. Questions about two matchers can be—

1. At some specified minimum False Reject Rate FRR, is matcher a more accurate (better) than matcher b?

2. At some specified False Accept Rate FAR, is matcher a equivalent in accuracy to matcher b, or not?

We concentrate on the latter question, where the False Accept Rate (security!) is specified. (The analysis is valid for all sorts of questions one might have about 1:1 matchers, though.)

Comparing matchers then is a matter of hypothesis testing at some fixed FAR:

$$\begin{aligned} H_o: &\ \text{matcher } a &=&\ \text{matcher } b, \\ H_a: &\ \text{matcher } a &\neq&\ \text{matcher } b. \end{aligned} \tag{16.14}$$

Thus the matchers operate at $\text{FAR}_a(t_a) = \text{FAR}_b(t_b) = \text{FAR}$ and the hypotheses are tested by comparing $\text{FRR}_a(t_a)$ and $\text{FRR}_b(t_b)$.

In the following we formulate statistical tests for deciding between hypotheses like (16.14). Classical contingency tests are one way of answering this question, but here we describe a method that derives the confidence intervals and uses these for decision making. We assume that the data, $\mathbf{X}_a$, $\mathbf{Y}_a$, $\mathbf{X}_b$ and $\mathbf{Y}_b$ are available, as in Section 16.3.1.

16.4.1 The FAR is known

First, for simplicity, suppose that we reliably know the thresholds that give the desired FAR for each of the matchers: $\text{FAR}_a(t_a) = \text{FAR}_b(t_b)$, the operating point of interest. The hypotheses of (16.14) then translate to

$$\begin{array}{llcl} H_o: & \text{FRR}_a(t_a) - \text{FRR}_b(t_b) & = & 0 \\ H_a: & \text{FRR}_a(t_a) - \text{FRR}_b(t_b) & \neq & 0. \end{array} \tag{16.15}$$

This situation is shown in Figure 16.4. Deciding (16.15) then amounts to computing a $(1-\alpha)\,100\%$ confidence interval for $\text{FRR}_a(t_a) - \text{FRR}_b(t_b)$ and testing whether zero is included in this interval.

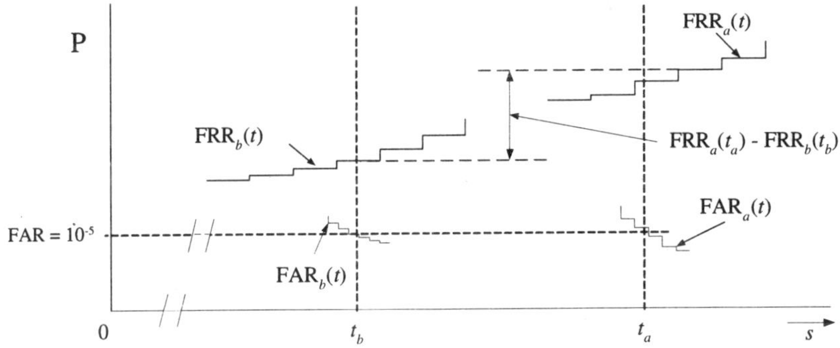

Figure 16.4: Matcher a and matcher b are compared by hypothesis testing $\text{FRR}_a(t_a)$ and $\text{FRR}_b(t_b)$. It is assumed here that $\text{FAR}(t_a) = \text{FAR}_b(t_b) = 10^{-5}$.

How do we compute this confidence interval? Here we are only concerned with the FRR of the matchers and for now we only need the match scores $\mathbf{X}_a$ and $\mathbf{X}_b$ of Table 16.4. These match scores are obtained from $\mathcal{P}$ subjects, and hence, both sets can be divided up into subset match scores $\mathcal{X}_{ai}$ and $\mathcal{X}_{bi}$, associated with the subjects $d_i, i = 1, ..., \mathcal{P}$. These sets of subsets can then be bootstrap sampled with replacement to obtain $(\mathbf{X}^\star_{aj}, \mathbf{X}^\star_{bj}), j = 1, ..., B$:

$$\begin{array}{lcl} \mathbf{X}_a & = & \{\mathcal{X}_{ai},\ i=1,...,\mathcal{P}\} \ \Rightarrow\ \{\mathbf{X}^\star_{a1},...,\mathbf{X}^\star_{aB}\}, \\ \mathbf{X}_b & = & \{\mathcal{X}_{bi},\ i=1,...,\mathcal{P}\} \ \Rightarrow\ \{\mathbf{X}^\star_{b1},...,\mathbf{X}^\star_{bB}\}. \end{array} \tag{16.16}$$

Key here is that the set of subjects $\{1, 2, ..., \mathcal{P}\}$ are sampled with replacement B times and that for each sampling the two bootstrap sets $(\mathbf{X}^\star_a, \mathbf{X}^\star_b)$ are constructed in the same fashion.

From this, B bootstrap estimates of $\mathrm{FRR}_a(t_a) - \mathrm{FRR}_b(t_b)$ can be computed as

$$\mathrm{DFRR}_j^\star = \mathrm{FRR}_{aj}^\star(t_a) - \mathrm{FRR}_{bj}^\star(t_b),\ j = 1, ..., B.$$

These estimates, again, can be ordered from small to large

$$\mathrm{DFRR}_{(j)}^\star,\ j = 1, ..., B,$$

and, in the usual fashion, the bootstrap confidence interval is obtained as

$$\Delta\,\mathrm{DFRR} = [\mathrm{DFRR}_{(k_1)}^\star, \mathrm{DFRR}_{(k_2)}^\star]. \tag{16.17}$$

The decision rule becomes

$$\begin{array}{ll} \text{Decide } H_o \text{ is true:} & \text{if } 0 \in \Delta\,\mathrm{DFRR}, \\ \text{Decide } H_a \text{ is true:} & \text{if } 0 \notin \Delta\,\mathrm{DFRR}. \end{array} \tag{16.18}$$

That is, the true value of $\mathrm{FRR}_a(t_a) - \mathrm{FRR}_b(t_b)$ is contained in $\Delta\,\mathrm{DFRR}$ with confidence $(1-\alpha)\,100\%$. Only if zero is not in this interval do we accept, with confidence $(1-\alpha)$, the alternative hypothesis that matcher a and matcher b have different FRR at operating point (specified by FAR), i.e., the matcher with lower observed FRR is *statistically significantly* better than the other.

16.4.2 Estimating the FAR first

Let us more realistically assume that the true operating thresholds t_a and t_b need to be somehow estimated as $\hat{t}_a$ and $\hat{t}_b$. These estimates are obtained *strictly* from the mismatch scores $\mathbf{Y}_a$ and $\mathbf{Y}_b$ of Table 16.4 as follows.

As in Section 16.3.2, we combine the sets of (16.11) as a set of pairs of subsets for each matcher:

$$\begin{array}{rcl} (\mathbf{X}_a, \mathbf{Y}_a) & = & \{(\mathcal{X}_{ai}, \mathcal{Y}_{ai}),\ i = 1, ..., \mathcal{P}\}, \\ (\mathbf{X}_b, \mathbf{Y}_b) & = & \{(\mathcal{X}_{bi}, \mathcal{Y}_{bi}),\ i = 1, ..., \mathcal{P}\}, \end{array} \tag{16.19}$$

Now for $k = 1, ..., B$ perform—

1. Generate two bootstrap sets of ordered pairs ($\mathbf{X}_a^\star$, $\mathbf{Y}_a^\star$), ($\mathbf{X}_b^\star$ and $\mathbf{Y}_b^\star$) by *simultaneously* and *identically* sampling with replacement $\mathcal{P}$ sets of ordered pairs from $(\mathbf{X}_a, \mathbf{Y}_a)$ and $(\mathbf{X}_b, \mathbf{Y}_b)$ of (16.19). The two bootstrap sets are

$$(\mathbf{X}_a^\star, \mathbf{Y}_a^\star) = \{(\mathcal{X}_{a1}^\star, \mathcal{Y}_{a1}^\star), ..., (\mathcal{X}_{a\mathcal{P}}^\star, \mathcal{Y}_{a\mathcal{P}}^\star)\}.$$

   and

$$(\mathbf{X}_b^\star, \mathbf{Y}_b^\star) = \{(\mathcal{X}_{b1}^\star, \mathcal{Y}_{b1}^\star), ..., (\mathcal{X}_{b\mathcal{P}}^\star, \mathcal{Y}_{b\mathcal{P}}^\star)\}.$$

2. We reorder $(\mathbf{X}_a^\star, \mathbf{Y}_a^\star)$ as

$$(\mathbf{X}_a^\star, \mathbf{Y}_a^\star)' = \{(X_{a(1)}^\star, Y_{a(1)}^\star), (X_{a(2)}^\star, Y_{a(2)}^\star), ..., (X_{a(\mathcal{P})}^\star, Y_{a(\mathcal{P})}^\star)\}$$

such that

$$Y^\star_{a(1)} \leq Y^\star_{a(2)} \leq ... \leq Y^\star_{a(k)} \leq ... \leq Y^\star_{a(\mathcal{P})},$$

from the lowest to the highest mismatch score.

Now with the requirement that the False Accept is FAR, we get threshold $\hat{t}_a = Y^\star_{a(k_o)}$ such that

$$k_o = \lfloor \mathcal{P} \times \text{FAR} + 1 \rfloor.$$

Construct set

$$\mathbf{X}^t_a = \{X^\star_i | (X^\star_i, Y^\star_i) \in (\mathbf{X}_a, \mathbf{Y}_a), Y^\star_i > \hat{t}_a\}$$

to estimate $\text{FRR}_a(\hat{t}_a)$.

3. Similarly, construct set $\mathbf{X}^t_b$ based on threshold $\hat{t}_b$ to estimate $\text{FRR}_b(\hat{t}_b)$.

4. Determine the test statistic $\text{DFRR}(\hat{t}_a, \hat{t}_b) = \text{FRR}_a(\hat{t}_a) - \text{FRR}_b(\hat{t}_b)$.

It is DFRR for which we need to generate B bootstrap estimates

$$\text{DFRR}^\star_k(\hat{t}_a, \hat{t}_b),\ k = 1, ..., B.$$

Ordering these estimates as $\text{DFRR}^\star_{(k)}(\hat{t}_a, \hat{t}_b)$ and following (16.17), we get confidence interval

$$\Delta\,\text{DFRR}(\hat{t}_a, \hat{t}_b) = [\text{DFRR}^\star_{(k_1)}(\hat{t}_a, \hat{t}_b), \text{DFRR}^\star_{(k_2)}(\hat{t}_a, \hat{t}_b)].$$

and the decision rule is the same as (16.18).

We apologize for the rather terse description of this confidence interval construction; it is a straightforward application of the subsets bootstrap of Chapter 15.

16.5 Biometric searching and sorting—the CMC

The Cumulative Match Curve (CMC) that was introduced in Chapter 6 is a performance measure for biometric identification systems (1: m search engines) that return sorted lists of candidates, specifically when the system always returns the identities associated with the K highest-scoring biometric samples from an enrollment database $\mathbf{M}$. The scores of the m biometric samples in $\mathbf{M}$ are usually computed with a biometric 1:1 matcher; that is, the m match scores of the input query biometric and the biometrics in $\mathbf{M}$ are computed and sorted. This is just *one* way to use a 1:1 matcher for searching. While the CMC is most often used for face recognition evaluation, it could be used as a performance measure for any biometric.

Assume that we have a (large) set of biometric samples with associated ground truth. Key to measuring a CMC curve is the assembly of two subsets of samples:

1. A *gallery* set $\mathcal{G}$. There are m biometrics in gallery set $\mathcal{G} = \{B_1, B_2, ..., B_m\}$; these are biometric identifiers of m *different* subjects $d_i, i = 1, ..., m$.

2. A *probe* set denoted as $\mathcal{Q}$ is a set or list of n "unknown" samples $(B'_1, B'_2, ..., B'_n)$ or $\{B'_\ell, \ell = 1, ..., n\}$, associated with the subjects $(d'_1, d'_2 ..., d'_n)$.
   The set $\mathcal{Q}$ can be from any set of individuals. However, usually probe identities are presumed to be in the gallery $\mathcal{G}$. The probe set may contain more than one biometric sample of a given person and need not contain a sample of each subject in $\mathcal{G}$.

Given a biometric $B'_\ell \in \mathcal{Q}$ and a biometric $B_i \in \mathcal{G}$, the output of the biometric matcher is a similarity score $s(B'_\ell, B_i)$. In order to estimate the Cumulative Match Curve, each probe biometric is matched to every gallery biometric and a total of $n \times m$ similarity scores

$$\mathcal{S}_\ell = \{s(B'_\ell, B_1), s(B'_\ell, B_2), ..., s(B'_\ell, B_m)\}, \;\; \ell = 1, ..., n$$

is computed, i.e., we have n sets $\mathcal{S}_\ell$ of m similarity scores each. The scores $s(B'_\ell, B_i), i = 1, .., m$ for each probe biometric B'_ℓ are ordered as

$$s(B'_\ell, B_{(1)}) \geq s(B'_\ell, B_{(2)}) \geq ... \geq s(B'_\ell, B_{(m)}) \tag{16.20}$$

and probe B'_ℓ is assigned the rank $k_\ell = k$ if the matching sample from $\mathcal{G}$ is $B_{(k)}$. Hence the rank estimate $\hat{r}_\ell = k_\ell$ of probe B'_ℓ is k if the matching gallery biometric is in the k-th location of the sorted list of (16.20), or equivalently, if the matching identity is in the k-th location of the associated list (or vector) of identities,

$$\begin{aligned} \mathbf{C}_m(B'_\ell;\, \mathcal{G}) &= (B_{(1)}, B_{(2)}, ..., B_{(k)}, ..., B_{(m)}) \\ &= (d_{(1)}, d_{(2)}, ..., d_{(k)}, ..., d_{(m)}), \end{aligned} \tag{16.21}$$

i.e., when $d'_\ell = d_{(k)}$. We denote this list as $\mathbf{C}_m(B'_\ell;\, \mathcal{G})$ because the sorted list depends on input query B'_ℓ and on the gallery $\mathcal{G}$.

16.5.1 Estimating the CMC

We have set $\mathbf{K}$ of n rank estimates $\{k_\ell; \ell = 1, ..., n\}$ (with $1 \leq k_\ell \leq m$), one estimate for each probe biometric B'_ℓ; each rank is defined *only if* the correct identity is in the ordered list of gallery biometrics $\mathbf{C}_m(B'_\ell;\, \mathcal{G})$ of (16.21).

Before going into the Cumulative Match Curve (CMC), let us first define the discrete rank probabilities $P(r), r = 1, ..., m$, of a biometric search engine. These probabilities associated with a search engine (or associated with the sorting capabilities of a match engine) are simply the probabilities, summing to 1, that the identity associated with a probe has rank r. The $P(r)$ are the true frequencies of occurrence, or discrete probabilities, of ranks $1 \leq r \leq m$ (Figure 16.5). Basically, $P(r), r = 1, ..., m$ could be any discrete Rank Probability Mass (RPM) function. However, a Rank Probability Mass function $P(r)$ as in Figure 16.5, with low average rank, is preferred.

Given the probe and gallery data, the $P(r)$ associated with a biometric matcher, are estimated by

$$\hat{P}(r = k) = \frac{1}{n}\,(\#\, k_\ell = k) = \frac{1}{n}\,(\#\, k_\ell \in \mathbf{K} = k), \;\; k = 1, ..., m. \tag{16.22}$$

The probability $\hat{P}(k)$ that a matching biometric has rank $r = k$ is estimated as the fraction of probe biometrics B'_ℓ for which $k_\ell = k$; $\hat{P}(1)$ is an estimate of the probability that the rank of any probe is 1, $\hat{P}(2)$ is an estimate of the probability that the rank is 2, and so on.

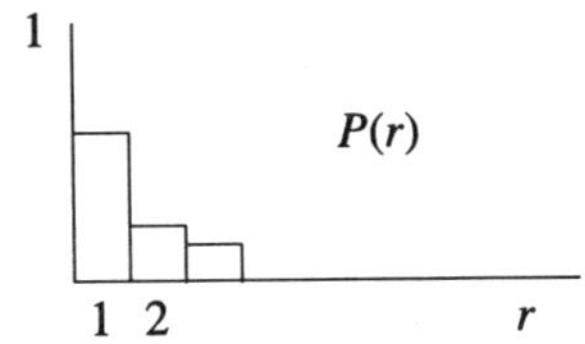

Figure 16.5: The true rank probabilities $P(r), r = 1, ..., m$ of a search engine, estimated by $\hat{P}(k)$ of Expression (16.22).

The Cumulative Match Curve estimates the distribution of the ranks $k_\ell, \ell = 1, ...n$ of probes $\{B'_1, B'_2, ..., B'_n\}$; $\text{CMC}(k)$ is the fraction of probe biometrics B'_ℓ that have rank $\hat{r}_\ell = k_\ell \leq k$. That is,

$$\text{CMC}(k) = \frac{1}{n}\,(\#\, k_\ell \leq k) = \frac{1}{n}\,(\#\, k_\ell \in \mathbf{K} \leq k) = \frac{1}{n} \sum_{\ell=1}^{n} \mathbf{1}\,(k_\ell \leq k). \qquad (16.23)$$

By definition, the true CMC is just

$$\text{CMC}(R) = \sum_{r=1}^{R} P(r); \quad R = 1, ..., m. \qquad (16.24)$$

The CMC estimate (written as $\text{CMC}(K)$) is the distribution of the estimated ranks $\hat{r}_\ell$, denoted as $\hat{r}_\ell = k_\ell$ and estimates the probability $Prob(r_\ell \leq K)$. The random variable r takes on discrete values $1, 2, ..., m$, determined by gallery $\mathcal{G}$ size m. Figure 16.6 shows a Cumulative Match Curve $\text{CMC}(k)$; a higher $\text{CMC}(1)$ and a steeper slope of the curve at $k = 1$ indicate a better 1:1 matcher to implement $1{:}\,m$ search through, for instance, sorting.

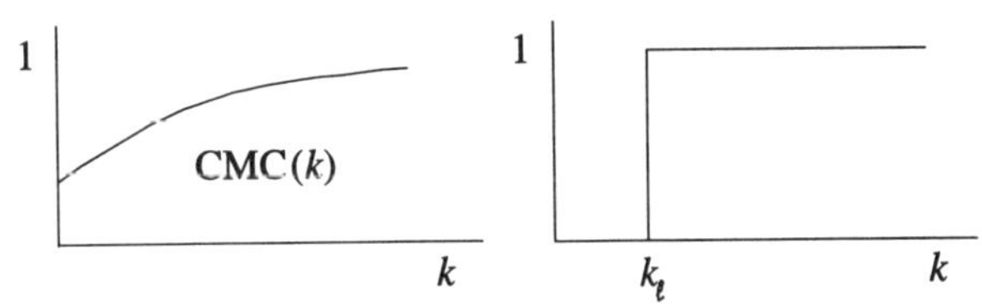

Figure 16.6: The CMC (left) is a weighted sum of step functions $\mathbf{1}(k_\ell \leq k)$ at $k = k_\ell$ as on the right.

Biometric data, organized as a probe $\mathcal{Q}$ and a gallery $\mathcal{G}$ ($\mathcal{G}$ mimics a "test" enrollment database M) are used to estimate the cumulative rank distribution of (16.23). Probes can be constructed in two ways:

1. *Closed universe:* Every probe biometric B'_ℓ is a biometric sample of a subject $d_i, i = 1, ..., n$ in the gallery $\mathcal{G}$ of size m.
2. *Open universe:* Not every probe biometric has to be a biometric of a subject in the gallery, i.e., there are biometrics in probe $\mathcal{Q}$ from outside the world of $\mathcal{G}$.

   That would mean that there are input biometric samples B'_ℓ that are *not* a sample from an identity in $\mathcal{G}$. Sample B'_ℓ then does not match any $B_i, i = 1, ..., m$.

Let us look at both cases.

16.5.2 Closed universe

Because of the closed-world assumption, each probe biometric B'_ℓ has a corresponding gallery biometric B_i and the correct match is always somewhere in the return vector of (16.21). This means that for each probe $\ell = 1, ..., n$, the rank r_ℓ, $1 \le r_\ell \le m$, is defined and can be determined from reordered list

$$\mathbf{C}_m(B'_\ell;\ \mathcal{G}) = (B_{(1)}, B_{(2)}, ..., B_{(m)}). \tag{16.25}$$

To identify an unknown subject d'_ℓ from the subjects $d_i, i = 1, ..., m$ in gallery $\mathcal{G}$, let the decision rule be to choose the highest-scoring candidate in vector $\mathbf{C}_m(B'_\ell;\ \mathcal{G})$ of (16.25) as the correct answer:

$$\text{Decide } d'_\ell \ = d_{(1)}, \tag{16.26}$$

the top answer from ordered vector $\mathbf{C}_m(B'_\ell;\ \mathcal{G}) = (d_{(1)}, d_{(2)}, ..., d_{(m)})$ or equivalently $\mathbf{C}_m(B'_\ell;\ \mathcal{G}) = (B_{(1)}, B_{(2)}, ..., B_{(m)})$.

Selecting the top gallery biometric $\mathbf{C}_1(B'_\ell;\ \mathcal{G}) = B_{(1)} \in \mathcal{G}$ as the correct match may appear to be a good identification strategy, because subject d'_ℓ is in gallery $\mathcal{G}$ and may be believed to be the most likely candidate on the top of the ranked list. The estimated rank probability $\hat{P}(r)$ of (16.22) is an estimate of the probability of correct identification, under this assumption. A subject cannot be falsely rejected and cannot be falsely accepted; he or she can *only* be misidentified.

Alternatively, to make a decision, one can construct a short vector (candidate list) $\mathbf{C}_K(B'_\ell;\ \mathcal{G})$ of identities of the K top-scoring gallery biometrics B_i. This list of identities associated with the top K gallery biometrics, $\mathbf{C}_K(B'_\ell;\ \mathcal{G}) = (B_{(1)}, B_{(2)}, ..., B_{(K)})^T$ (or equivalently, $\mathbf{C}_K(B'_\ell;\ \mathcal{G}) = (d_{(1)}, d_{(2)}, ..., d_{(K)})$) comprises the K most likely identities.

Now, suppose the gallery $\mathcal{G}$ is some "most wanted" list and the decision rule is

$$\begin{array}{ll} \text{Decide } d'_\ell \text{ is "\textit{On list:}"} & \text{if } d'_\ell \in \mathbf{C}_K(B'_\ell;\ \mathcal{G}), \text{ or equivalently,} \\ \text{Decide } d'_\ell \text{ is "\textit{On list:}"} & \text{if rank of } d'_\ell \le K. \end{array} \tag{16.27}$$

That is, a correct decision is deemed to have been made if the true identity of unknown subject d'_ℓ is in the candidate list of length K. This is the *negative identification* problem: "Is subject d'_ℓ on some list?" Of course, the closed world assumes that *every* subject d'_ℓ that is authenticated is on this "wanted" list.

This assumption will be relaxed in Section 16.5.4.

As is the case when we only consider the top candidate on the list (16.26), also for $K > 1$, a subject d'_ℓ cannot be falsely rejected and cannot be falsely accepted, he or she can only be mistakenly left off the candidate list. Again, only identification errors can be made, with probability

$$Prob\,(d'_\ell \text{ is misidentified}) \;=\; Prob\,(d'_\ell \text{ is misidentified; } K) \;=\; \hat{P}_F(K). \qquad (16.28)$$

This error probability depends first of all on candidate list size K and is lower for larger K. It further depends on the gallery size m and is higher for larger m. A third influence is the quality of the match engine, or the distributions of match score and non-match scores. The probability of correct identification then is

$$Prob\,(d'_\ell \text{ is identified}) \;=\; Prob\,(d'_\ell \text{ is identified; } K) \;=\; \text{CMC}(K) \;=\; 1 - P_F(K),$$

where $P_F(K)$ is the probability of misidentification of subject $d'_\ell \in \mathcal{G}$ of (16.28). The Cumulative Match Curve is an estimate of the probability of correct identification as a function of candidate list size K identification

$$\text{CMC}(K) \;=\; \sum_{k=1}^{K} \hat{P}(k), \;\; K = 1, ..., m. \qquad (16.29)$$

This curve $\text{CMC}(K)$ converges to 1 when K approaches m. The faster the convergence, the more likely it is that the correct matches are in the candidate lists for a given biometric matcher.

16.5.3 The CMC, FAR/FRR relation

Even though the decision rule of (16.27) does not use any thresholds, the match engine produces scores $s(B'_\ell, B_i)$, and we can relate the CMC to the False Accept and False Reject rates of the underlying 1:1 biometric matcher used in the $1\!:\!m$ search engine, by realizing that on every search the match score of the correct template can be interpreted as a "virtual threshold" on the non-match scores $s(B'_\ell, B_i)$, $d'_\ell \neq d_i$ from the rest of the database. (Note that this relation between the CMC and the FAR/FRR only holds when the $1\!:\!1$ matcher is used for sorting the gallery scores.)

Each probe $\mathcal{Q}$ is a biometric B'_ℓ with a corresponding biometric B_i in gallery $\mathcal{G}$. Matching probe biometric B'_ℓ to a gallery of size m is equivalent to drawing of m random variables:

1. One match score $X = s(B'_\ell, B_i)$ for $B_i \in \mathcal{G}$ with B'_ℓ and B_i from the same identity.
2. A set of $(m-1)$ mismatch scores $\{Y_1, ..., Y_{m-1}\}$, i.e., $\{s(B'_\ell, B_i), i = 1, ..., (\ell - 1), (\ell + 1), ..., m\}$, where the B_i are in the gallery but not including biometric B_i matching to B'_ℓ.

To look at this, we turn to the rank probabilities of (16.22) first to simplify the analysis

$$\hat{P}(k) \;=\; Prob\,(\text{rank } B'_\ell \text{ is } k), \;\; k = 1, ..., m.$$

These can be determined using joint probabilities $p(x, y_1, ..., y_{m-1})$ of the m random variables $\{s(B'_\ell, B_i), i = 1, ..., m\}$.

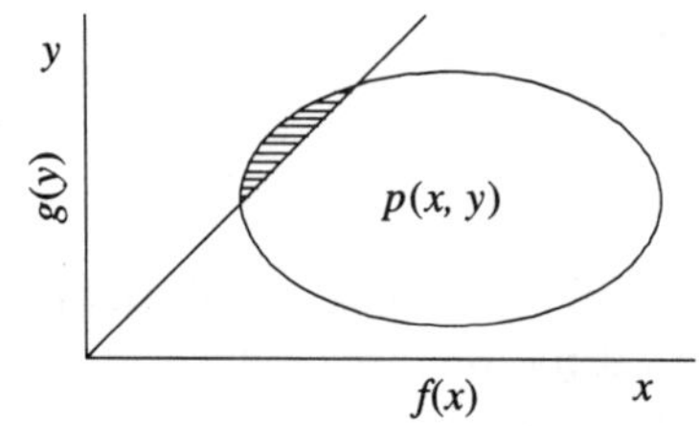

Figure 16.7: The probability of misidentification when gallery size is $m = 2$.

Gallery size $m = 2$

Pick a gallery size of $m = 2$ for the moment; this can be depicted in the xy plane. Any probe biometric B'_ℓ with $\ell = 1$ will result in a pair of random variables (X, Y), with $X = s(B'_\ell, B_1)$ a match score and $Y = s(B'_\ell, B_2)$ a non-match score. The match score is from probability density function (PDF) $f(x)$; the non-match scores are from PDF $g(y)$, each associated with a 1:1 matcher. There exist only two rank probabilities

$$\begin{aligned} P(1) &= Prob\,(\text{rank } B'_\ell \text{ is } 1) = Prob\,(X \geq Y), \\ P(2) &= Prob\,(\text{rank } B'_\ell \text{ is } 2) = Prob\,(Y > X). \end{aligned}$$

Let the joint probability density for a probe (X, Y) be $p(x, y) = f(x)\,g(y)$, assuming independence of the match and non-match score (see Figure 16.7). The rank probability $P(1)$ is the integral of $p(x, y)$ over the area $x > y$ and is the probability of correct identification. Rank probability $P(2) = 1 - P(1)$, the probability of misidentification, is the integral of $p(x, y)$ over the area $x < y$. These two rank probabilities can be written down.

First look at the probability of correct identification, $P(1)$. In Figure 16.7, this is the probability that match score $X > Y$. This is the probability mass for $x > y$

$$\begin{aligned} P(1) &= Prob\,(X > Y) = \int\!\!\int_{x>y} p(x, y)\,\mathrm{d}y\,\mathrm{d}x \\ &= \int_{x=0}^{\infty} f(x) \int_{y=0}^{x} g(y)\,\mathrm{d}y\,\mathrm{d}x = \int_{y=0}^{\infty} g(y) \int_{x=y}^{\infty} f(x)\,\mathrm{d}x\,\mathrm{d}y \\ &= \int_{x=0}^{\infty} f(x)\,[1 - \mathrm{FAR}(x)]\,\mathrm{d}x = \int_{y=0}^{\infty} g(y)\,[1 - \mathrm{FRR}(y)]\,\mathrm{d}y \qquad (16.30) \end{aligned}$$

The interpretation of (16.30) is shown in Figure 16.8. Given any match score $X = x$, the probability that $y < x$ is the probability that y is not a False Accept when adjusting threshold t_o to declare score x a match, i.e., is $[1 - \mathrm{FAR}(x)]$. Conversely, given any y, the probability that $x > y$ is $[1 - \mathrm{FRR}(y)]$, the probability that x is not a False Reject at threshold y.

What is the probability that the correct identity has rank 2, i.e., what is $P(2)$? From Figure 16.7 it is seen that the probability that the non-match similarity score Y is larger

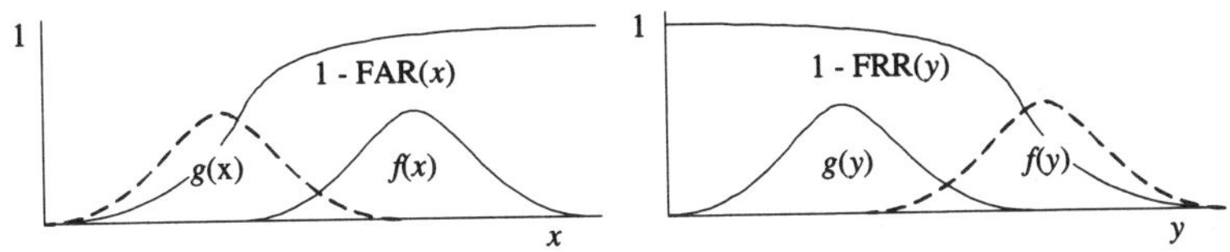

Figure 16.8: For gallery size $m = 2$, the probability of correct identification is the integral of $f(x)[1 - \mathrm{FAR}(x)]$ and $g(y)[1 - \mathrm{FRR}(y)]$ as in (16.30).

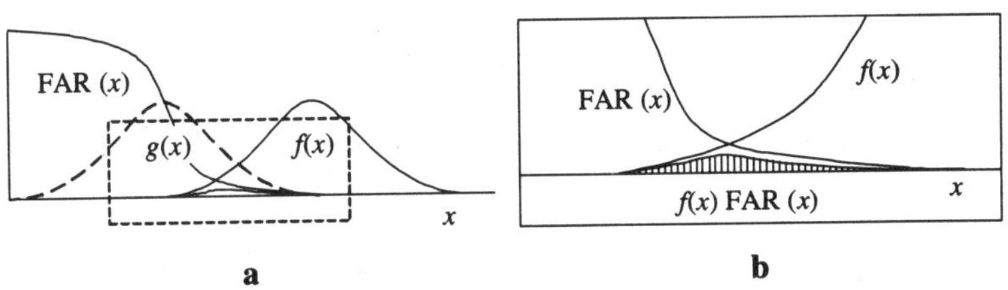

Figure 16.9: For gallery size $m = 2$, the probability of misidentification is the integral of $f(x)\,\mathrm{FAR}(x)$.

than X, i.e., the integral of $p(x, y)$ over the shaded region $y > x$, is

$$P(2) = 1 - P(1) = Prob\,(X < Y) = \int\int_{x<y} p(x,y)\,\mathrm{d}y\,\mathrm{d}x$$

$$= \int_{x=0}^{\infty} f(x)\,\mathrm{FAR}(x)\,\mathrm{d}x = \int_{y=0}^{\infty} g(y)\,\mathrm{FRR}(y)\,\mathrm{d}y. \tag{16.31}$$

Interpretations of these probabilities are as above. The first integral of (16.31) is shown in Figure 16.9a. Given some match score $X = x$, the probability that a non-match score y is larger than x is $\mathrm{FAR}(x)$, the probability of a False Accept at threshold x. The shaded area in the detail of Figure 16.9b is the probability of incorrect identification; this is the integral $f(x)\,\mathrm{FAR}(x)$ over all possible match scores x.

The second integral of (16.31) is an integral of possible values for mismatch scores y. For any y the probability that match score x is less than y is the probability of a False Reject at threshold y; the integral over $g(y)\,\mathrm{FRR}(y) = P(2)$.

Gallery size $m > 2$

When $m > 2$, a probe Q is the drawing of *one* random match score X and the drawing of $(m-1)$ random non-match scores $Y_i,\ i = 1, ..., m-1$.

Let us first examine rank probability $r = 1$, i.e., $P(1)$. This is $Prob$ (rank X is 1), hence the probability that $X > Y_1, X > Y_2, ..., X > Y_{m-1}$. Again assuming independence, the joint probability density of $(x, y_1, y_2, ..)$ is

$$p(x, y_1, ..., y_{m-1}) = f(x) \prod_{i=1}^{m-1} g(y_i).$$

The rank probability $P(1)$ is the integral of $p(x, y_1, y_2, ..)$, as in Figure 16.7, but over a hyper region this time:

$$\begin{aligned} P(1) &= \int_x \int_{y_1<x} \cdots \int_{y_{m-1}<x} p(x, y_1, ..., y_{m-1})\, \mathrm{d}x\, \mathrm{d}y_1 \cdots \mathrm{d}y_{m-1} \\ &= \int_{x=0}^{\infty} f(x) \left(\prod_{i=1}^{m-1} \int_{y=0}^{x} g(y_i)\, \mathrm{d}y_i \right) \mathrm{d}x. \end{aligned}$$

Using (16.30), this becomes

$$P(1) = \int_{x=0}^{\infty} f(x)\, [1 - \mathrm{FAR}(x)]^{m-1}\, \mathrm{d}x. \tag{16.32}$$

Figure 16.10 compares $P(1)$ for $m = 2$ of (16.30) with the $P(1)$ for $m > 2$ of (16.32). The probability $[1 - FAR(x)]^{m-1} \ll [1 - FAR(x)]$ for large m and $Prob$ (rank X is 1) obviously becomes less and less for larger m.

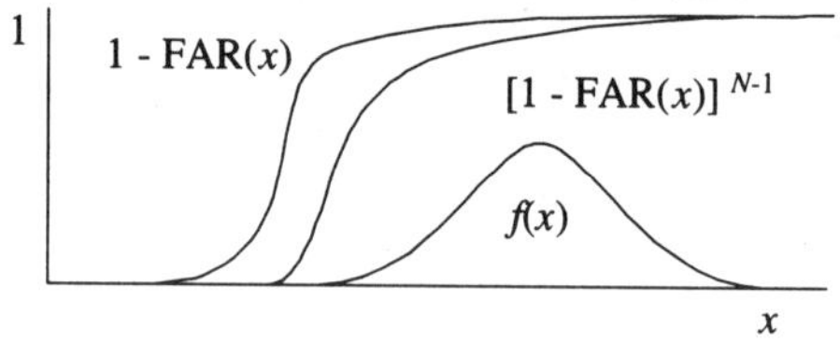

Figure 16.10: The probability $P(1)$ that match score X is assigned rank 1 goes down for increasing m.

By extending the above, we are now ready to construct $P(r)$, the probability that the "some" correct identity ends up in the r-th position of the return vector $\mathbf{C}_m$. The probability that rank X is r is the probability that $(r - 1)$ of the $(m - 1)$ non-match scores $Y_1, ..., Y_{m-1}$ are greater than X. This is given by

$$P(r) = \binom{m-1}{r-1} \int_x \int_{y_1 \cdots y_{k-1} > x} \int_{y_r \cdots y_{m-1} < x} p(x, ..., y_{m-1})\, \mathrm{d}x \cdots \mathrm{d}y_{m-1},$$

accounting for the number of ways $(r - 1)$ non-match scores can be selected from $(m - 1)$. Using (16.30) and (16.31), this becomes

$$\binom{m-1}{r-1} \int_0^{\infty} \left(\prod_{i=1}^{r-1} \int_{y=0}^{x} g(y_i)\, \mathrm{d}y_i \right) f(x) \left(\prod_{i=k}^{m-1} \int_{y=x}^{\infty} g(y_i)\, \mathrm{d}y_i \right) \mathrm{d}x.$$

And we have

$$P(r) = \binom{m-1}{r-1} \int_0^\infty [\mathrm{FAR}(x)]^{k-1} f(x) [1 - \mathrm{FAR}(x)]^{m-r} \, dx. \tag{16.33}$$

The interpretation of (16.33) is as above. Given some match score $X = x$ with probability density $f(x)$, there are $(r-1)$ mismatch scores Y_ℓ greater than x with probability $[\mathrm{FAR}(x)]^{r-1}$. There are an additional $(m-r)$ mismatch scores Y_ℓ that are less than x with probability $[1 - \mathrm{FAR}(x)]^{r-1}$.

An approximate interpretation of (16.33) is the following. If we assume that the match score PDF $f(x)$ is relatively narrow and can be well approximated by an impulse function $\delta(x - \hat{x})$, with $\hat{x}$ the expected match score of B'_ℓ, (16.33) becomes

$$P(r) \approx \binom{m-1}{r-1} [\mathrm{FAR}(\hat{x})]^{r-1} [1 - \mathrm{FAR}(\hat{x})]^{m-r}. \tag{16.34}$$

This is the probability that $(r-1)$ non-match scores Y are greater than $\hat{x}$, *and* $(m-r)$ non-match scores Y are less than $\hat{x}$.

Remember, the Cumulative Match Curve CMC(K) of (16.36) is an estimate of the probability of correct identification as a function of candidate vector $\mathbf{C}_K$ size K. A subject d' is "correctly identified" if the corresponding identity in the gallery has rank K or less, and following (16.24), we get for the true CMC

$$\begin{aligned} \mathrm{CMC}(K) &= \sum_{r=1}^{K} P(r) \\ &= \sum_{r=1}^{K} \binom{m-1}{r-1} \int_0^\infty [\mathrm{FAR}(x)]^{r-1} f(x) [1 - \mathrm{FAR}(x)]^{m-r} \, dx. \end{aligned} \tag{16.35}$$

When using (16.34), this becomes approximately

$$\mathrm{CMC}(K) \approx \binom{m-1}{r-1} \sum_{r=1}^{K} [\mathrm{FAR}(\hat{x})]^{r-1} [1 - \mathrm{FAR}(\hat{x})]^{m-r}, \quad K = 1, ..., m. \tag{16.36}$$

The probability that the true rank r of the correct identity $\leq K$ is the sum of the true individual probabilities that rank is r from $r = 1$ to $r = K$.

Returning the top K candidates is rarely a practical way of running an identification system, since it always yields an ambiguous answer, and the candidate list is of fixed size, regardless of how poor those matches may be. For the human expert it is more effective for them to examine only (but all) those candidates that are close to the probe rather than the procrustean strategy of a fixed-size candidate list. It may be argued that some limit must be imposed on the size of the returned candidate list, but if the list of candidates exceeding the threshold is too long to be dealt with, it cannot be denied that the biometric is too weak for the task, and imposing some limit on the list length will result in a penalty in reliability. Selectivity-reliability curves are to be preferred as measures of identification system performance.

There is one issue with the relation Expression (16.36) between the CMC of a 1:m search engine and the FAR/FRR of a 1:1 match engine. Given the way the CMC is computed by *sorting* $s(B'_\ell, B_i)$ as in (16.20) is only *one* particular way of implementing 1:m search based an a 1:1 search engine. In Section 16.6, we look ar searching as a sequence of pairwise comparisons of (B_i, B_j) with respect to input query B'_ℓ.

16.5.4 Opening up the closed world

What happens if the probe data $\mathcal{Q}$ and gallery data $\mathcal{G}$ are not closed? Assume that next to the $n_{known} = m$ identities that are known to the 1:m identification system through enrollment database M, there are an additional $n_{unknown}$ biometrics of unknown subjects d' in the probe. That is, $n_{unknown}$ imposter biometrics are added to the probe $\mathcal{Q}'$ of new size $n_{known} + n_{unknown} = m + n_{unknown}$.

In this case, misidentifications occur *and* they occur for *each* of the $n_{unknown}$ unknown biometrics in new probe $\mathcal{Q}'$ simply because these imposters are always wrongly identified. The result is that the rank probabilities and the CMC are weighted by the prior probability that $d' \in \mathcal{G}$ (the original gallery, database $\mathbf{M}$)

$$Prob\,(d \in \mathcal{G}) = \frac{n_{known}}{n_{known} + n_{unknown}} = \frac{m}{m + n_{unknown}}.$$

The true rank probabilities $P'(r)$ and the probability of correct identification $P(K) = \text{CMC}(K)$ are lower. We have

$$P'(r) = \frac{n_{known}}{n_{known} + n_{unknown}} \times P(r) = \frac{m}{m + n_{unknown}} \times P(r), \quad r = 1, ..., m,$$

and an estimate of the CMC in operational conditions is

$$\text{CMC}'(K) = \frac{m}{m + n_{unknown}} \times \text{CMC}(K), K = 1, ..., m$$

with $\text{CMC}(K)$ given in (16.35).

The rank probability estimates and the Cumulative Match Curve estimates have no extra information to offer over the closed-world scenario because $P'(r)$ and $\text{CMC}'(K)$ simply incorporate prior probabilities that a subject d' is in gallery $\mathcal{G}$ and hence is known. While the CMC is a useful characterization of a system in controlled evaluations, since it does not take into account the imposters in the "open universe" [164] that are inevitable in any real application, it is only a partial characterization of the performance of a system.

Percentage rank distribution

If we define $\rho = 100 \times K/m$ as the percentage rank and substitute $K = m\,\rho/100$ into (16.29), then $\text{CMC}(\rho)$ is the percentage rank distribution. The variable ρ is still a discrete random variable, but the percentage rank distribution is normalized on a scale $\rho \in [0, 100]$. This allows for the comparison of CMCs for different gallery sizes m. Since the proportion of the non-matching scores that exceed the matching score is independent of the database size, as m increases, increasing K in proportion (evaluating at constant ρ) would give us the same chance of finding the correct candidate in the top K.

However, no matter what is used, the cumulative rank distribution or the Rank Probability Mass (RPM) distribution, the correct identification rate estimates $\mathrm{CMC}(K) = 1 - \hat{P}_F(K)$ are deceptively high when the open-world assumption is used. Phillips in [163] offers an analysis of the CMC and Receiver Operator Characteristics (ROC) curve. Here the CMC is related to the FAR and FRR by exploring the duality between identification and verification.

16.6 Biometric searching and ranking

Chapter 6 and Section 16.5 had some discussion on rank-order statistics; in this section we tentatively explore the issue a little more. Rather than looking at $1\colon m$ search as m operations on the database **M**, we postulate the notion of a rank engine and look at biometric search as a sequence of possibly m^2 ranking operations on pairs (B_i, B_j) of enrolled biometric identifiers in **M**. That is, all biometric identifiers $B_i, B_j \in \mathbf{M}, i \neq j$ are ranked with respect to input query B'_ℓ and exchange places, if necessary, to obtain a reordered vector $\mathbf{C}_m(B'_\ell;\ \mathbf{M})$. Note that a ranking engine is presented here as *one* potential search mechanism going beyond m times 1:1 matching; more generic search frameworks are subject of future research.

16.6.1 Matching versus searching

Enrollment is a prerequisite to be able to model $1\colon m$ biometric search probabilistically, be it large scale, or any scale. Till now, we mentioned enrollment databases **M** or galleries $\mathcal{G}$ of size m, but we perhaps did not really define what we mean by *biometric search.* And we perhaps did not stress enough the differences between biometric matching and biometric searching, which are used for biometric verification and identification, respectively—

1. *Matching* is intended to make a hard decision. Two templates B', B are input to the biometric match engine and a similarity score $s = S(B', B) \geq 0$ is the output. Based on this score, a "YES/NO" decision is made and *two* possible errors, a False Match and a False Non-Match, can be made. Characteristic of a match engine is that *a threshold value t_o is needed* to make the match decision.

2. *Searching* is meant to find close matches in the database **M** of enrolled identities.

   The database **M** consists of m biometric templates $(B_1, B_2, ..., B_m)^T$ corresponding to m distinct identities $(d_1, d_2, ..., d_m)^T$. A query sample $B'_\ell = f(\mathcal{B}'_\ell)$ corresponding to identity d'_ℓ is presented to the search engine and a short candidate list,

$$\begin{aligned} \mathbf{C}_K(B'_\ell;\ \mathbf{M}) &= (B_{(1)}, B_{(2)}, ..., B_{(K)})^T \\ &= (d_{(1)}, d_{(2)}, ..., d_{(K)})^T, \end{aligned} \tag{16.37}$$

   of K identities in **M** is returned as output. The candidates $B_{(k)}$ are ranked $k = 1, ..., K$ according to some measure of "closeness" to B'_ℓ. We denote this candidate list as $\mathbf{C}_K(B'_\ell;\ \mathbf{M})$ to indicate that the list greatly depends on enrollment database **M**.

Note that the parameter K does not have to be fixed; in fact, it is probably best to compute K after the $(B_1, B_2, ..., B_m)^T$ have been ranked. Characteristic of biometric searching is that *no particular threshold value* t_o *is needed* to perform a biometric search optimally. This is, for example, shown in Figure 6.3 in Chapter 6 where two enrolled biometric templates are compared without using some *a priori* fixed threshold t_o. We therefore tentatively claim that in order to perform biometric search, biometric rank capabilities between all pairs, or subsets of pairs, (B_i, B_j) of enrolled templates need to be developed.

Searching is often defined as a closed-world problem in that query biometric B'_ℓ is assumed to be represented in database **M**, i.e., $d'_\ell \equiv d_j$ such that $d_j \in \mathbf{M}, 0 \le j \le m$ (we will relax this assumption later). Here with B'_ℓ we mean another sample of the biometric $\mathcal{B}_\ell$, which for now is assumed to be in database **M**. The expectation of a biometric search is then, of course, that the *true* identity of query biometric B'_ℓ is present in the candidate list; and, if at all possible, the true identity $\mathcal{B}_\ell$ should be number 1 in $\mathbf{C}_K(B'_\ell;\ \mathbf{M})$, i.e., $d_{(1)} = d'_\ell$. This might not be the case, and as such, a biometric search engine can *only* make one type of error. An error is made when the true identity $\mathcal{B}_\ell$ is *not* ranked number 1 or if the true identity is not in $\mathbf{C}_K(B'_\ell)$. This is called a misidentification error.

16.6.2 Biometric search engine

There are two things that should be remembered when comparing searching to matching—

1. A match engine can make two types of mistakes, i.e., a *False Accept* and a *False Reject*.

   A search engine, as defined by (16.37), can only make one type of mistake, *a misidentification*.

2. Match decisions are based on a continuous random variable s, the match scores $s(B', B)$ between templates or samples.

   Search decisions (of any size database **M**) are based on a discrete random variable r (rank) that can take on m values. After all, each biometric $B_i, i = 1, ..., m$ in **M** is assigned a unique rank in each candidate list $\mathbf{C}_K(B'_\ell)$, where B'_ℓ is the query biometric. The rank $k = k_\ell$ of the correct identity $d'_\ell = d_{(k)}$ in list (16.37) is an estimate of true rank r of database entry $B_{(k)}$ with respect to real-world biometric $\mathcal{B}'_\ell$. If $k > K$, with K, the length of the candidate list, B'_ℓ is misidentified.

However, $1\!:\!m$ search cannot be modeled without the notion of enrollment—

Enrollment: Given m subjects $d_i, i = 1, ..., m$, acquire samples from biometrics $\mathcal{B}_i, i = 1, ..., m$. For each $\mathcal{B}_i$ a corresponding B_i is constructed and enrolled in database **M**, which is really just a vector

$$\mathbf{M} = (B_1, B_2, ..., B_m)^T. \tag{16.38}$$

In its simplest form, a biometric database is just an ordered list or a vector, if you will, for example, ordered according to the chronology of enrollment of the subjects $d_i, i = 1, ..., m$.

Some subsystem must have carefully taken biometric samples from our m subjects and somehow must have designed the biometric representations $(B_1, B_2, ..., B_m)$ in enrollment database $\mathbf{M}$.

After this enrollment procedure, each biometric search of a query biometric $\mathcal{B}'_\ell$ against $\mathbf{M}$ can be considered as an operation generating a reordering of the list $\mathbf{M}$. Given some real-life input $\mathcal{B}_\ell$, some biometric representation B'_ℓ of this input is constructed, and the output of the query is, as already indicated in (16.37), the vector

$$\mathbf{C}_m(B'_\ell;\ \mathbf{M}) \ = \ (B_{(1)}, B_{(2)}, ..., B_{(m)})^T. \tag{16.39}$$

This biometric search is just a reordering, or ranking, of the biometric list $\mathbf{M}$. This ranking is a function of the query biometric B'_ℓ and the biometrics in $\mathbf{M}$, as indicated by the notation $\mathbf{C}_m(B'_\ell;\ \mathbf{M})$ of (16.39). The expectation is that sample biometrics $B_{(i)} \in \mathbf{C}_K$ for small i are similar to sample B'_ℓ in appearance.

This is the expectation of a search engine, But can this achieved in reality? And how would this ranking be done optimally? To answer these questions, we propose a rank engine.

16.6.3 Rank engine

As we argued above, to be able to search, one needs to be able to rank biometric pairs and one needs a rank engine. When introducing a match engine that operates with a threshold t_o in Chapter 5, we also briefly introduced a rank engine that does not use any threshold, but we will do it more formally here. This allows us to model the errors that a search engine can make in Section 16.6.4.

A functional view of a rank engine is given in Figure 16.11. A rank engine takes in essence three biometrics, $\mathcal{B}_\ell$, $\mathcal{B}_i$, and $\mathcal{B}_j$, as input and decides if $\mathcal{B}_\ell$ is closer in appearance to $\mathcal{B}_i$ than to $\mathcal{B}_j$. The biometrics $\mathcal{B}_i$ and $\mathcal{B}_j$ are sensed and sampled during enrollment; at that stage the biometric machine representations B_i and B_j are constructed. At query time, a biometric $\mathcal{B}_\ell$ is presented to the rank engine (Figure 16.11). The query biometric is sensed, sampled, and developed into a machine representation B'_ℓ.

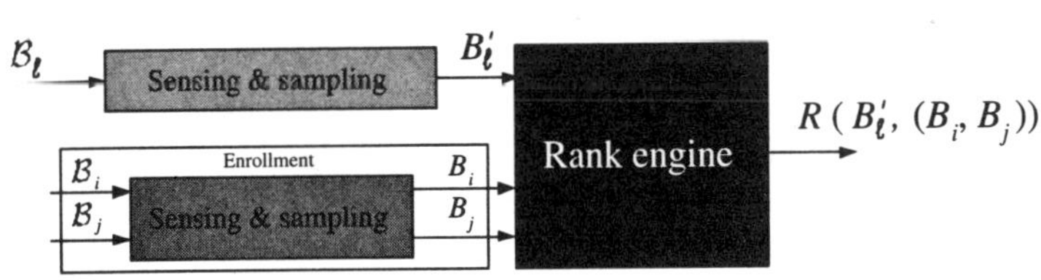

Figure 16.11: A rank engine has the capability to rank enrolled biometric identifiers B_i and B_j with respect to a query biometric identifier B'_ℓ.

The rank engine itself on the right in Figure 16.11 is just a black box that takes as input

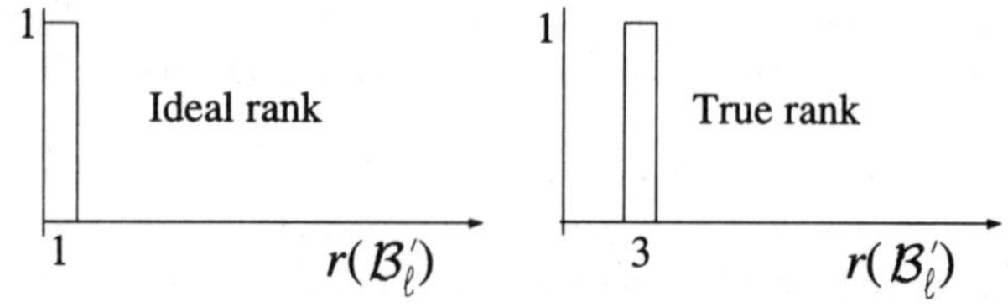

Figure 16.12: The true $r(\mathcal{B}_i)$ when $\mathcal{B}_\ell$ and $i = \ell$ is not necessarily 1.

B'_ℓ, B_i, and B_j, and has as ideal output

$$R(B'_\ell, (B_i, B_j)) = \begin{cases} 1 & \text{if} \quad r(B_i) \leq r(B_j), \\ 0 & \text{if} \quad r(B_i) > r(B_j), \end{cases} \tag{16.40}$$

where $r(B_i)$ and $r(B_j)$ are the ranks of B_i and B_j with respect to B'_ℓ, which are estimates of the ranks of B_i and B_j with respect to real-world biometric $\mathcal{B}_\ell$, any sample of the true biometric, respectively. If in (16.40) the ranks of every B_i and B_j are determined based on scoring of a 1:1 matcher, the application of (16.40) m^2 times is the same as applying m 1:1 match operations on the elements $\{B_i, i = 1, ..., m\}$ in **M**. However, by defining a pairwise ranking, the ranking of most pairs (B_i, B_j) might be irrelevant and the ranking of other pairs could be determined by the 1:1 matcher. For some pairs, $R(B'_\ell, (B_i, B_j))$, e.g., when B_i and B_j are "close" in some sense, the ranking could be computed using not just a 1:1 matcher, but rankers that operate on the three sets of feature representation could be used, e.g., the fingerprint minutiae of fingerprint biometrics, B'_ℓ, B_i, and B_j.

Then, ideally after a search with query sample B'_ℓ is completed, for each ordered pair $(..., B_{(i)}, ..., B_{(i)}, ...)$, with $j > i$, in the ranked list

$$\mathbf{C}_m(B'_\ell;\ \mathbf{M}) = (B_{(1)}, B_{(2)}, ..., B_{(m)}), \tag{16.41}$$

we should have $R(B'_\ell, (B_{(i)}, B_{(j)})) = 1$, meaning that B'_ℓ is closer to $B_{(i)}$ than to $B_{(j)}$. Moreover, if the real-world $\mathcal{B}_\ell \equiv \mathcal{B}_i$, we should have $\forall\ j \neq i$: $R(B'_\ell, (B_i, B_j)) = 1$, and $B_{(1)}$ should equal B_i. Further, this should be the case for every $i = 1, ..., m$.

Clearly, this is only achievable for databases **M** that can be ordered for each instance of B'_ℓ so that $B_{(1)}$ is associated with the true biometric $\mathcal{B}_i = \mathcal{B}_\ell$, i.e., with the correct identity rank 1. This should be the case for all query biometric samples $B'_\ell = f(\mathcal{B}_i)$ from all enrolled identities $\mathcal{B}_i \in \mathbf{M}$, $i = 1, ..., m$. This, of course, is rarely possible in real situations and search engines make mistakes in that the correct identity is not always rank 1. For example, the true, ideal (but not very realistic) rank probability of $\mathcal{B}_i \in \mathbf{M}$ for query $B'_\ell = f(\mathcal{B}_i)$ could be $Prob(\text{rank } \mathcal{B}_i = 1) = 1$. On the other hand, the true (but not very realistic) rank probability of $\mathcal{B}_i \in \mathbf{M}$ for query $B'_\ell = f(\mathcal{B}_i)$ could also be, e.g., $Prob(\text{rank } \mathcal{B}_i = 3) = 1$ (see Figure 16.12). Or, in more realistic situations, we will have $E(r_i) = 1$ or $E(r_i) = 3$.

16.6.4 Search errors

To model the search errors, we have to determine what the random variable is. In the case of a match engine, there exists a (typically, continuous) random variable s, called score. In contrast to score s, rank r is a discrete random variable, simply because of the inherent finiteness m of the probabilistic model for r. The probability of ranks of a $1{:}m$ search engine is characterized as a probability mass distribution $P(r)$, defined for $r = 1, ..., m$.

Searching is repeated ranking

The actual implementation of (16.39) could be the repeated application of 1:2 *ranking* on pairs (B_i, B_j) in $\mathbf{M}$. For each pair (B_i, B_j)—

$$\begin{array}{lll} \text{if} & R(B'_\ell, (B_i, B_j);\ \mathbf{M}(m)) = 1 & \text{do nothing,} \\ \text{if} & R(B'_\ell, (B_i, B_j);\ \mathbf{M}(m)) = 0 & \text{swap } B_i \text{ and } B_j, \end{array} \tag{16.42}$$

till no more changes can be made. Searching is then like sorting and the algorithm of (16.42) is an inefficient $O(m^2)$ bubble sort. To reorder $\mathbf{M}$ (a list) as in (16.42), $O(m^2)$ operations are needed. More efficient $O(m \log m)$ algorithms might be developed.

By extensive pre-processing of $\mathbf{M}$, the algorithmic complexity of (16.42) might become linear in m or even sub-linear in m so that the $1{:}m$ search system is scalable. Pre-computing the rank functions $R(B'_\ell, (B_i, B_j);\ \mathbf{M}(m))$ may involve the complete database $\mathbf{M}$, as is in indexing approaches to biometric search [75]. Extensive training of $R(B'_\ell, (B_i, B_j);\ \mathbf{M}(m))$ seems to be the approach taken, for example, in speaker identification systems [150].

The most important thing to point out is, that a $1{:}m$ search engine cannot be developed without the enrollment of m subjects in database $\mathbf{M}$. If one has access to database $\mathbf{M}$, the rank functions $R(B'_\ell, (B_i, B_j);\ \mathbf{M}(m))$ can be pre-computed and perhaps more efficient search algorithms can be designed.

A search engine makes rank errors

Remembering the search list (16.39),

$$\begin{aligned} \mathbf{C}_K(B'_\ell; \mathbf{M}) &= (B_{(1)}, B_{(2)}, ..., B_{(K)})^T \\ &= (d_{(1)}, d_{(2)}, ..., d_{(K)})^T, \end{aligned}$$

we see that the output of a large-scale search of database $\mathbf{M}$ is a reordered list or vector. The output is always a list (vector) and is *not* some sort of distribution—there is only one biometric $B_{(k)}$ at location k. This is because we are dealing with a *finite world* or usually called a *closed world* (see Chapter 6 for a precise description of a closed world). At this point, we want to emphasize that because m, the number of identities in $\mathbf{M}$, *is finite*, things like $r(B'_\ell)$ are *discrete* and are *finite* too—no matter how large ℓ is. Therefore, we are dealing with a discrete random variable $r(B'_\ell)$, which is estimated as k when $d'_\ell = d_{(k)}$ in the ranked return list; i.e., the subject associated with B'_ℓ has rank k.

So we have established that scale m is finite, no matter how large the scale of the biometric identification problem. This is *not* an assumption, this is simply a property of the problem. We also have other properties of the problem that seemingly are assumptions and contradictions:

1. Each identity $d_i, i = 1, ..., m$ is enrolled *only* once.

2. This only means that the biometrics *uniqueness property* introduced in Chapter 1 is satisfied and there are no B_1 and B_2 in $\mathbf{M}$ so that $B_1 = f(\mathcal{B})$ and $B_2 = f(\mathcal{B})$.

3. For a particular input biometric $B'_\ell = f(\mathcal{B}_i), \mathcal{B}_i \in \mathbf{M}$, let us assume that we have (*a*) $R(B'_\ell, (B_i, B_j); \mathbf{M}) = 0$ for $j = 1, ..., k-1$; (*b*) $R(B'_\ell, (B_i, B_j); \mathbf{M}) = 1$ for $j = k, ..., m$; This means that $k - 1$ biometric identifiers are ranked higher than the matching biometric B_i. The rank of matching biometric B'_ℓ is then k.

4. Each biometric query $B'_\ell = f(\mathcal{B}_i), \mathcal{B}_i \in \mathbf{M}$ does have a true rank in the reordered list
$$\mathbf{C}(\mathcal{B}_\ell;\ \mathbf{M}) = (B_{(1)}, B_{(2)}, ..., B_{(m)}) = (B_{(r)}, r = 1, ..., m).$$
There exists a true r such that $d'_\ell = d_{(r)}$. This is simply the rank of $B_i = f(\mathcal{B}_i) \in \mathbf{M}$. However, put a little more precisely, we have no unique r (except for in rare cases, as in Figure 16.13). We have a distribution $P_\ell(r)$, the true rank of query B'_ℓ in the reordered list.

   The random variable $r(B'_\ell)$ is the expected rank $r(\mathcal{B}_i)$ in the candidate list for the query samples $B'_\ell = f(\mathcal{B}_i \in \mathbf{M})$. If on average $(k-1)$ rank-order errors are made, we have $E(r) = k$ for some $k \geq 1$ and k is not equal to 1 per se.

The perfect search engine that always returns $k = 1$ for every $B'_\ell = f(\mathcal{B}_i \in \mathbf{M}), i = 1, ..., m$ simply does not exist in real situations. The true underlying Rank Probability Mass (RPM) distributions for many $\mathcal{B}_\ell \in \mathbf{M}$ will be like the one on the right in Figure 16.13; few $\mathcal{B}_\ell$ will have the ideal RPM as on the left; moreover, we will not have $P(1) = 1$ or $P(3) = 1$, but we will have distributions around $r = 1$ and $r = 3$.

16.6.5 Probability of correct identification

Recall that the output of a large scale search of database $\mathbf{M}$, is a reordered list (16.39)
$$\mathbf{C}(B'_\ell;\ \mathbf{M}) = (B_{(1)}, B_{(2)}, ..., B_{(m)}).$$
This is the output of a search engine and only under ideal circumstances, for all $B'_\ell = f(\mathcal{B}_i \in \mathbf{M})$, the rank of the desired enrolled biometrics template in $\mathbf{C}(B'_\ell;\ \mathbf{M})$ is 1. Under not so ideal but still true circumstances, the true underlying rank $r(B'_\ell) = r(f(\mathcal{B}_\ell)) = r(\mathcal{B}_i) = r \geq 1$ (as in Figure 16.13). However, we cannot compute $r(\mathcal{B}_i)$ of some true biometric $\mathcal{B}_i$ in $\mathbf{M}$, we can only obtain an estimate $k_i = \hat{r}(\mathcal{B}_i)$; or more precisely, we can estimate a probability mass distribution for $r(\mathcal{B}_i)$.

If we sample $\mathcal{B}_i$ L times and obtain samples $B_i(j)$, $j = 1, ..., L$, we can measure the ranks for these samples as a set $\mathcal{K}_i = \{k_i(1), k_i(2), ..., k_i(L)\}$. From this set a statistic can be computed that expresses the quality of the search engine for a specific $\mathcal{B}_i$ in database $\mathbf{M}$. This statistic is the probability mass function of $B_i \in \mathbf{M}$ determined by rank criteria (16.40) and database $\mathbf{M}$.

If we obtain L samples from each $\mathcal{B}_i$, $i = 1, ..., m$ each, we have a set of L ranks $\mathcal{K}_i$ for each $B_i, i = 1, ..., m$ in $\mathbf{M}$. By taking the union of all $\mathcal{K}_i$,
$$\mathbf{K} = \bigcup_{i=1}^{m} \mathcal{K}_i, \tag{16.43}$$

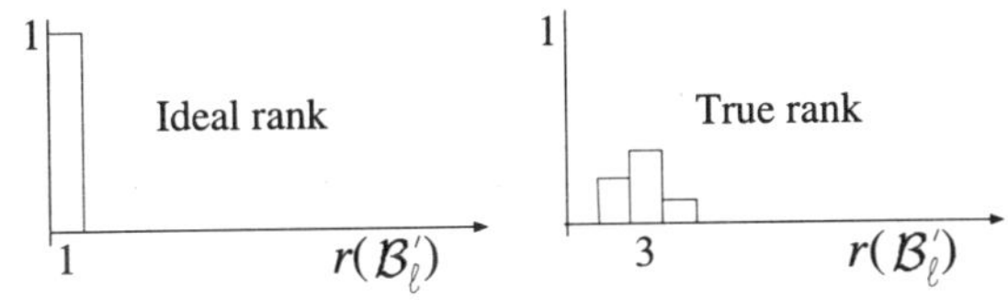

Figure 16.13: Of course, the estimated $\hat{r}(\mathcal{B}_2)$ when $\mathcal{B}_\ell \equiv \mathcal{B}_2$ does not necessarily have $E(r) = 1$.

we have a set of ranks from which statistics that express the overall quality of a search engine can be computed. This statistic is the overall probability mass function of search engine, and is determined by rank criteria (16.40) and data set $\mathbf{M}$.

Probability rank mass function

As already hinted at before, we need a rank-order statistic to express the "quality" of large-scale identification system performance. For each query biometric samples from $\mathcal{B}_i \in \mathbf{M}$ the Rank Probability Mass, denoted as $P_i(r)$, is defined as

$$\textit{Prob}\,(\text{rank of } \mathcal{B}_i \equiv r) \;=\; P_i(r) = \; P(r, \mathcal{B}_i;\; \mathbf{M}), \quad r = 1, ..., m. \tag{16.44}$$

This is just the probability that an identity d_i associated with query B_i ends up with rank r in vector $\mathbf{C}_m$. The true $P_i(r)$ of (16.44) can only be estimated as $\hat{P}_i(k) = P(k, \mathcal{B}_i; \mathbf{M})$ as follows.

Suppose we have access to real-world biometric $\mathcal{B}_i$ enrolled in $\mathbf{M}$ and can sample $\mathcal{B}_i$ at will L times. For each sample $B_i(j), j = 1, ..., L$, we have an estimate $k_i(j)$. We can sample $\mathcal{B}_i$ over and over again (L times) and we have a sample of the distribution $P_j(r)$ denoted as above $\mathcal{K}_i = \{k_i(1), k_i(2), ..., k_i(L)\}$, and we have an empirical Rank Probability Mass function (RPM estimate)

$$\hat{P}_i(k) = \hat{P}(k, B_i;\; \mathbf{M}) = \frac{1}{L}\,(\#\, k_i(j) \in \mathcal{K}_i \text{ that equals } k), \quad k = 1, ..., m. \tag{16.45}$$

In the estimate of (16.45), we use a set of ranks associated with biometric $B_i(j)$, $\mathcal{K}_i$. A total set of ranks $\mathbf{K}$ using all B_i in $\mathbf{M}$ is given in (16.43). Using this set of ranks, we get

$$\hat{P}(k) = \frac{1}{m}\sum_{i=1}^{m} \hat{P}_i(k) = \frac{1}{Lm}(\#\, k_i(j) \in \mathbf{K} \text{ that equals } k), \quad k = 1, ..., m. \tag{16.46}$$

So, there are a lot of things that we can do, using the search engines' PRM curve estimates. For example, it is now possible to devise estimators, for say $E(r_i)$, the mean of the probability mass function of (16.44). Confidence intervals for estimates of rank-order statistics can be obtained by partitioning the set $\mathbf{K}$ as we did with match scores $\mathbf{X}$ and mismatch scores $\mathbf{Y}$ in Chapter 15 and applying the subsets bootstrap.

A key point to mention here is that we have a statistic, the Rank Probability Mass (RPM) estimate of (16.45). This statistic, as shown, models the behavior of a search engine. A rank-order search result $\mathcal{K}_i = \{k_i(1), k_i(2), ..., k_i(L)\}$ is just a set of numbers $1 \leq k_i(j) \leq m$, and each $k_i(j)$ is associated with the database entry i, i.e., an identity ID_i or d_i. Statistics of these sets of ranks and subsets of these ranks can be estimated. The estimation and comparison of such statistics can be developed with the material presented in this book.

Further study of such rank-order statistics remain an important research topic.

16.7 ROC versus CMC

In Chapter 13 a number of publicly available biometric sample databases were enumerated. These databases can be used to objectively compare different match engines for the different biometric identifiers.

In general, these test databases are fully annotated as to what the identities or identifiers are of the subjects whose samples have been acquired. This means that sets of match scores $\mathbf{X}$ and sets of mismatch scores $\mathbf{Y}$ can be easily computed. Using these scores, the $\text{FRR}(T)$ and the $\text{FAR}(T)$ can be estimated using the empirical distributions of (5.8) and (5.9). From these estimates, an ROC curve can be constructed and used to evaluate and compare matchers. These type of tests are technology evaluations, as described in Chapter 7. In such situations, it should be noted, however, that the generated False Accepts may not represent the "true" False Accepts occurring in real scenarios. These False Accepts represent just random incidences that occur when cross-matching samples that are acquired from different biometrics $\mathcal{B}$. The resulting ROC curves of a particular matcher should be interpreted for what they are. The False Accepts are the result of "zero-effort forgeries" as defined by the signature recognition community.

Testing a biometric authentication system in an operational evaluation (see Chapter 7) is a different story. In this case, some errors will never be detected and hence cannot be measured.

1. In the case of a positive authentication system (verification, identification), the False Accept of a subject d may never be detected. If subject d is erroneously accepted, the only person that knows about this error is the subject d, and if the subject does not report the False Accept, the tester will never know. Hence, the only data that are available are the False Rejects—assuming of course that the rejected subject notifies the system operators. The False Accepts cannot be reliably detected and therefore the FAR cannot be estimated!

2. On the other hand, in the case of a negative authentication or screening, a False Negative might never be noticed. If an undesirable subject d is for some reason not detected, the subject d is again the only person that knows about this error, and may never report the occurrence of the False Negative error.

   Also for this authentication protocol one of the error conditions cannot possibly be always detected. Therefore, the FNR of screening applications cannot be measured.

Hence, in an operational evaluation of a biometric installation, in reality only part of the truth data is known:

1. When testing a positive authentication system, only a set of match scores $\mathbf{X}$ can be obtained. These are the scores of genuine subjects that are rejected, the False Rejects.

2. When testing a negative authentication system, on the other hand, only a set of mismatch scores $\mathbf{Y}$ can be measured. These are the scores of innocent subjects that are erroneously detected as members of the screening database $\mathbf{N}$, the False Positives.

Therefore, unlike the case of technology evaluations, for operational evaluation only part of the ground truth is available.

In such cases, obviously one cannot compute ROC curves and the only thing that can be done is to resort to measuring something like a Cumulative Match Curve (CMC) statistic of a matcher (see Chapter 6, 16.5, and 16.6. Repeating (16.35), we have

$$\mathrm{CMC}(K) = \sum_{r=1}^{K} \binom{m-1}{r-1} \int_0^{\infty} [\mathrm{FMR}(x)]^{r-1}\, f(x)\, [1 - \mathrm{FMR}(x)]^{m-r}\, \mathrm{d}x.$$

This expression contains the False Match Rate $\mathrm{FMR}(T)$ (FAR, FPR) and the probability density function $f(x)$ of match scores, which determines (if known) the False Non-Match rate $\mathrm{FNMR}(T)$ (FRR, FNR). In that sense, the CMC of (16.7) is a statistic based on both error rates, while only scores associated with one error type are available. Rank-order statistics can be used to express the quality of a large-scale match result. This statistic, as shown, models the behavior of a search engine. Given rank-order search results $\mathbf{K} = \{\mathcal{K}_1, \mathcal{K}_2, ..., \mathcal{K}_m\}$, where each $\mathcal{K}_\ell$ is associated with a database identity, statistics for sets of these ranks and subsets of these ranks can be estimated. Further research on such rank-order statistics estimation remains an important topic for further research.

16.7 XDC VERSUS CMC

Hence, in an operational ... [illegible] ... is known.

[illegible]

[illegible]

... the use of ... [illegible] ... available.

[illegible]

17

What's next?

Biometrics are just a concept, plus lots of data, and way too much confusion. We have seen several different biometrics in a variety of stages of technological development, from obsolescence (Bertillon) through commercial product (e.g., iris) to speculative research (odor) [155]. We have tried to show that each biometric has its own advantages and we can predict that technology advances and medical discoveries will lead to development of new means of biometric identification in the future. In fact, almost any physical property of the human body (density, reflectance, absorbtion, emission, chemical composition), if it can be defined and measured with sufficient precision, could act as a biometric, particularly when it can be considered as a spatial variable or repeating temporal signal. Such signals are prone to difficulties of normalization, susceptibility to long-term variations, and non-individual factors; this is the case with all the established biometrics. Biometrics therefore presents some very difficult problems and is nowhere near the stage at which it is currently portrayed in movies. The technological developments in biometrics are a tiny piece of the technological puzzle that can increase safety and security in everybody's day-to-day life.

Technological advances will no doubt engender a panoply of potential biometrics. Yet those which are used in practice will be determined by commercial forces and suitability to different situations rather than simply feasibility, which has been the principal limitation hitherto. Moreover, legislation and legalities involved in resolving disputes are likely to be an important factor in determining which biometrics are successful in the long run, and perhaps rightfully so. The technologies that have human experts who can justify and defend a "biometric match" will fare better in this respect.

Automated biometric systems will be used for large-scale authentication. In this regard, there are two serious pitfalls that should be kept in mind at all times and avoided:

- All too often, biometrics are introduced into an application as an afterthought. Thorough integration of the biometric subsystem with more traditional authentication protocols is essential.

- Biometric authentication does not guarantee, should not guarantee, and cannot guarantee, 100 percent certainty of the matching decision. Yet what biometrics loses in the way of certainty, it gains in the way of tight linkage between identities. That is, while automated authentication using *non-biometric* credentials is achieved through exact matching (and hence provides 100 percent accuracy), the credential may actually be in the possession of an impersonator.

Based on these observations it is clear that the field of information security research will, sooner or later, have to be extended to encompass biometrics, bringing about a tighter coupling between physical and logical security. The biometric linkage between identities, is not really that reliable when it comes to real-life applications, and it will never be 100 percent. The capabilities of cross-linking identities through biometrics surely does not pose problems that are insuperable and can be addressed with traditional information security.

17.1 Recommendations

Aside from these caveats, in this book we have identified a number of important issues bearing on the selection of a biometric. We summarize these as follows:

- *Realistic error estimates*

  In the marketing material of biometric systems, all too often error rates are not reported, poorly reported; or, worse, the systems are claimed to be 100 percent accurate. In laboratory environments one may achieve close to 100 percent accuracy when good-quality biometric samples are used. For applications in the field, where the conditions are far from ideal, unavoidable noise and other ambient conditions will adversely affect the system performance and any system might be doomed to failure even before its inception.

  When describing system performance, accuracy should be reported as at least two numbers: the False Accept and False Reject Rates. Further, an estimate of the Failure to Enroll (FTE) should be given. Confidence intervals should be supplied to give an indication of the significance of these estimates. Along with these numbers, some feel for the database quality should be given. For example, at a minimum, it should be stated how the samples were obtained and some sample biometrics should be shown.

  Special attention should always be given to the way the False Accept Rate is measured. Is the quoted False Accept Rate a zero-effort False Accept Rate, or is it the so-called true False Accept Rate—and what does that mean then?

- *Identification versus verification*

  Although in this book we defined both verification and identification as the execution of an authentication protocol, the two authentication methods are very different. Verification is a match of one biometric with only 1 biometric template. Identification, on the other hand, involves a search against a database of m identities that have been enrolled for some reason (either civilian or criminal) *and* the m biometric identifiers are (centrally) accessible. However, the step from 1:1 verification to 1:m identification is not as straightforward, as is commonly believed (even when $m = 2$) and needs to be better understood.

  The False Accept Rate for an identification system is roughly m times the False Accept Rate of the 1:1 verification system for that particular biometric. This means that in a positive identification system the equivalent biometric 1:1 matcher has to run at extremely low False Accept Rates to ensure reliable person identification. Thus, if security is of paramount importance, it is often inadvisable to rely solely on biometric identification in these sorts of applications.

Biometric screening, or negative identification, is also by definition a $1\!:\!m$ search problem if the biometric is used as the only "credential." For these large, pure biometric screening protocols it is desirable to have a low False Positive Rate, which, again, will be roughly m times the False Positive (Accept) Rate of the 1:1 system. If this is not the case, too many subjects will be incorrectly identified as, for instance, a member of the "most-wanted" database for each database inquiry. Handling and dismissing these exceptions can be expensive.

- *Enrollment databases*

  Enrollment is an important aspect of any biometric authentication system. The search capabilities of a $1\!:\!m$ search engine critically depend of the quality of the enrollment database $\mathbf{M}$. When a search engine is to be developed around a legacy biometric database $\mathbf{M}'$ of m individuals, this can impose unrealistic requirements on the biometric match engines. On the other hand, when designing such a system from "scratch," the design and population of the database $\mathbf{M}$ are more flexible. However, this should be kept in mind from day 1 of system design and during the entire lifespan of the system.

  This is, of course, equally valid for 1:1 authentication systems with m members. The overall quality of the central or distributed biometric database is of extreme importance. Any poor enrollment is (1) an immediate security problem, and (2) an immediate decrease in overall system performance.

- *System evaluation procedures*

  Procedures for evaluating biometric authentication systems are not very well defined nor very well understood. In general, biometrics system performance evaluations are of three types: technology evaluations, scenario evaluations, and operational evaluations.

  Technology evaluations are the type of evaluations that have been in use in the speech and document retrieval areas for quite a while now. Some organization, often a government agency, releases databases of training and test data at some point, and test participants submit their algorithms within some period of time after the release of the test data. The results are then compared over these pre-collected training and test data sets. Sometimes these type of tests on annotated database are called *benchmarks*.

  A scenario evaluation is the testing of biometric authentication where the complete end-to-end system is installed at some test facility. These test facilities are often at some national research lab commissioned to do the comparative testing. The databases that result from such a process are typically not as clean or complete as those in a technical evaluation. Problems with enrollment and imaging conditions become more evident in this type of evaluation.

An operational evaluation involves performance measurement in a real environment with real users. The system should be tested in a conditions similar to those an ATM faces in the real world. That is, the system should be tested for environmental robustness, response to unenrolled persons, and attacks of different types. In such a scenario, the most important error condition, (a False Accept by a deliberate impostor; a False Negative through good disguisement) may never be detected.

Technical and scenario evaluations rely on the selection of volunteers, which can have unknown repercussions on the test results with respect to the system's actual performance (as measured in operational evaluation). However, assuming that care is taken in building a database representative of the user population and that the inevitable collection problems can be engineered away or statistically modeled better, technology evaluations are a reasonable (and relatively cheap) way to compare the care accuracy numbers of different biometric matchers.

- *Probabilistic answers and costs*

  Biometrics brings the need to analyze security in probabilistic terms. Automated biometric authentication does not provide a crisp *"YES/NO"* answer, but rather a probabilistic answer. Thus things like cost analysis can be used during authentication—the cost of a False Accept can be different from the cost of a False Reject.

  From a security point of view, one of the greatest strengths of biometrics is non-repudiation. The user transmits biometric information to the authentication system that is unique to the user, and the user cannot deny that the application was used or accessed by her/him. This, in itself, adds deterrence over the traditional authentication techniques. But again, when biometrics are used, non-repudiation is guaranteed only within certain probabilistic bounds.

  More interaction between researchers in biometrics and researchers in security is needed here. After all, biometrics are related to authentication protocols, which are traditional security topics. Non-repudiation is also a well-defined and desired computer security property of authentication systems [208]. In particular, there is a need to develop probabilistic frameworks to analyze both the enrollment stage (which may be done with non-biometric identifiers) and the biometric authentication stage of end-to-end systems. That is, the certainty of authentication should depend on the match probability at authentication time plus the certainty of identification at enrollment time.

17.2 Current issues

We are just now starting to see the deployment of medium- to large-scale biometrics installations. This means a number of systems issues will become increasingly important in the near future (next five years).

- *Enrollment policies*

  This is perhaps the most ignored aspect of biometric authentication systems. Enrollment (and de-enrollment) should generally be implemented according to well-specified policies, at least from the point of view of the public. Construction of screening databases, of course, is largely done by federal agencies that maintain criminal databases. It is also important to keep in mind that the authentication system is only as accurate as the accuracy of the enrollment. Biometric sample collection during enrollment should be conducted very carefully by qualified personnel in order to ensure good overall results. Database maintenance is an ongoing, never-ending concern, because degrading of the enrollment database inevitably degrades a system's performance. When this occurs, it might be better to just "freeze" a legacy biometric system and start a new system with more robust matching technology.

- *Interface usability*

  The second-most-ignored aspect of biometric authentication systems, and we probably have not stressed this enough in this book, is the user interface. The weakest link in the design of most systems is where the user interacts with the system; biometric authentication systems are of course no exception. Much work is done on this aspect at all the field sites where biometric authentication systems are actually being installed; unfortunately, this has not resulted in any comprehensive description of the state in the art in this area.

  Despite this awareness, many existing biometric authentication systems are particularly awkward to interact with. This is often the result of attempts to control the quality of the acquired biometric sample by asking the human to adapt to the input device. Instead, biometric user interfaces need to be as easy and robust as those of ATMs. Problems with user-friendliness may impact the performance of biometric authentication not only in terms of Failure to Enroll (FTE) and Failure to Use (FTU), but also in terms of False Accept and False Reject Rate. However there have been no quantitative studies done by user interface researchers to quantify these degradations. The user interface area is problematic as it is—simply because biometric researchers do not seem to feel this is a high-priority issue.

  The biometrics research community is well advised to study the interaction of users with sensory devices; after all, it will be the public (users) who will eventually accept a biometric device or not—and much is determined here by its usability.

- *Biometric system testing*

  Currently, in the biometric research community, two seemingly incompatible biometric evaluation statistics are used: (i) ROC-based statistics using matcher scores and (ii) statistics like the CMC that use *ranks* of identities The CMC is a rank-based statistic that handles a discrete, finite number m of ranks. This is just what is needed because a $1{:}\,m$ search is a finite problem with ranked lists as output.

  An obvious way to rank identities is to sort similarity scores in a non-increasing fashion. What the CMC, as it is used, expresses is the "sorting" capabilities of a biometric matcher. Note that sorting is a special case of ranking or reordering.

  The statistics ROC and CMC are probabilistically compatible and there is use for these statistics in both system training and system testing. The precise relation and uses of these statistics is a wide-open biometric research area, which we only touched on a little in this book.

- *Modular biometric system design*

  Biometric systems are systems that are enrolled (trained to recognize a biometric) using biometric data. These systems are also designed, tested, and trained using biometric data. If enrollment needs to be fully completed before deployment, we have the following situation. After enrolling identities, a biometric authentication system is "exposed" to the real world. This means that every biometric component of the system is exposed. This calls for the design of biometric systems in the form of modules that perform functions on biometric templates. The accuracy predictions and accuracy estimates are much easier to do in such a modular fashion.

- *Overall system security*

  Security (safety) is the protection of a service or system against threats. There is both the threat of a violent physical attack on the system, and the threat of certain people impersonating the users of the system (or other individuals) to somehow compromise the application or the integrity of its users.

  Defense against such attacks could well become a convoluted process of security measures and countermeasures, as is the case today in message authentication in the area of computer security. One avenue is a protocol involving multiple authentication methods (combinations of possessions, knowledge, or biometrics). Here biometric authentication systems can be inherently more secure than legacy authentication systems. This is because there is, in theory, a more secure (non-repudiable) linking of individuals to the more common and accepted identity databases (passports, birth certificates, etc.).

  Many questions about *how* to make biometric authentication work without creating additional security loopholes still remain unanswered. At present little work is being done in this area. While there are the beginnings of "Protection Profiles" in common criteria certification schemes, in general there is not much mention of biometrics in the security literature.

- *Privacy of data*

  The possibility of infringing on civil liberties in biometric authentication systems is very real. Beyond traditional privacy issues such as having a biometric sample taken in the first place, integration of voluntary verification systems and involuntary screening systems raises further concerns. Credibly ensuring privacy can be a "make-or-break" issue when installing a biometrics system in some venues.

  In fact, the privacy issues are important, and it should be clear that research in biometric privacy is a very important topic in biometrics.

17.3 The future

While the future (determined by evolving public opinion through debate) is always hazy, we will venture to make a few general predictions about what the field of biometrics will look like (say, ten years from now). Note that these are merely the authors' opinions and do not necessarily reflect the position of IBM as a corporation.

- *Biometric plurality*

  We have seen throughout the book that no single biometric has clear attributes that will guarantee its ascendancy over the others. Each has strengths and weaknesses and the different characteristics mean that we can expect all the major biometrics to remain viable in at least some area. Fingerprints, faces, and signatures will all continue to be used as biometrics in legacy systems. Clearly face recognition will be widely tested in surveillance applications as the technology continues to mature (although it remains to be seen how soon its effectiveness will justify widespread deployment).

  Speaker identification will become widely used as the only biometric for the growing market in pure telephony applications. The advantage of pervasive and remote authentication makes it attractive for financial applications. However, when real security is essential these advantages might be liabilities simply because the existing infrastructure is hard to secure.

  Identification: It seems that in the short term, spending on biometrics will continue to be dominated by large-scale civil government and criminal installation, and for these fingerprint must retain its dominance. It is possible that there will be increasing penetration by iris for applications that do not rely on legacy data, and the use of face recognition where filtering and binning make its lower accuracy acceptable (plus, of course, the potential for covert uses). For very large scale systems the possibility of using more than one (or two) fingerprints per person will still favor the choice of fingerprints.

  Verification: In this market, the future seems much less clear, with diversity being preserved for a number of reasons:

  - more competing technologies;
  - fewer legacy systems;
  - capricious public acceptance, subject to future fads and media coverage;
  - the gradual installation of many small systems.

  Here even the weakness of some biometrics can be seen as a virtue since the consequent lack of scalability means that templates cannot be used for unauthorized searches against large databases, or other such abuses.

  Even with this uncertainty we can hazard a few guesses. Fingerprint recognition clearly has a number of advantages because of its maturity (with high accuracy and low-cost algorithms and sensors). But public acceptance could go either way—encouraging acceptance as being a reliable, well-understood mechanism for identification, or discouraging it with criminal associations and fear of "function-creep." The outcome will be based on public opinion, as it should be. Face recognition will likely become pre-eminent for small-scale cooperative authentication, such as personalization of electronic devices, because of its passive, non-contact sensing by a small, cheap, and multi-use sensor. This makes face recognition also very attractive in special environments like people's homes.

- *Single modality systems*

  While multi-modal fusion can in theory provide improved security, it requires two (or more) sets of input devices, feature extractors, and databases. These costs and the integration issues are likely to outweigh the benefits of such systems if an equivalent single modality can be shown to have adequate performance. It is only when biometrics with high discriminative power are unavailable for some reason (e.g., user objections or difficulty of acquisition) that combinations of weaker biometrics will be used. Fusion of multiple samples of the *same* type of biometric (like the fingerprints above) is much more likely. Still, there may be some utility in falling back to a weaker biometric when a stronger one is unavailable. For instance, one might use fingerprints as the primary authentication method, but when exceptions occur, run face recognition on the individual as well [150].

- *Low-impact biometrics*

  There is likely to be an increasing interest in easy-to-acquire biometrics. For instance, a number of new technologies for identifying a person from a distance are being researched (particularly for deployment in government facilities). While we expect several orders of improvements in performance of these research systems, they probably will never achieve the level of accuracy of more cooperative or invasive biometrics. Such minimally intrusive systems are very attractive because they make applications simpler for people to use and hence accept as a regular part of their lives.

  Another, more sinister, motivation is the use of this type of system for covert surveillance of various kinds. Even if individuals' movements are not tracked by police organizations, there is still the possibility of pervasive, annoying customized advertising as seen in the movie *Minority Report* (e.g., "Good afternoon, Mr. Yakamoto. How did you like that three-pack of tank tops you bought last time you were in?").

- *Cross-linked databases*

  There are currently a large number of non-biometric governmental databases—FBI, DEA, Immigration, state driver's licenses, sex offender lists, etc. And there are already moves to link these together, as well as to supplement them with information obtained from commercial databases such as airline passenger lists. This will only increase in the future and, in the eyes of many, such linkage is justified on the grounds of national security. Since one of the basic strengths of biometrics is its uniqueness, this will likely be exploited to the full extent possible to further establish and confirm links among distributed databases. While it may be prudent to watch/observe the activities of certain targeted individuals very carefully, governments must also zealously guard against abuses of the system. To this end, some consensus needs to be established as to which databases are useful and reasonable for linking (e.g., perhaps not video rentals or medical records). Correspondingly, technical means need to be developed for actively preventing the use of any which are deemed irrelevant or too invasive.

- *Legal wrangling*

  As biometrics enters the mainstream of society, numerous privacy issues arise. These have substantial implications for civil, commercial uses of biometrics. Possible legislation may include things like the right to opt out of any biometric identification scheme (e.g., credit cards). This means there have to be suitable exception handling mechanisms, and the costs of these processes need to be seriously considered. Another possible complication could occur if some jurisdictions mandate that no biometric data may ever be stored in a database, except possibly for matters of national interest (e.g., voter registration). This would rule-out "pay with your thumb" schemes for commerce that depend on large-scale customer databases (although the use of smartcards to store templates locally may alleviate this to some extent). While these constraints are only speculation, the time to lobby *for or against* such policies is probably now, before such systems become widespread.

  No advancement in biometrics will ever put an end to this legal wrangling because this is just the normal process of forming public opinion and public acceptance. All this may make biometrics either relevant or irrelevant. In the latter case, it might be only the biometric community who will be to blame.

17.4 A final word

Biometrics has many aspects, many of which we have attempted to review, or at least explain in this book. Our intent was to allow the question "*Which biometric is best?*" to be answered in a better-informed, detailed way using this book. While the answer may be only tentative, and surely not definitive, at least the question can be phrased a little better by asking: "Which biometric is best, given that the requirements for the application are specified as follows ..."

We are left to say that biometrics is in principle a well-defined, but fundamentally an ill-posed, problem. However, there is a perception that biometrics is a mature technology ready to be deployed in security applications and this warrants attention. A lot of attention in normal times is good. Too much attention in abnormal times may not be *that* good because it may not always be serious, or serious enough. There is nothing *not* serious about security; yet security is a nebulous concept, because it is surely less than 100 percent. Questions like *How much less?* and *At what cost?* then become very relevant and the answer should be simple: e.g., *99.99 percent at one cent per incident.* But such answers are based on probabilities and assumptions. These assumptions are often not clearly defined, modeled, trained, and stated; and such answers might even be perpetuating biometric myths and misinterpretations. Biometric identifiers are not sufficiently distinctive to be 100 percent reliable, except in very rare instances. Biometrics, by its very definition is not foolproof, but it raises privacy and safety concerns simply because it does work to a certain extent and can still be much improved. This explains all the attention that biometrics is getting. But real issues like *security* and *privacy* deserve more attention, and biometrics is just *one* piece of the puzzle. Biometrics does offer security, but surely less than 100 percent.

In writing this book we tried to explain that biometrics is important and we discovered that it is a very fascinating science; and as such it deserves all the attention it has received.

There are many uses of biometrics that will improve overall public safety. Of course, there are abuses of biometric technologies; however, there is ample time to address these in serious ways. Whether we have expressed all of this well, we will of course know only with feedback from the readers. Therefore, the authors would like our readers to contact us at the publisher's address about any remaining unanswered concerns.

References

[1] AAMVA Standards Working Group. AAMVA Standard for the Driver License / Identification Card 2000. Technical Report AAMVA DL/ID-2000, The American Association of Motor Vehicles Administrators, June 2000.

[2] R. Achs, R.G. Harper, and N.J. Harrick. Unusual dermatoglyphics associated with major congenital malformations. *New England Journal of Medicine*, 275:1273–1278, December 1966.

[3] ACLU Reports. Drawing a blank: Tampa police records reveal poor performance of face-recognition technology. January 2002.

[4] F.M. Alkoot and J. Kittler. Improving the performance of the product fusion strategy. In *Proc. 15th International Conference on Pattern Recognition, Barcelona*, volume 2, pages 164–167, 2000.

[5] American National Standards Institute (ANSI). Biometric Information Management and Security. Technical Report X9.84-2001, http://www.x9.org/books.html, 2001.

[6] H.S.M. Beigi, S.H. Maes, U.V. Chaudhari, and J.S. Sorensen. IBM model-based and frame-by-frame speaker recognition. In *Speaker Recognition and its Commercial and Forensic Appications*, Avignon, April 1998.

[7] P.H. Belhumeur, J.P. Hespanha, and D.J. Kriegman. Eigenfaces vs. Fisherfaces: Recognition using class specific linear projection. *IEEE Transanctions on Pattern Analysis and Machine Intelligence*, 19(7):711–720, July 1997.

[8] F. Bergadano, D. Gunetti, and C. Picardi. User authentication through keystroke dynamics. *ACM Transactions on Information and System Security (TISSEC)*, 5(4):367–397, November 2002.

[9] J. Bery. The history and development of fingerprinting. In H.C. Lee and R.E. Gaensslen, editors, *Advances in Fingerprint Technology*, pages 1–38. CRC Press, Boca Raton, FL, 1994.

[10] W. Bicz, Z. Gurnienny, and M. Pluta. Ultrsound sensor for fingerprints recognition. In *Proc. of SPIE, Vol. 2634, Optoelectronic and electronic sensors*, pages 104–111, June 1995.

[11] BioAPI Consortium. *BioAPI Specification Version 1.1*. The BioAPI Consortium, March 2001.

[12] Biometric Systems Lab. *HaSIS — A Hand Shape Identification System.*

[13] Biometric Systems Lab, Pattern Recognition and Image Processing Laboratory, and U.S. National Biometric Test Center. FVC2002: Fingerprint verification competition.

[14] Biometrics Working Group. Best practices in testing and reporting performance of biometric devices. http://www.afb.org.uk/bwg/bestprac.html, 2000.

[15] D.M. Blackburn, M. Bone, and P.J. Phillips. FRVT 2000: Facial recognition vendor test. Technical report, DoD Counterdrug Technology Development Office, Defence Advance Research Project Agency, National Institute of Justice, Dahlgren, VA; Crane, IN; Arlington, VA, December 2000.

[16] R.M. Bolle, J.H. Connell, N. Haas, R. Mohan, and G. Taubin. Veggievision: A produce recognition system. In *Proc. Third IEEE Workshop on Applications of Computer Vision*, pages 244–251, Sarasota, FL, December 1996.

[17] R.M. Bolle, J.H. Connell, A. Hampapur, E. Karnin, R. Linsker, G.N. Ramaswamy, N.K. Ratha, A.W. Senior, J.L. Snowdon, and T.G. Zimmerman. Biometric technologies ... emerging into the mainstream. Technical Report RC22203 (W0110-041), IBM Research Division, Yorktown Heights, NY, October 2001.

[18] R.M. Bolle, J.H. Connell, S. Pankanti, and N. Ratha. On the security of biometrics authentication. IBM Technical Report, 2002.

[19] R.M. Bolle, J.H. Connell, and N.K. Ratha. Biometric perils and patches. *Pattern Recognition*, (12):2727–2738, December 2002.

[20] R.M. Bolle, N. Ratha, and J.H. Connell. Biometric authentication: Security and privacy. In *Proc. 1st Workshop on Pattern Recognition in Information Systems, PRIS 2001*, pages 2–11. ICEIS PRESS, July 2001.

[21] R.M. Bolle, N.K. Ratha, and S. Pankanti. Evaluating authentication systems using bootstrap confidence intervals. In *Proceedings of AutoID'99*, pages 9–13, Summit, NJ, October 1999.

[22] R.M. Bolle, N.K. Ratha, and S. Pankanti. Evaluation techniques for biometrics-based authentication systems (FRR). In *Proc. 15th Int. Conf. on Pattern Recognition*, pages 835–841, September 2000.

[23] R.M. Bolle, N.K. Ratha, and S. Pankanti. Error analysis of pattern recognition systems – The subsets bootstrap. *Computer Vision and Image Understanding*, To appear, 2003.

[24] F.A. Bouchier, J.S. Ahrens, and G. Wells. Laboratory evaluation of the IriScan prototype biometric identifier. Technical Report SAND96-1033 RS-8232-2/960378, Sandia National Laboratories, Albuquerque, NM, April 1996.

[25] V. Bouletreau, N. Vincent, R. Sabourin, and H. Emptoz. Handwriting and signature: One or two personality identifiers? In *Proc. of the 14th International Conference on Pattern Recognition*, volume II, pages 1758–1760, Brisbane, Austria, August 1998.

[26] C.M. Brislawn, J.N. Bradley, R.J. Onyshczak, and T. Hopper. The FBI compression standard for digitized fingerprint images. In *Proc. of SPIE*, volume 2847, pages 344–355, August 1996.

[27] R.R. Brooks and S.S. Iyengar. *Multi-sensor Fusion: Fundamentals and Applications with Software*. Prentice-Hall, Upper Saddle River, NJ, 1997.

[28] R. Brunelli and T. Poggio. Face recognition: Features versus templates. *IEEE Transanctions on Pattern Analysis and Machine Intelligence*, 15(10):1042–1052, October 1993.

[29] M. Burge and W. Burger. Ear biometrics. In A. Jain, R. Bolle, and S. Pankanti, editors, *Biometrics, Personal Identification in Networked Society*, pages 273–285. Kluwer Academic Publishers, Boston, MA, 1999.

[30] M.J. Burge and W. Burger. Ear biometrics in computer vision. In *Proceedings of the International Conference on Pattern Recognition*, pages 826–830, 2000.

[31] J. Campbell. Speaker recognition. In A.K. Jain, R.M. Bolle, and S. Pankanti, editors, *Biometrics: Personal Identification in Networked Society*, pages 165–190. Kluwer Academic Press, Boston, MA, 1999.

[32] J.P. Campbell and D.A. Reynolds. Corpora for the evaluation of speaker recognition systems. In *Proceedings of the IEEE International Conference on Acoustics, Speech, and Signal Processing*, volume 2, pages 829–832, 1999.

[33] J.P. Campbell (Ed.). NIST 1999 Speaker Recognition Workshop. *Digital Signal Processing*, 10(1–3), January/April/July 2000.

[34] R. Cappelli, A. Lumini, D. Maio, and D. Maltoni. Fingerprint classification by directional image partitioning. *IEEE Transanctions on Pattern Analysis and Machine Intelligence*, 21(5):402–421, May 1997.

[35] CBEFF Technical Development Team. Common Biometric Exchange File Format (CBEFF). Technical Report NISTIR 6529, The National Institute of Standards and Technology (NIST), January 2001.

[36] D. Charlton (Ed.). A 'pointless' exercise. *Fingerprint Whorld*, 21(107):19, January 2002.

[37] R. Chellappa, C.L. Wilson, and S. Sirohey. Human and machine recognition of faces: A survey. *Proceedings of the IEEE*, 83(5):705–740, May 1995.

[38] A.M. Choudhary and A.A.S. Awwal. Optical pattern recognition of fingerprints using distortion-invariant phase-only filter. In *Proc. of SPIE, Vol. 3805, Photonic devices and algorithms for computers*, pages 162–170, October 1999.

[39] J. Clark and A. Yuille. *Data Fusion for Sensory Information Processing Systems.* Kluwer Academic Publishers, Boston, MA, 1990.

[40] R. Clarke. Human identification in information systems: Management challenges and public policy issues. *Information Technology & People*, 7(4):6–37, December 1994.

[41] W.G. Cochran. *Sampling Techniques.* Wiley Series in Probability and Mathematical Statistics. John Wiley & Sons, New York, 3rd edition, 1977.

[42] Criminal Justice Information Services (CJIS). Electronic fingerprint transmission specification. Technical Report CJIS-RS-0010 (V7), Criminal Justice Information Services Division, Washington, D.C., January 1999.

[43] R. Curbelo. Noisy fingerprint identification by artificial neural networks. In *Proc. of SPIE, Vol. 3728, 9th Workshop on Virtual Intelligence/ Dynamic Neural Networks*, pages 432–449, March 1999.

[44] DARPA. *Human ID at a Distance (HumanID).* http://www.darpa.mil/iao/HID.htm.

[45] Daubert Update. *Latent Print Examination, Fingerprints, Palmprints and Footprints.* http : //onin.com/fp/.

[46] J. Daugman. Recognizing persons by their iris patterm. In A.K. Jain, R.M. Bolle, and S. Pankanti, editors, *Biometrics: Personal Identification in Networked Society*, pages 103–122. Kluwer Academic Press, Boston, MA, 1999.

[47] J.G. Daugman. High confidence visual recognition of persons by a test of statistical independence. *IEEE Transanctions on Pattern Analysis and Machine Intelligence*, 15(11):1148–1161, November 1993.

[48] J.G. Daugman and G.O. Williams. A proposed standard for biometric decidability. In *CardTechSecureTech*, pages 223–234, Atlanta, GA, 1996.

[49] DBPP. Biometric Device Protection Profile (BDPP). Technical Report Draft Issue 0.28, UK Government Biometrics Working Group, September 2001.

[50] Digital Descriptor Systems, Inc. *Non-contact fingerprint scanner.* http://www.ddsi-cpc.com/productsmain.htm.

[51] G. Doddington, W. Liggett, A. Martin, M. Przybocki, and D. Reynolds. Sheep, goats, lambs and wolves: A statistical analysis of speaker performance. In *Proceedings of IC-SLD'98, NIST 1998 speaker recognition evaluation*, Sydney, Australia, November 1998.

[52] J.G.A. Dolfing. *Handwriting Recognition and Verification.* PhD thesis, University of Eindhoven, Eindhoven, the Netherlands, 1998.

[53] R. Donovan. *Trainable Speech Synthesis.* PhD thesis, Cambridge University, Engineering Department, Cambridge, UK, 1996.

[54] C. Dorai, N. Ratha, and R.M. Bolle. Detecting dynamic behavior in compressed fingerprint videos: Distortion. In *Proc. IEEE Computer Vision and Pattern Recognition*, pages 320–326, June 2000.

[55] B. Duc, E.S. Bigün, J. Bigün, G. Maître, and S. Fischer. Fusion of audio and video information for multi modal person authentication. *Pattern Recognition Letters*, 18(9):835–843, 1997.

[56] R.O. Duda and P.E. Hart. *Pattern Classification and Scene Analysis*. John Wiley & Sons, Inc., New York, 1973.

[57] A.L. Duwaer. *Data processing system with a touch screen and a digitizer tablet, both integrated in one input device*. US Patent No. 5,231,381, 1993.

[58] G.J. Edwards, C.J. Taylor, and T.F. Cootes. Interpreting faces using active appearance models. In *Third International Conference on Automatic Face and Gesture Recognition*, pages 300–305, Nara, Japan, April 1998.

[59] B. Efron. Bootstrap methods: Another look at the Jackknife. *Ann. Statistics*, 7:1–26, 1979.

[60] Equinox. http://www.equinoxsensors.com/. New York, NY; Baltimore, MD.

[61] R.H. Ernst. *Hand ID system*. US Patent No. 3,576,537, 1971.

[62] B. Germain et al. Issues in large scale automatic biometric identification. In *IEEE Workshop on Automatic Identification Advanced Technologies*, pages 43–46, Stony Brook, NY, November 1996.

[63] European ACTS projects. *M2VTS Project: MULTI-MODAL BIOMETRIC PERSON AUTHENTICATION*. http://www.tele.ucl.ac.be/PROJECTS/M2VTS/.

[64] EyeDentify Europe N.V. http://www.eye-dentify.com/. Wommelgem, Belgium.

[65] B. Fader. Note: Apply moisterizer only after gaining access. *New York Times, February, 24*, page C5, 2003.

[66] FBI, U.S. Department of Justice, Washington, D.C. 20402. *The Science of Fingerprints, Classification and Uses*, 1984.

[67] Federal Bureau of Investigations. *WSQ gray-scale Fingerprint Image Compression Specification*, 1993.

[68] J. Ferryman, editor. *Performance Evaluation of Tracking and Surveillance*. 2001 IEEE Conference on Computer Vision and Pattern Recognition, Kauai, Hawaii, December 2001.

[69] J. Ferryman, editor. *Third International Workshop on Performance Evaluation of Tracking and Surveillance Systems*. 2002 European Conference on Computer Vision, Copenhagen, Denmark, June 2002.

[70] Fingerprint Data Interchange Workshop. Summary of the 1998 NIST Fingerprint Data Interchange Workshop, September 1998.

[71] D.T. Follette, E.B. Hultmark, and J.G. Jordan. Direct optical input system for fingerprint verification. *IBM Technical Disclosure Bulletin*, (74C 00989), April 1974.

[72] S. Furui. Recent advances in speaker recognition. In J. Bigün, G. Chollet, and G. Borgefors, editors, *Audio- and Video-based Biometric Person Authentication*, volume 1206 of *Lecture Notes in Computer Science*, pages 237–252. Springer-Verlag, Heidelberg, Germany, 1997.

[73] F. Galton. *Memories of My Life*. Methuen, London, 1908.

[74] M.D. Garris and R.M. McCabe. *NIST special database 27: Fingerprint Minutiae from Latent and Matching Tenprint Images*. Advanced Systems Division, Image Recognition Group, National Institute for Standards and Technology, June 2002.

[75] R. Germain. Large scale systems. In A.K. Jain, R.M. Bolle, and S. Pankanti, editors, *Biometrics: Personal Identification in Networked Society*, pages 311–326. Kluwer Academic Press, Boston, MA, 1999.

[76] R.S. Germain, A. Califano, and S. Colville. Fingerprint matching using transformation parameter clustering. *IEEE Computational Science and Engineering*, pages 42–49, Oct-Dec 1997.

[77] H.P. Graf. Sample-based synthesis of talking heads. In *Proc. IEEE ICCV Workshop Recognition, Analysis, and Tracking of Faces and Gestures in Real-Time Systems*, pages 3–7, Vancouver, BC, July 2001.

[78] T.J. Grycewicz. Techniques to improve binary joint transform correlator performance for fingerprint recognition. *Optical Engineering*, 38(1):114–119, January 1999.

[79] Y. Hamamoto. A Gabor filter-based method for identification. In L.C. Jain, U. Halici, I. Hayishi, S.B. Lee, and S. Tsutsui, editors, *Intelligent Biometric Techniques in Fingerprint and Face Recognition*, pages 137–151. CRC Press, Boca Raton, FL, 1999.

[80] E.M. Hamann, H. Henn, T. Schack, and F. Seliger. Securing e-business applications using smart cards. *IBM Systems Journal*, 40(3):635–647, 2001.

[81] W.L. Harkness. Properties of the extended hypergeometric distribution. *Annals of Mathematical Statistics*, 36(3):938–945, 1965.

[82] N.J. Harrick. Techniques to improve binary joint transform correlator performance for fingerprint recognition. *Applied Optics*, 33:2774, 1962.

[83] Harrick Scientific. Data sheet 8: Frustrated reflection fingerprinting, *circa* 1970.

[84] L.P. Heck and M. Weintraub. Handset-dependent background models for robust text-independent speaker recognition. In *Proceedings of the IEEE International Conference on Acoustics, Speech, and Signal Processing*, April 1997.

[85] R. Hill. Retina identification. In A.K. Jain, R.M. Bolle, and S. Pankanti, editors, *Biometrics: Personal Identification in Networked Society*, pages 123–142. Kluwer Academic Press, Boston, 1999.

[86] J. Holmes, L. Wright, and R. Maxwell. A performance evaluation of biometric identification devices. Technical Report SAND91-0278/UC-906, Sandia National Laboratories, Albuquerque, NM; Livermore, CA, June 1991.

[87] L. Hong and A. Jain. Multimodal biometrics. In A.K. Jain, R.M. Bolle, and S. Pankanti, editors, *Biometrics: Personal Identification in Networked Society*, pages 327–344. Kluwer Academic Press, Boston, MA, 1999.

[88] R.A. Huber and A. Headrick. *Handwriting Identification: Facts and Fundamentals.* CRC Press LCC, Boca Raton, FL, April 1999.

[89] HumanScan GmbH. BioID http://www.bioid.com/.

[90] Immigration and Naturalization Services. *INS Passenger Accelerated Service System (INSPASS).* http://www.ins.usdoj.gov/graphics/lawenfor/bmgmt/inspect/inspass.htm.

[91] Intel Corporation. *Intel^{RT} Common Data Security Architecture (CDSA)*, 2002.

[92] I/O Software Inc. *Biometric Application Programming Interface (BAPI).* http://www.iosoftware.com/products/licensing/bapi/glossary.htm, 2002.

[93] D.K. Isenor and S.G. Zaky. Fingerprint identification using graph matching. *Pattern Recognition*, 19(2):113–122, 1986.

[94] I.H. Jacoby, A.J. Giordano, and W.H. Fioretti. *Personnel Identification Apparatus.* US Patent No. 3,648,240, 1972.

[95] A. Jain, L. Hong, S. Pankanti, and R. Bolle. On-line identity authentication system using fingerprints. *Proceedings of the IEEE*, 85:1365–1388, September 1997.

[96] A.K. Jain, R.M. Bolle, and S. Pankanti (Eds.). *Biometrics: Personal Identification in Networked Society.* Kluwer Academic Publishers, Boston, MA, 1999.

[97] A.K. Jain, R.M. Bolle, and S. Pankanti (Eds.). Introduction to biometrics (Chapter 1). In A.K. Jain, R.M. Bolle, and S. Pankanti, editors, *Biometrics: Personal Identification in Networked Society*, pages 1–41. Kluwer Academic Publishers, Boston, MA, 1999.

[98] A.K. Jain, L. Hong, and R.M. Bolle. On-line fingerprint verification. *IEEE Transanctions on Pattern Analysis and Machine Intelligence*, 19(04):302–313, April 1997.

[99] A.K. Jain, L. Hong, and S. Pankanti. Biometrics identification. *Communications of the ACM*, 43(2):91–98, 2000.

[100] A.K. Jain, S. Prabhakar, and L. Hong. A multichannel approach to fingerprint classification. *IEEE Transanctions on Pattern Analysis and Machine Intelligence*, 21(4):348–359, April 1999.

[101] A.K. Jain, S. Prabhakar, L. Hong, and S. Pankanti. FingerCode: A filterbank for fingerprint representation and matching. In *Proc. of IEEE Computer Vision and Pattern Recognition 1999*, volume 2, pages 187–193, 1999.

[102] A.K. Jain, A. Ross, and S. Pankanti. A prototype hand geometry-based verification system. In *2nd IEEE International Conference on Audio- and Video-based Biometric Person Authentication*, pages 166–171, Washington, DC, March 1999.

[103] B. Javidi and J. Wang. Position-invariant two-dimensional image correlation using a one-dimensional space integrating optical processor: Application to security verification. *Optical Engineering*, 35(9):2479–2486, Sept. 1996.

[104] S. Jung, R. Thewes, T. Scheiter, K.F. Gooser, and W. Weber. A low-power and high-performance CMOS fingerprint sensing and encoding architecture. *IEEE Journal of Solid-state Cicuits*, 34(7):978–984, July 1999.

[105] Justice Blackmun. *William Daubert, et ux., etc., et al., Petitioners v. Merrell Dow Pharmaceuticals, Inc.* Supreme Court of the United States, Washington, DC 20543, 1993.

[106] T. Kanade. *Picture Processing System by Computer Complex and Recognition of Human Faces*. PhD thesis, Dept. of Information Science, Kyoto University, 1973.

[107] I. Kansala and P. Tikkanen. Security risk analysis of fingerprint based verification in PDAs. In *Proc. IEEE AutoID 2002*, pages 76–82, Tarrytown, NY, March 2002.

[108] K. Karhunen. Uber lineare Methoden in der Warscheinlichkeitsrechnung. In *Ann. Acad. Sci. Fennicae, ser A1, Math. Phys.*, volume 37, 1946.

[109] C. Kaufman, R. Perlman, and M. Spencer. *Network Security, Private Communication in a Public World*. Prentice Hall PTR, Upper Saddle River, NJ, 1995.

[110] R. Khanna. Systems engineering for large scale fingerprint systems. In N.K. Ratha and R.M. Bolle, editors, *Automatic Fingerprint Recognition Systems*. Springer, US, Cambridge, MA, 2003.

[111] Patrick J. Kiger. Alphonse Bertillon and the science of criminal identification. http://www.discovery.com/stories/deadinventors/deadinventors.html.

[112] S. King, H. Harrelson, and G. Tran. Testing iris and face recognition in a personnel identification application. In F.L. Podio and Dunn J.S, editors, *Proceedings of the Biometrics Consortium Conference*. NIST, US Department of Commerce, Cristal City, VA, February 2002.

[113] M. Kirby and L. Sirovich. Application of the Karhunen-Loève procedure for the characterization of human faces. *IEEE Transanctions on Pattern Analysis and Machine Intelligence*, 12(1):103–108, January 1990.

[114] J. Kittler and F.M. Alkoot. Relationship of sum and vote fusion strategies. In J. Kittler and F. Roli, editors, *Multiple Classifier Systems 2001*, volume 2096 of *Lecture Notes in Computer Science*, pages 339–348. Springer-Verlag, Heidelberg, Germany, July 2001.

[115] J. Kittler, M. Hatef, R.P.W. Duin, and J. Matas. On combining classifiers. *IEEE Transanctions on Pattern Analysis and Machine Intelligence*, 20(3):226–239, 1998.

[116] J. Kittler, Y.P. Li, J. Matas, and M.U. Ramos Sánchez. Lip-shape dependent face verification. In J. Bigün, G. Chollet, and G. Borgefors, editors, *Audio- and Video-based Biometric Person Authentication*, volume 1206 of *Lecture Notes in Computer Science*, pages 61–68. Springer-Verlag, Heidelberg, Germany, March 1997.

[117] A. Kong, A. Griffith, D. Rhude, G. Bacon, and S. Shahs. Department of Defence & Federal Biometric System Protection Profile for Medium Robustness Environments. Technical Report Draft Version 0.02, US Department of Defence, March 2002.

[118] J. Koolwaaij. *Automatic Speaker Verification in Telephony: A Probabilistic Approach*. PhD thesis, University of Nijmegen, Nijmegen, the Netherlands, September 2000.

[119] S. Kotz and S. Nadarajah. *EXTREME VALUE DISTRIBUTION: Theory and Applications*. World Scientific Publishing Co., Inc., River Edge, NJ, 2000.

[120] H.C. Lee and R.E. Gaensslen (Eds.). *Advances in Fingerprint Technology*. CRC Press, Boca Raton, FL, 1994.

[121] L.L. Lee, T. Berger, and E. Aviczer. Reliable on-line human signature verification systems. *IEEE Transanctions on Pattern Analysis and Machine Intelligence*, 18(6):643–647, June 1996.

[122] S.-H. Lee, S.-Y. Yi, and E.-S. Kim. Fingerprint identification by use of volume holographic optical correlator. In *Proc. of SPIE, Vol. 3715, Optical Pattern Recognition*, pages 321–325, March 1999.

[123] R.Y. Liu and K. Singh. Moving blocks Jackknife and Bootstrap capture weak dependence. In R. LePage and L. Billard, editors, *Exploring the Limits of the Bootstrap*, pages 225–248. John Wiley & Sons, Inc., New York, 1992.

[124] M.M. Loève. *Probability Theory*. Van Nostrand, Princeton, NJ, 1955.

[125] J. Luettin, N.A. Thacker, and S.W. Beet. Speaker identification by lipreading. In *Proc. of the 4th International Conference on Spoken Language Processing (ICSLP'96), Vol. 1*, 62–65, 1996.

[126] Lumidigm, Inc. http://www.lumidigm.com/. Albuquerque, NM.

[127] S.H. Maes, J. Navratil, and U.V. Chaudhari. Converstaional speech biometrics. In J. Liu and Y. Ye, editors, *E-Commerce agents. Marketplace Solutions, Security Issues, and Supply Demands*, pages 166–179. Springer-Verlag, Berlin Heidelberg, 2001.

[128] J.-F. Mainguet, M. Pegulu, and J.B. Harris. Fingerchip^{TM}: Thermal imaging and finger sweeping in a silicon fingerprint sensor. In *Proc. of AutoID 99*, pages 91–94, October 99.

[129] D. Maio and D. Maltoni. Direct gray-scale minutiae detection in fingerprints. *IEEE Transanctions on Pattern Analysis and Machine Intelligence*, 19(1):27–40, January 1997.

[130] D. Maio, D. Maltoni, R. Cappelli, J.L. Wayman, and A.K. Jain. FVC2000: Fingerprint verification competition. *IEEE Transanctions on Pattern Analysis and Machine Intelligence*, 24(3):402–412, 2002.

[131] T. Mansfield, G. Kelly, D. Chandler, and J. Kane. Biometric product testing final report. Technical Report CESG Contract X92A/4009309, Centre for Mathematics and Scientific Computing, National Physics Laboratory, Middlesex, UK, March 2001.

[132] T. Mansfield and J. Wayman. Best practices in testing and reporting performance of biometric devices, For biometrics working group. Technical Report Issue 2 Draft 9, Centre for Mathematics and Scientific Computing, National Physics Laboratory, Middlesex, UK, February 2002.

[133] T. Matsumoto, H. Matsumoto, K. Yamada, and S. Hoshino. Impact of artificial "gummy" fingers on fingerprint systems. In *Proceedings of SPIE Vol. #4677, Optical Security and Counterfeit Deterrence Techniques IV*, pages 244–251, January 2002.

[134] S.J. McPhee, M.A. Papadakis, L.M. Tierney, and R. Gonzales, editors. *Current medical diagnosis and treatment.* Appleton and Lange, Stamford, CT, 1997.

[135] M.H. Metz, Z.A. Coleman, N.J. Phillips, and C. Flatow. Holographic optical element for compact fingerprint imaging system. In *Proc. of SPIE, Vol. 2659, Optical security and counterfeit deterrance techniques*, pages 141–151, 1996.

[136] B. Miller. Vital signs of identity. *IEEE Spectrum*, 31(2):22–30, 1994.

[137] B. Moayer and K.S. Fu. A syntactic approach to fingerprint pattern recognition. *Pattern Recognition*, 7:1–23, 1975.

[138] B. Moayer and K.S. Fu. A tree system approach for fingerprint pattern recognition. *IEEE Trans. on Computers*, C-25(3):262–274, 1976.

[139] R.T. Moore. Automatic fingerprint identification systems. In H.C. Lee and R.E. Gaensslen, editors, *Advances in Fingerprint Technology*, pages 163–191. CRC Press, Boca Raton, FL, 1994.

[140] M.E. Munich and P. Perona. Camera-based ID verification by signature tracking. In *Proceedings of the European Conference on Computer Vision*, pages 782–796, 1998.

[141] R.N. Nagel and A. Rosenfeld. Computer detection of freehand forgeries. *IEEE Transactions on Computers*, 26(9):895–905, September 1977.

[142] V.S. Nalwa. Automatic on-line signature verification. *Proceedings of the IEEE*, 85(2):215–239, February 1997.

[143] S. Nanavati, M. Thieme, and R. Nanavati. *Biometrics: Identity Verification in a Networked World.* John Wiley & Sons, New York, March 2002.

[144] National Research Council. DNA technology in forensic sciene. Technical report, National Academy Press, Washinghton D.C., 1992.

[145] National Research Council. The evaluation of forensic DNA evidence. Technical report, National Academy Press, Washinghton D.C., 1996.

[146] J. Navratil, U.V. Chaudhari, and G.N. Ramashamy. Speaker verification using target and background dependent linear transforms and muti-system fusion. In *Proc. EUROSPEECH 2001*, November 2001.

[147] Net Nanny Software International, Inc. http://www.biopassword.com/.

[148] J. Neter and W. Wasserman. *Applied linear statistical models.* Richard D. Irwing, Inc., Homewood, IL, 1974.

[149] NIST. American national standard for information systems – data format for the interchange of fingerprint, facial, and scar mark and tattoo (smt) information, ansi-itl 1-2000 (nist special publication 500-245), September 2000.

[150] NIST Speech Group. *NIST Year 2003 Speaker Recognition Evaluation Plan.* http://www.nist.gov/speech/tests/spk/2003/doc/2003-spkrec-evalplan-v2.2.pdf, 2003.

[151] NIST/Biometrics Consortium Biometrics Interoperability, Assurance, and Performance Working Group. *Biometric Application Programming Interface (API) for Java Card.* See BCWG_JCBiometricsAPI_v01_1.pdf at http://www.javacardforum.org/Documents/Biometry/, 2002.

[152] M.S. Nixon, J.N. Carter, D. Cunado, P.S Huang, and S.V. Stevenage. Automatic gait recognition. In A.K. Jain, R.M. Bolle, and S. Pankanti, editors, *Biometrics: Personal Identification in Networked Society*, pages 231–248. Kluwer Academic Press, Boston, MA, 1999.

[153] M.S. Obaidat and B. Sadoun. Keystroke dynamics based authentication. In A.K. Jain, R.M. Bolle, and S. Pankanti, editors, *Biometrics: Personal Identification in Networked Society*, pages 213–229. Kluwer Academic Press, Boston, 1999.

[154] M.S. Obaidat and B. Sadoun. Keystroke dynamics based authetication. In A.K. Jain, R.M. Bolle, and S. Pankanti, editors, *Biometrics: Personal Identification in Networked Society*, pages 213–229. Kluwer Academic Press, Boston, 1999.

[155] The US Army Research Office. *Odor Type Detection Program.* http://www.aro.army.mil/research/odortypedetection.pdf, 2002.

[156] L. O'Gorman. Seven issues with human authentication technologies. In *Proc. IEEE AutoID 2002*, pages 185–186, Tarrytown, NY, March 2002.

[157] R.D. Olsen. Identification of latent prints. In H.C. Lee and R.E. Gaensslen, editors, *Advances in Fingerprint Technology*, pages 163–191. CRC Press, Boca Raton, FL, 1994.

[158] S. Pankanti, S. Prabhakar, and A.K. Jain. On the individuality of fingerprints. In *Proceedings of the IEEE Conference on Computer Vision and Pattern Recognition*, pages I:805–812, Kauai, Hawaii, December 2001.

[159] P.S. Penev. *Local Feature Analysis: A Statistical Theory for Information Representation and Transmission.* PhD thesis, The Rockefeller University, 1998.

[160] K.C. Persaud, D.-H. Lee, and H.-G. Byun. Objective odour measurements. In A.K. Jain, R.M. Bolle, and S. Pankanti, editors, *Biometrics: Personal Identification in Networked Society*, pages 251–270. Kluwer Academic Press, Boston, 1999.

[161] W.W. Peterson, T.G. Birdsall, and W.C. Fox. The theory of signal delectability. *Transactions of the IRE*, PGIT-4:171–212, April 1954.

[162] C.P. Pfleeger. *Security in Computing*. Prentice Hall PTR, Upper Saddle River, NJ, 1996.

[163] P.J. Phillips. On performance statistics for biometric systems. In *Proceedings of IEEE AutoID 1999*, pages 111–116, October 1999.

[164] P.J. Phillips, P. Grother, R. Micheals, D.M. Blackburn, T Elham, and J. Mike Bone. FRVT 2002: Facial recognition vendor test. Technical report, DoD Counterdrug Technology Development Office, Defence Advance Research Project Agency, National Institute of Justice, Dahlgren, VA; Crane, IN; Arlington, VA, April 2003.

[165] P.J. Phillips, H. Moon, P.J. Rauss, and S.A. Rizvi. The FERET September 1996 database and evaluation procedure. In J. Bigün, G. Chollet, and G. Borgefors, editors, *Audio and Video-based Biometric Person Authentication*, volume 1206 of *Lecture Notes in Computer Science*. Springer-Verlag, Heidelberg, Germany, April 1997.

[166] P.J. Phillips, H. Moon, S.A. Rizvi, and P.J. Rauss. The FERET evaluation methodology for face-recognition algorithms. *IEEE Transanctions on Pattern Analysis and Machine Intelligence*, 22(10):1090–1104, October 2000.

[167] P.J. Phillips, S. Sarkar, I. Robledo, P. Grother, and K.W. Bowyer. The gait identification challenge problem: Data sets and baseline algorithm. In *Proc. of the International Conference on Pattern Recognition, Vol. 1*, 385–388, August 2002.

[168] R. Plamondon and G. Lorette. Automatic signature verification and writer identification — The state of the art. *Pattern Recognition*, 22(2):107–131, 1989.

[169] D.M. Politis. Computer-intensive methods in statistical analysis. *IEEE Signal Processing*, 15(1):39–55, January 1998.

[170] H.V. Poor. *An Introduction to Signal Detection and Estimation*. Springer-Verlag, New York, 1988.

[171] J.F. Porter. *On the 30 error criterion*. Unpublished but described in J.L. Wayman, editor, *National Biometric Test Center Collected Works*, National Biometric Test Center, University of San Jose, CA, 1997-2000.

[172] S. Prabhakar and A.K. Jain. Decision-level fusion in fingerprint verification. *Pattern Recognition*, 35(4):861–874, 2002.

[173] F.J. Prokoski and R. Riedel. Infrared identification of faces and body parts. In A.K. Jain, R.M. Bolle, and S. Pankanti, editors, *Biometrics: Personal Identification in Networked Society*, pages 191–212. Kluwer Academic Press, Boston, 1999.

[174] M. Przybocki and A. Martin. *The 1999 NIST Speaker Recognition Evaluation Speaker Detection and Speaker Tracking*. EUROSPEECH 99 6th European Conference on Speech Communication and Technology, Budapest, Hungary, September 1999.

[175] G.N. Ramaswamy. Conversational biometrics: The future of personal identification. Technical report, IBM Research Division, Yorktown Heights, NY, September 2001.

[176] A. Ranalli. Fingerprint matching via spatial correlation with regional coherence. In *Proc. of SPIE, Vol. 2932, Human detection and positive identification: Methods and technologies*, pages 161–167, January 1997.

[177] C.V.K. Rao. *Pattern Recognition Techniques Applied to Fingerprints*. PhD thesis, Linköping University, Sweden, 1977.

[178] N. Ratha and R. Bolle. Smartcard based authentication. In A. Jain, R. Bolle, and S. Pankanti, editors, *Biometrics, Personal Identification in Networked Society*, pages 369–384. Kluwer Academic Publishers, Boston, MA, 1999.

[179] N. Ratha, J.H. Connell, and R.M. Bolle. An analysis of minutiae matching strength. In J. Bigun and F. Smeraldi, editors, *Proceedings 3rd IEEE International Conference on Audio- and Video-Based Biometric Person Authentication*, pages 223–228. Springer Verlag, Heidelberg Berlin, June 2001.

[180] N.K. Ratha, S. Chen, and A.K. Jain. Adaptive flow orientation based texture extraction in finger print images. *Pattern Recognition*, 28(11):1657–1672, November 1995.

[181] N.K. Ratha, J.H. Connell, and R.M. Bolle. Enhancing security and privacy in biometrics-based authentication systems. *IBM Systems Journal*, 40(3):614–634, 2001.

[182] N.K. Ratha, J.H. Connell, and R.M. Bolle. Biometrics break-ins and band-aids. *Pattern Recognition Letters*, 24(13):2105–2113, September 2002.

[183] N.K. Ratha, K. Karu, S. Chen, and A.K. Jain. A real-time matching system for large fingerprint database. *IEEE Transanctions on Pattern Analysis and Machine Intelligence*, 18(8):799–813, Aug. 1996.

[184] N.K. Ratha, V.D. Pandit, R.M. Bolle, and V. Vaish. Robust fingerprint authentication using local structural similarity. In *Fifth IEEE Workshop on Applications of Computer Vision*, pages 29–34, December 2000.

[185] D. A. Reynolds. Automatic speaker recognition: Current approaches and future trends. In *Proc. IEEE AutoID 2002*, pages 103–108, Tarrytown, NY, March 2002.

[186] D.A. Reynolds. Comparison of background normalization methods for text independent speaker verification. In *Proc. of the European Conference on Speech Technology*, pages 963–966, Rhodes, 1995.

[187] D.A. Reynolds. The effects of handset variability on speaker recognition performance: Experiments on the Switchboard Corpus. In *Proc. of the IEEE International Conference on Acoustics, Speech, and Signal Processing*, pages 113–116, Atlanta, GA, May 1996.

[188] A.T. Rivers. *Modeling Software Reliability During Non-Operational Testing*. PhD thesis, North Carolina State University, 1998.

[189] D. Roberge, C. Soutar, and B.V.K. Kumar. Optimal trade-off filter for the correlation of fingerprints. *Optical Engineering*, 38(1):108–113, January 1999.

[190] A.E. Rosenberg and S. Parthasarathy. Speaker background models for connected digit password speaker verification. In *Proc. of the Int. Conf. on Acoustics, Speech, and Signal Processing*, pages 81–84, Atlanta, GA, 1996.

[191] E. Rosenberg, J. DeJong, C.-H. Lee, B.H. Juang, and F.K. Soong. The use of cohort normalized scores for speaker verification. In J. Ohala, editor, *Proceedings of the 1992 International Conference Spoken Language Processing*, volume 1, pages 599–602. University of Alberta, Alberta, CA, 1992.

[192] T. Rowley. Silicon fingerprint readers: A solid state approach to biometrics. In *Proc. of the CardTech/SecureTech, Orlando, FL*, pages Vol. 1, 152–159, Washington D.C., May 1997.

[193] T. Ruggles. Comparison of biometric techniques. Technical report, The California State Legislature, http : //biometric-consulting.com/bio.htm, April 1996. Revised May 8, 2001.

[194] A. Samal and P.A. Iyengar. Automatic recognition and analysis of human faces and facial expressions: A survey. *Pattern Recognition*, 25(1):65–77, 1992.

[195] F. S. Samaria and A. C. Harter. Parameterisation of a stochastic model for human face identification. In *Proceedings of 2nd IEEE Workshop on Applications of Computer Vision, Sarasota FL*, pages 138–142, December 1994.

[196] J. Scheeres. *Airport Face Scanner Failed.* http://www.wired.com/news/privacy/0,1848,52563,00.html, May 2002.

[197] B. Schneier. The uses and abuses of biometrics. *Communications of the ACM*, 42(8):136, 1999.

[198] A. Senior. A combination fingerprint classifier. *IEEE Transanctions on Pattern Analysis and Machine Intelligence*, 23(10):1165–1174, 2001.

[199] A.W. Senior. Recognizing faces in broadcast video. In *IEEE International Workshop on Recognition, Analysis, and Tracking of Faces and Gestures in Real-Time Systems*, pages 105–110, September 1999.

[200] D. Setlak. Fingerprint sensor having spoof reduction features and related methods. US Patent Number: 5,953,441, September 1999.

[201] D.P. Sidlauskas. *3D hand profile identification apparatus.* US Patent No. 4,736,203, 1988.

[202] C. Simon and I. Goldstein. A new scientific method of identification. *New York State Journal of Medicine*, 35(18), September 1935.

[203] L. Sirovich and M. Kirby. Low-dimensional procedure for the characterization of human faces. *J. Optical Society of America*, 4:519–524, 1987.

[204] D.A. Socolinsky, L.B. Wolff, J.D. Neuheisel, and C.K. Eveland. Illumination invariant face recognition using thermal infrared imagery. In *Proc. IEEE Computer Vision and Pattern Recognition, Vol. 1,* 527–534, 2001.

[205] M.K. Sparrow and J. Penelope. A topological approach to the matching of single fingerprints: development of algorithms for use on rolled impressions. Technical Report Special Publication 500-126, National Bureau of Standards, 1985.

[206] R.W. Sproat. *Multilingual Text-to-Speech Synthesis: The Bell Labs Approach,* Lucent Technologies Staff, Bell Laboratories, Lucent Technologies, Murray Hill, NJ, USA. Kluwer Academic Publishers, Boston, MA, October 1997.

[207] S.N. Srihari, S.-H. Cha, H. Arora, and S. Lee. Individuality of handwriting. *Journal of Forensic Sciences*, 47(4):1–17, July 2002.

[208] W. Stallings. *Network and Internetwork Security.* Prentice Hall, Englewood Cliffs, NJ, 1995.

[209] G.C. Stockman, S. Kopstein, and S. Benett. Matching images to models for registration and object detection via clustering. *IEEE Transanctions on Pattern Analysis and Machine Intelligence*, 4(3):229–241, May 1982.

[210] A. Stoianov, C. Soutar, and A. Graham. High-speed fingerprint verification using an optical correlator. *Optical Engineering*, 38(1):99–107, January 1999.

[211] D.A. Stoney. A systematic study of epidermal ridge minutiae. *Journal of Forensic Sciences*, 32:1182–1203, 1987.

[212] J.D. Stosz and L.A. Alyea. Automated system for fingerprint authentication using pores and ridge structures. In *Proc. of SPIE, Vol. 2277, Automatic system for the identification and inspection of humans*, pages 210–223, October 1994.

[213] D.L. Swets and J. Weng. Using discriminant eigenfeatures for image retrieval. *IEEE Transanctions on Pattern Analysis and Machine Intelligence*, 18(8):831–836, August 1996.

[214] I.M. Tarbell. Identification of criminals: The scientific method used in France. http://chnm.gmu.edu/courses/magic/plot/bertillon.html.

[215] C.J. Tilton. An emerging biometric standard. *IEEE Computer Magazine,* Special Issue on Biometrics, 1:130–135, February 2001.

[216] M. Turk and A. Pentland. Eigenfaces for recognition. *Journal of Cognitive Neuro Science*, 3(1):71–86, 1991.

[217] University of South Florida. http://marathon.csee.usf.edu/GaitBaseline/. Tampa, FL.

[218] U.S. Department of Justice. *Solicitation: Forensic Friction Ridge (Fingerprint) Examination Validation Studies.* National Institute of Justice, Office of Science and Technology, Washington, DC 20531, March 2000.

[219] B. Victor, K.W. Bowyer, and S. Sarkar. An evaluation of face and ear biometrics. In *Proceedings of the International Conference on Pattern Recognition*, pages I:429–432, August 2002.

[220] Visionics Corporation, Inc.; now Identix, Inc. http://www.visionics.com/. Jersey City, NJ.

[221] R. Wang, T.J. Hua, J. Wang, and Y.J. Fan. Combining of Fourier transform and wavelet transform for fingerprint recognition. In *Proc. of SPIE, Vol. 2242, Wavelet Applications*, pages 260–270, March 1994.

[222] C.I. Watson. *NIST special database 10: Supplemental Fingerprint Card Data for NIST Special Database 9.* Advanced Systems Division, Image Recognition Group, National Institute for Standards and Technology, February 1993.

[223] C.I. Watson. *NIST special database 14: Fingerprint Card Pairs 2*. Advanced Systems Division, Image Recognition Group, National Institute for Standards and Technology, February 1993.

[224] C.I. Watson. *NIST special database 4: 8-bit Gray scale Images of Fingerprint Image Groups*. Advanced Systems Division, Image Recognition Group, National Institute for Standards and Technology, February 1993.

[225] C.I. Watson. *NIST special database 9: Mated Fingerprint Card Pairs*. Advanced Systems Division, Image Recognition Group, National Institute for Standards and Technology, February 1993.

[226] C.I. Watson. *NIST special database 24: NIST Digital Video of Live-scan Fingerprint Database*. Advanced Systems Division, Image Recognition Group, National Institute for Standards and Technology, February 1998.

[227] J.L. Wayman. A scientific approach to evaluating biometric systems using mathematical methodology. In *Proceedings of CardTech/SecureTech.*, pages 477–492, Orlando, FL, May 1997.

[228] J.L. Wayman. Error rate equations for the general biometrics system. *IEEE Automation and Robotics Magazine*, 6(1):35–48, March 1999.

[229] J.L. Wayman. *National Biometric Test Center Collected Works*. National Biometric Test Center, San Jose, CA, August 2000.

[230] J.H. Wegstein and J.F. Rafferty. Matching fingerprints by computer. Technical Report 466, National Bureau of Standards, 1969.

[231] D. Welsh and K. Sweitzer. *Presented at CardTech/SecureTech.*, May 1997.

[232] R.P. Wildes. Iris recognition: An emerging biometric technology. *Proceedings of the IEEE*, 85(9):1348–1363, September 1997.

[233] R.P. Wildes, J.C. Asmuth, G.L. Green, S.C. Hsu, R.J. Kolczynski, J.R. Matcy, and S.E. McBride. A machine-vision system for iris recognition. *Machine Vision and Applications*, 9:1–8, 1996.

[234] C.L. Wilson, C.I. Watson, and E.G. Paek. Combined optical and neural fingerprint matching. In *Proc. of SPIE, Vol. 3073, Optical Pattern Recognition*, pages 373–382, March 1997.

[235] L. Wiskott and C. von der Malsburg. Recognizing faces by dynamic link matching. In *Proceedings of the International Conference on Artificial Neural Networks*, pages 347–352, 1995.

[236] I. Witten, A. Moffat, and T. Bell. *Managing Gigabytes: Compressing and Indexing Documents and Images*. Van Nostrand Reinhold, New York, NY, 1994.

[237] H.J. Wolfson and Y. Lamdan. Geometric hashing: A general and efficient model-based recognition scheme. In *IEEE Proc. First International Conf. on Computer Vision*, pages 238–249, 1988.

[238] T. Worthington, T. Chainer, J. Wilford, and S. Gunderson. IBM dynamic signature verification. *Computer Security*, pages 129–154, 1985.

[239] M.Y.-S. Yao, S. Pankanti, N. Haas, N. Ratha, and R.M. Bolle. Quantifying quality: A case study in fingerprints. In *Proc. IEEE AutoID 2002*, pages 126–131, Tarrytown, NY, March 2002.

[240] J.R. Young and H.W. Hammon. Automatic palmprint verification study. Technical Report RADC-TR-81-161 Final Technical Report, Rome Air Development Center, June 1981.

[241] R. Zunkel. Hand geometry based authentication. In A.K. Jain, R.M. Bolle, and S. Pankanti, editors, *Biometrics: Personal Identification in Networked Society*, pages 87–102. Kluwer Academic Press, Boston, MA, 1999.

Index